herbarium

This stunning book of photographs by Robyn Stacey, one of this country's finest photographers, is the first of its kind. Stacey, along with essayist, Ashley Hay, throws open the closed doors of the National Herbarium of New South Wales at the Royal Botanic Gardens, Sydney, to reveal the secret history of Australia's flora. *Herbarium* tells fascinating stories about the nature of collecting, those who collected, what they collected and when, and also provides the scientific background to each of the specimens. A list of botanical notes provides a unique link between the specimens and their collectors. The exquisite photographs of the botanical specimens – some extremely rare – comprise a collection of extraordinary beauty.

Robyn Stacey is a Senior Lecturer in the School of Communication, Design and Media at the University of Western Sydney. She is one of Australia's most acclaimed photographers and has exhibited her photographic work in numerous solo and group exhibitions in Australia and internationally since the mid-1980s to high critical acclaim. Her work is represented in the National Gallery of Australia, all state galleries in Australia and Art Bank, as well as in university and private collections, both here and abroad.

Ashley Hay has written two books of narrative non-fiction, *The Secret: The Strange Marriage of Annabella Milbanke and Lord Byron* (Duffy and Snellgrove, 2000), and *Gum: The Story of Eucalypts and Their Champions* (Duffy and Snellgrove, 2002). She has also published essays and short fiction and works for *The Bulletin*.

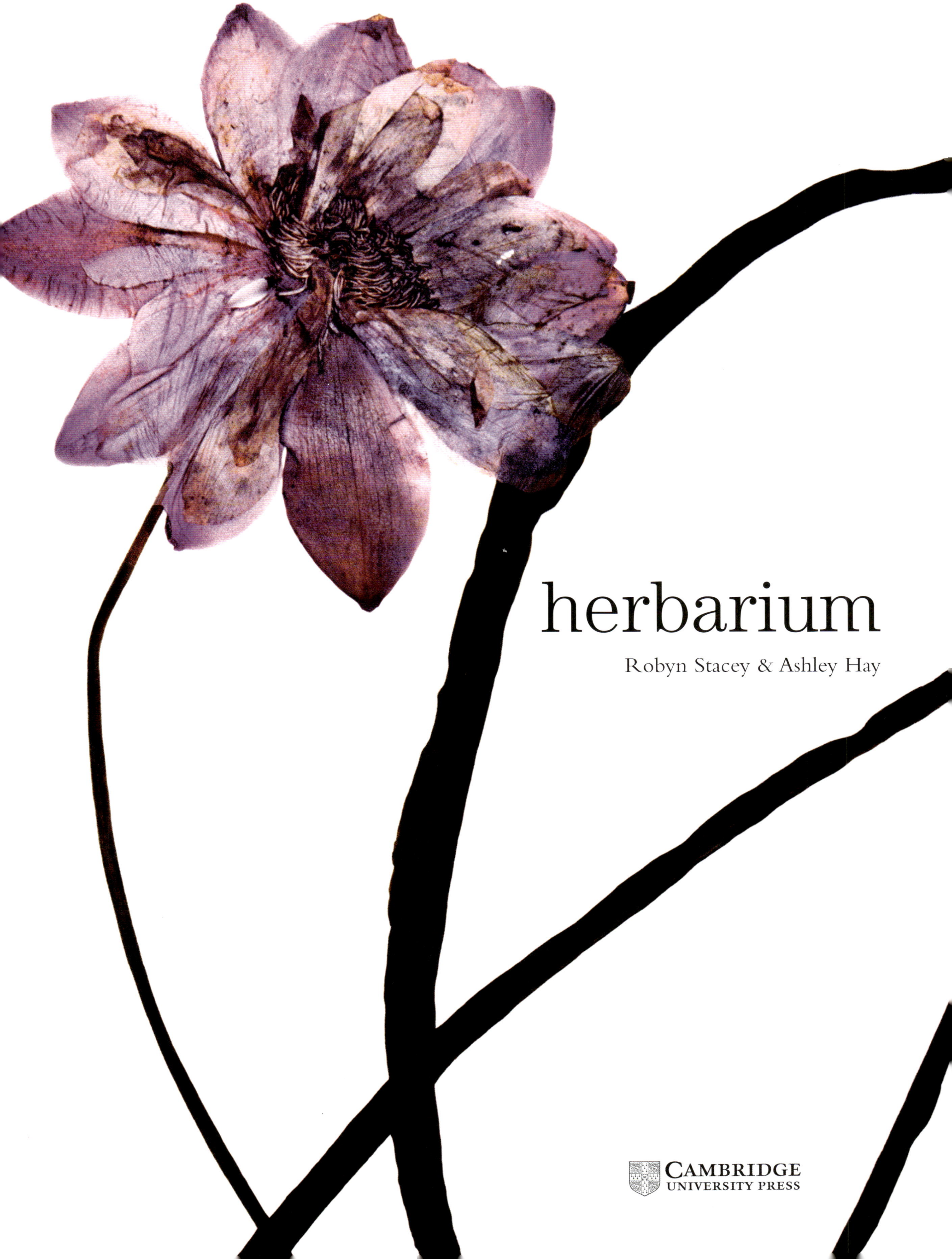

herbarium

Robyn Stacey & Ashley Hay

CAMBRIDGE
UNIVERSITY PRESS

For John, Helen and Allan

PUBLISHED BY THE PRESS SYNDICATE OF THE UNIVERSITY OF CAMBRIDGE
The Pitt Building, Trumpington Street, Cambridge, United Kingdom

CAMBRIDGE UNIVERSITY PRESS
The Edinburgh Building, Cambridge CB2 2RU, UK
40 West 20th Street, New York, NY 10011–4211, USA
477 Williamstown Road, Port Melbourne, VIC 3207, Australia
Ruiz de Alarcón 13, 28014 Madrid, Spain
Dock House, The Waterfront, Cape Town 8001, South Africa

http://www.cambridge.org

First published by Cambridge University Press 2004

Printed in Australia by BPA Print Group

Typeface Snell Roundhand Script, Bembo (*Adobe*) 11/15 pt *System* QuarkXPress®

A catalogue record for this book is available from the British Library

National Library of Australia Cataloguing in Publication data

Stacey, Robyn.
Herbarium

Bibliography.
Includes index.
ISBN 0 521 60392 7 (Slipcase).

ISBN 0 521 84277 8 (hbk).

1. National Herbarium of New South Wales. 2. Botany – Australia – History. 3. Plant collecting – Australia – History. 4. Plant collectors – Australia – History. 5. Botanical Specimens – Collection and preservation – Australia. I. Hay, Ashley. II. Title.

580.740994

ISBN 0 521 60392 7 hardback (slipcased)

This is one of a limited edition of 300 copies.

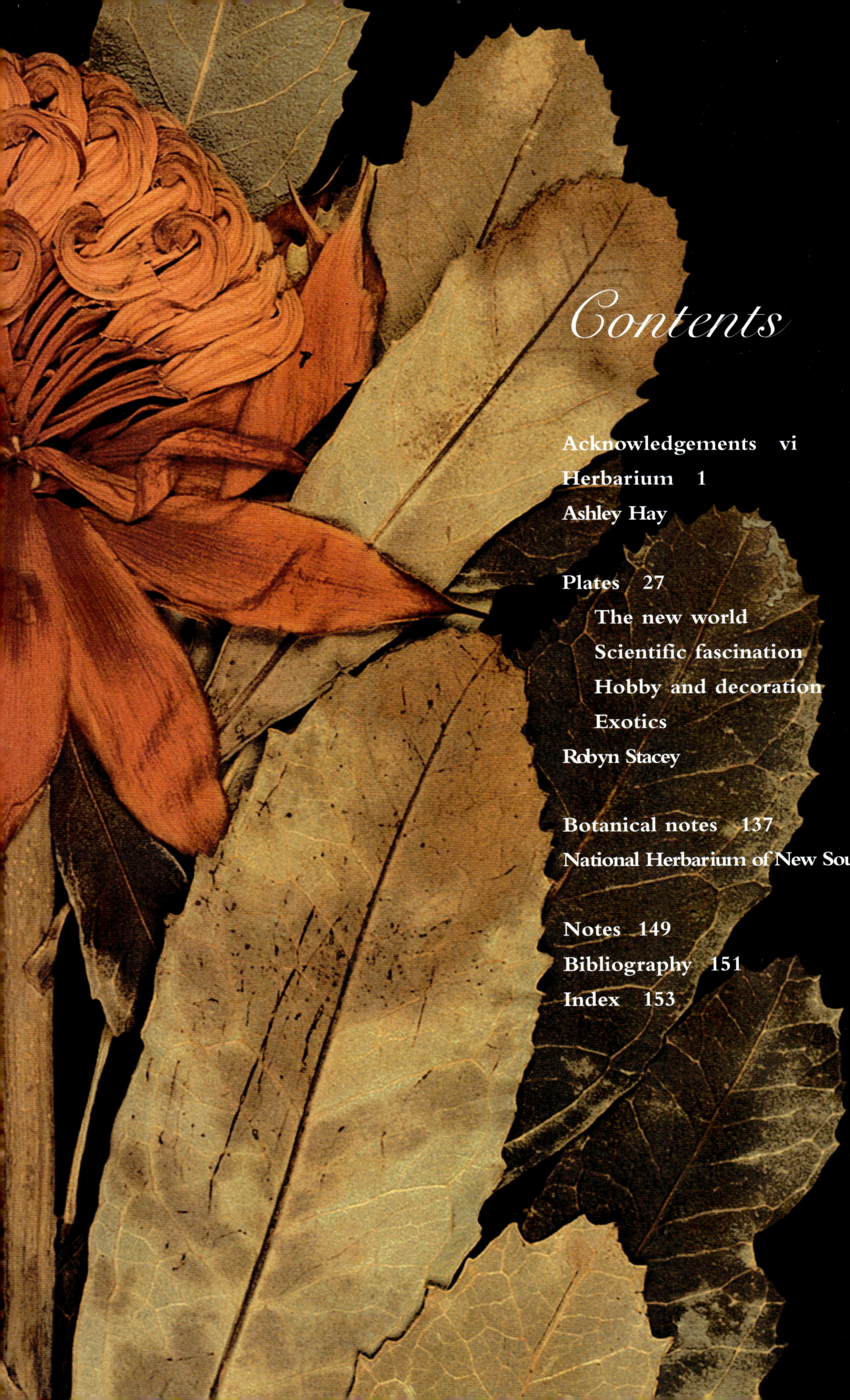

Contents

Acknowledgements

Thanks to Frank Howarth, who was Director (1996–2003) of what is now the Royal Botanic Gardens Trust, for initially proposing the idea of the book and for his enthusiasm for the project, and to Dr Tim Entwisle (current Director) for his ongoing commitment and support.

Ashley Hay for her generosity of spirit and for making the stories and history of the Herbarium come to life in such a vivid and poetic way.

Melissa Fraser for her invaluable professionalism at the picture editing stage and for her brilliant design acumen in realising the final outcome.

To all the Herbarium staff at the Royal Botanic Gardens, Sydney. Books like this develop over time and with input from many sources. During the three years I spent photographing in the Herbarium all the staff were very generous with their expertise and time. In particular I would like to thank Karen Wilson whose vast knowledge of the history of the Herbarium is only matched by her botanical expertise. Louisa Murray, the Herbarium's collections co-ordinator, for her support throughout the project, and particularly for her time and effort in supplying the scientific information for the botanical notes. To Dr Phillip Kodela and Katherine Downs for their time and patience in helping me source the specimens and for providing botanical information. Dr Peter Weston for all his help and scientific expertise with the orchids and Dr Alan Millar for his extensive knowledge about seaweeds and the relevant collectors. To Dr Barbara Briggs (currently Honorary Research Associate, formerly Senior Assistant Director (Scientific) 1972–1997) for the botanical chats and insights she provided. To Dr Barry Conn (Manager, Plant Diversity) for his support and oversight of the process in the final stages.

Thanks to Leonie Stanberg, Clare Herscovitch, Peter Jobson, Zonda Erskine and Julie Taylor for their time and assistance in sourcing specimens. And thanks to Jaime Plaza who kindly lent me equipment and was very generous with access to facilities and to Miguel Garcia for all his assistance in the library over the years.

Kim Armitage for being a great Commissioning Editor and to all the team at Cambridge for their commitment and belief in the book.

Cameron Hall and Craig O'Brien at Trannys who ensured that the scanning and colour management of the image files were superb.

Professor Jann Conroy, Associate Professor Robert Spooner-Hart, and the Centre for Horticulture and Plant Sciences at the University of Western Sydney for their ongoing and generous support of the book.

Stills Gallery for their encouragement and enthusiasm for the project, from the first Hot House exhibition in 2001 to the publication of *Herbarium* in 2004.

Robyn Stacey

Thanks to Robyn Stacey for the invitation to write something to sit beside the richness of her images.

To staff (past and present) at the Botanic Gardens Trust – Frank Howarth, Tim Entwisle, Karen Wilson, Barbara Briggs, Peter Weston and Andrew Orme – for their time, generosity and encouragement – and particularly to Miguel Garcia and Judy Blood in the library, and to Louisa Murray, the Herbarium's collections' co-ordinator, for all assistance. Thanks, too, to Lionel Gilbert whose many volumes drawing on the vast archives of the library at the Royal Botanic Gardens, Sydney, make it feel as if the first huge cull of material has been made for you.

To Jill Chapman in the archives of the Powerhouse Museum, Sydney.

To Vicky Noble and Judith Magee at the Natural History Museum, London, for time with Banks' and Solander's herbarium specimens and with Sydney Parkinson's extraordinary drawings of them.

To Kim Armitage and the production staff of Cambridge University Press, to Malory Weston – and to Roger McDonald, tangentially, for introducing me to them.

To Garry Linnell and Kathy Bail at *The Bulletin* for time and understanding.

And to Gail MacCallum, Steve Offner and Nigel Beebe for their different sorts of patience with words.

Ashley Hay

All the staff at the National Herbarium of New South Wales (Botanic Gardens Trust, Sydney) but in particular, those who wrote text, and the following staff who assisted Robyn Stacey and Ashley Hay in carrying out the project: Barry Conn, Katherine Downs, Miguel Garcia, Phillip Kodela, Linn Linn Lee, Louisa Murray, Julie Taylor, Peter Weston, Karen Wilson and Peter Wilson.

For fern identification – Peter Bostock, Principal Botanist, Queensland Herbarium, Queensland Environmental Protection Agency.

National Herbarium of New South Wales

Cambridge University Press and the authors wish to thank the Centre for Horticulture and Plant Sciences at the University of Western Sydney for their financial support in the form of a grant to Robyn Stacey. We also wish to thank the Friends of the Gardens, Sydney, for their generous support.

The Press and the authors also wish to express their appreciation to a number of individuals. Thank you to Louisa Murray, who wrote the majority of the entries in the botanical notes, and Dr Alan Millar for his contribution to the notes. We are also grateful to Clare Herscovitch, Dr Phillip Kodela, Bob Makinson, Dr Barry Conn, Leonie Stanberg, Karen Wilson, Jan Allen and John Thomson who also provided entries.

Finally, we would like to acknowledge all staff at the National Herbarium of New South Wales, Royal Botanic Gardens, Sydney, especially the Director, Dr Tim Entwisle, and staff: Miguel Garcia, Linn Linn Lee, Katherine Downs, Phillip Kodela, Karen Wilson, Peter Weston, and Peter Wilson. Thank you.

Cambridge University Press

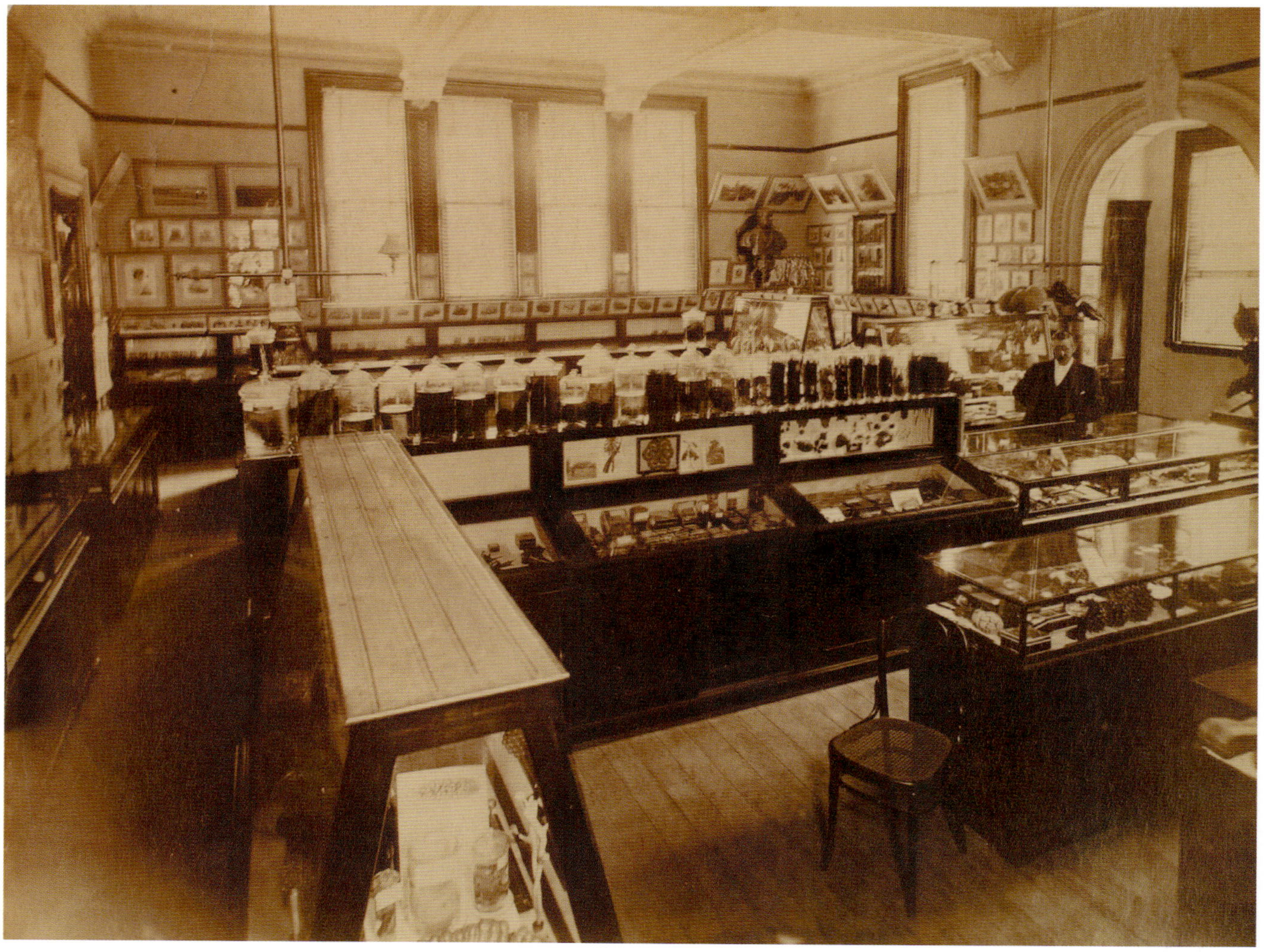

The original Botanical Museum, attached to the Herbarium, c. 1900

Herbarium

Whereas the richest garden only contains but a few thousand species growing at any one time, a herbarium may contain tens or hundreds of thousands of species . . .

J. H. Maiden, 1899

You can imagine it as a vast filing cabinet, one that's large enough to walk around inside. These are big, high rooms, one on top of the other and filled with corridors of shelves stacked ceiling-high with red plastic boxes. It's dim in here, and quiet, and the air is soft with the leftover smell of naphthalene. This is the National Herbarium of New South Wales at the Royal Botanic Gardens, Sydney, repository for more than a million plant specimens – not just from every corner of this continent but from all over the world.

A three-dimensional reference work, herbaria aren't just for carefully worded definitions of everything from trees to lichen; they hold actual pieces of them. The Latin name for these collections, *hortus siccus*, is literally a dried garden, and each pressed specimen that's mounted on a sheet of paper is inscribed with what the plant is, who collected it, when, and where.

The ideal garden in many ways, a herbarium needs no water, no pruning, no fertiliser. It defies seasons, climate, geography and even time itself, accommodating the newly found spiky green leaves of Australia's famous dinosaur-tree, the 40-metre Wollemi Pine, as easily as a tiny starburst of whortleberry cactus picked in Mexico in the nineteenth century, or a banksia taken during European botany's first glimpse of eastern Australia in 1770. It's a place where buds from Australia's south coast can sit with buds from its north, alpine flora with algae, plants from this country with those from anywhere else.

Botany, a science of comparison, would be impossible without this archive; botanists decide what a plant is by understanding what it isn't. Confronted with a branch they've never seen before, they work through the species stored in a herbarium's drawers – plants already defined and named – to judge the similarities and differences between the newcomer and the rest of the floral world. A plant pressed onto a sheet has no official place until it's named and classified, either as something already known or, if it matches nothing that any herbarium holds, as a new thing altogether.

Specimen names have their own layers of language. Take a piece of white stringybark. A eucalypt, its genus is *Eucalyptus*. This species, in particular, is *Eucalyptus globoidea*. Like all eucalypts, it belongs in the Myrtaceae family, some of whose members produce valuable things like cloves, allspice, guavas. In the Herbarium in Sydney's Gardens, there's a piece of this eucalypt that's valuable above all others. Collected in 1901 near Berrima in New South Wales, it's the type specimen, the single sample selected and preserved to be the permanent reference point for this species. This one is the holotype, the one sample designated by the botanist who first classified it as the 'original' for the name of the white stringybark. There are isotypes too, duplicates of that holotype taken from specimens collected at the same time from the same plant, which are often sent to other herbaria to provide a reference point for their collections.

In a way, this fragment of white stringybark in its drawer in Sydney stands for all white stringybarks. It defines *E. globoidea*.

But to collect something, to classify it and name it, is by no means to reach the end of its story. Amongst all the foliage in herbaria, botany wages fierce battles. Just as new names are created, old classifications are reconsidered, struck out and replaced. This botanist regards that species as the same as another, and changes its name. Years pass; another botanist looks at the same specimens and sees not similarities but differences: the name changes again. This genus is split. That species turns out to contain three distinct things. Handfuls of other species are scooped up together and called the same name. Some specimen sheets are cramped with these revisions.

The mosaic of the planet's plants shifts and grows constantly as botanical knowledge changes with new eras, new workers – new methods and insights. And herbarium collections reflect all this movement.

Diverse by definition, a herbarium can be anything from a small collection – one box holding a few things gathered by a single person – to sprawling institutions that draw on myriad people across generations of time, aiming for as many things from as many places as possible. Sydney itself may be known to hold more than a million plants, but there's no estimate of the number of collectors responsible for them all. It would run into the thousands.

And beneath the sheer weight of so many specimens and the catalogue of growth and place that they create is the moment when each one was picked. One person was walking the 'slow and roundabout' walk of a plant-hunter: maybe they were a scientist focused on a single genus, a professional employed to collect this species one week and another the next, or an enthusiast delighted by collecting on any walk they took. Whatever the case, they saw something, stopped, and cut a stem, selecting one sample of that fern, that shrub, that tree to stand for its whole breed. They pressed it, dried it, and made it part of a collection. In these stories of hunting and gathering, of so many people unable to walk through nature without pausing and picking, there's an edge of compulsion: that human urge to collect that, beyond science, might apply to postcards or teapots as much as to plants.

Lift the name of any collector from any herbarium specimen sheet. Magnify it with some of their history, their anecdotes, coincidences and disasters. And Australian botany – the stories behind this huge dried garden – comes up into view like a photograph emerging from the blank of its paper.

History

Species: Viola banksii K. R. Thiele & Prober
Common name: A Native Violet
Family: VIOLACEAE
Collector: J. Banks & D. Solander
Locality: Botany Bay, New South Wales
Date: May 1770

The world begins with water and gas spinning on its globe. Then, 3460 million years ago, photosynthesis – that clever connection between chlorophyll and sunlight – begins. The first land plants grow from 450 million years ago. Ancient algae evolve into mosses; fungi move across the most inhospitable surfaces; some mosses creep, becoming the ancestors of ferns. From 410 million years ago come Australia's first vascular plants. There are conifers, and then, 200 million years on again, the flowering plants that thicken and prevail on the face of the earth.

Time passes. The continents move and split, and rainforests and oceans cover Australia. Twenty-five million years ago, the weather changes, and the country's famous drought-resistant plants – acacias, banksias, eucalypts – begin to dominate the landscape. The rich velvety gold of a banksia's flower breaks in among the greens, the greys, the browns of the land's foliage. The soft white blossom of a scribbly gum bursts in late spring, touching the sky above its pretty bark.

If the count could ever be completed, Australia's flora may be found to nurture more than 25,000 species. Some, conservative in their evolution, remain fairly unchanged across millions of years; others diversify in and adapt to different environments. In either case, the complete suite of Australia's flora and fauna is one of the oldest and most distinct on the earth.

Against the time span of their evolution, the activity of collecting plants stretches as far as civilisation. By the third century BC, Theophrastus was already arguing that botany should move beyond medicinal and other practical uses to 'consider the distinctive characters and general nature of

plants from the standpoint of their morphology, their behaviour in the face of external conditions, their mode of generation and their whole manner of living'. As for the need to get into the field and study plants where they grew, it was Pliny who praised experience as 'the best teacher', criticising schools where 'it is more agreeable to sit on benches ... than to go out into deserted places and look for different herbs at each season of the year'.

By the sixteenth century, Italy's great medical cities – Pisa and Padua in particular – were reviving botanical activity, just as its great commercial cities – Venice, Florence, Genoa, Milan – had revolutionised science in the fifteenth. Beyond the established canon of medicinal and agricultural plants on which people had always focused, botanists began the larger task of counting and naming all the vegetation around them.

The catalogue of the floral world had begun. 'Not a hundredth part of the herbs existing in the whole world was described' by the Greeks, wrote one physician from Ferrara, 'but we add more every day'. Amongst all this activity, in the early decades of the 1500s, a man called Luca Ghini found a way to make plants both portable and immortal.

A driving force behind the establishment of Pisa's botanic gardens, one of the renaissance world's first, Ghini had created there a harbour of living plants for quick reference. Sometime in the first decades of the sixteenth century, he also began to preserve plants by placing them between sheets of paper, under the pressure of some weight, until they were dried and could be mounted on card. Here was the first recorded *hortus siccus*, the first 'dried garden'. Here were the first herbarium specimens, allowing piles of plant portions to be kept for easy reference or, even better, exchanged for those collected by other people in other places.

At first, this culture of collecting reflected the world as it was revealed by thrusts of exploration, trade and colonisation. An influx of strange and exotic specimens – plants, shells, animals, anything offered up by somewhere else – had wealthy European gentlemen constructing special cupboards in which to show off these amazing treasures from the ends of the earth. As Francis Bacon put it, these were cabinets where 'whatsoever singularity chance and the shuffle of things hath produced; whatsoever Nature hath wrought in things that want like may be kept'. The plants they held were trophies, precious objects, status symbols, haphazardly gathered and arranged, and almost works of art themselves in their value and display. Between the mid-1500s and the beginning of the eighteenth century, collections of such 'rare, exceptional, extraordinary, exotic and monstrous things' flourished.

Of course, the people who regarded their *hortus siccus* as something to cherish and flaunt were a long way from the scholars who would later relish such sheets as a means of explaining and naming the world. But the herbarium specimens of each were the same, as was the desire to possess them.

These two motivations – the compulsion to possess precious things, and the compulsion to collect in the name of advancing science – collided spectacularly in the life of Joseph Banks, the man who would catapult Great Britain to the forefront of botanical conquest. As a boy, he bought specimens from women who collected the berries and leaves, herbs and roots, used in pharmacy, and by the time he reached his twenties, his herbarium already bulged with the collections of more than a decade.

Although he went to Oxford, Banks, like many gentlemen of his age, didn't take a degree. But his commitment to botany was demonstrated by his paying a Cambridge lecturer to travel down and tutor him (Oxford's professor of botany had delivered only one lecture in 35 years). After some time at university, Banks' attention turned to his Grand Tour, a requisite part of life for any rich young man. Most went to Europe, which made Banks scoff: 'Every blockhead does that; my Grand Tour shall be one around the whole globe'.

He would accompany the expedition of Lieutenant James Cook to the South Seas, with a private party of eight including Daniel Solander and another naturalist, two artists, and four servants, plus dogs and luggage. This included reams of paper for drying and storing plant specimens (including proof pages of Milton's *Paradise Lost*); more than 100 books (including the great botanical works); material for his artists (to capture the plants at their freshest); an underwater telescope, bottles, barrels, nets, hooks, wax, bug catchers, even a guitar. Observers described it as 'the Argonautic Expedition for the Study of Nature'. 'No people,' said one, 'ever went to sea better fitted out for the purpose of Natural History'.

The pursuit of that purpose often required some ingenuity. Denied permission to land in Brazil, for example, Banks and Solander slunk ashore at night to hunt their plants, while other specimens were sent aboard labelled 'grass' and ostensibly for the ship's seafaring livestock.

Under sail, the herbarium had its own routine. 'Now do I wish that our friends in England could by the assistance of some magical spying glass take a peep at our situation,' Banks wrote. 'Dr Solander setts at the Cabbin table describing, myself at my Bureau journalizing, between us hangs a large bunch of sea-weed, upon the table lays the wood and barnacles ...'

The plants were taken to the captain's cabin (given over to Banks' entourage), sketched by the artists, and then carefully pressed and dried. In the afternoons, there was time to study specimens, to elaborate on the quick descriptions that had been attached to them as they were picked, and to consult the travelling library for synonymous plants from elsewhere. (Solander's thoughts on which plant fitted which family, what species resembled what, were so changeable as he moved through this completely foreign flora that he swapped bound notebooks, in which the pages couldn't be rearranged, for a system of loose sheets of paper – still known as 'Solander slips' – that could be shuffled and reorganised as it became obvious that *this* belonged in *this* family, not *that* one, or that *that* sat somewhere in *this* genus, not anywhere near the first slot he'd suggested.)

During any landfall, Banks' journal hummed with references to 'botanizing', the plants at one spot on Australia's east coast so thrilling him that he convinced Cook to alter its proposed name. Stingray Bay became Botany Bay instead.

And there, from the first opportunity, he was 'into the woods and ... found many plants'. Out along a river, he was stopped by the small purple points of some native violets and picked a posy of their flowers. So plentiful was the new and strange flora of this place that by the fifth, and rainy, morning he was 'well contented to find an excuse for staying on board to examine them a little'. When the afternoon cleared, however, it was straight back to 'our old occupation of collecting, in which we had our usual good success'.

'Our collection of Plants was now grown so immens[e]ly large,' Banks discovered before finishing a week in New South Wales, 'that it was necessary that some extr[a]ordinary care should be taken of them least they should spoil in the books' in which they were being dried. 'I therefore devoted this day to that business and carried all the drying paper, near 200 Quires of which the larger part was full, ashore and spreading them upon a sail in the sun kept them in this manner expos[e]d the whole day, often turning them and sometimes turning the Quires in which were plants inside out. By this means they came on board at night in very good condition.'

The young Joseph Banks sat in the sun on the shore of Botany Bay, the richness of his treasure-trove of plants – from Brazil, from Tahiti, from New Zealand, from the waters of the Pacific, from this one tiny piece of an immense new continent – surrounding him as he prepared them for his herbarium.

Arriving home in England in 1771, the *Endeavour* was heralded by London's newspapers as having 'made a voyage around the world, and touched at every coast and island where it was possible to get on shore, to collect every species of plant and other rare productions in nature'. It had brought home more than 30,000 specimens of over 3,600 different species, almost half of which had never been seen in Europe.

England was captivated – all these new and gorgeous things. It was as if the biggest cabinet of curiosities had landed in its midst. People went to parties hoping that Banks and Solander might arrive with an account of their voyage, 'which I am told,' said one lady, 'is very amusing'. Banks commissioned a portrait of himself surrounded by the finest prizes of his mighty 'grand tour': a ceremonial cloak, a wooden club, and a herbarium sheet at his feet.

In his capacious house on New Burlington Street, he converted several rooms into a global cabinet, open to anyone who was interested. 'Here is ... a large collection of insects, several fine specimens of the bread and other fruits,' one visitor sighed, '... together with a compleat *hortus siccus* of all the plants collected in the course of the voyage ... What raptures must they have felt to land upon countries where every thing was new to them! whole forests of nondescript trees clothed with the most beautiful flowers and foliage ... I could be extravagant upon this topic ...'

If the function of a herbarium is to collect and preserve plants in one place, it's the scientific work of classification that transforms it from a curiosity to a resource: botanical rendering has no value, in the world of science, until it's been published. A fine collector (even before his voyage he'd been known to accumulate everything from china and caricatures to a list of how much his friends weighed), Banks appeared to be a man whose publications would be forever delayed.

The leading natural historians of the day couldn't wait for this next scientific step to be made. One confessed he was 'almost entirely' deprived of sleep by the fact that the herbarium's classifications hadn't been finalised, written up, and published. 'Consider my friend,' he wrote to a fellow scholar, 'if these treasures are kept back what may happen to them. They may be devoured by vermin of all kinds. The house where they are lodged may be burnt. Those destined to describe them may die.'

As Solander did, unexpectedly, in 1782, after years of puzzling at his sheets of pressed plants and notes. This bottlebrush, he wondered, did it belong to the genus *Metrosideros* he'd devised for some trees in Tahiti? Was that gum tree the same genus again? He never unravelled the answer, and Banks published neither this material, nor any other scientific papers on his collections. The huge herbarium moved house, from New Burlington Street to Soho Square, still open to any gentleman of science who wanted to consult it. But Banks had another collection project underway, and it came with the patronage of the King, George III.

Both passionate about botany, George III and the young botanist formed a bond over the Royal Botanic Gardens at Kew, which Banks had visited for the first time in 1771 and called 'the finest … in Europe'. Quickly offered its honorary directorial position, Banks embraced it with a new proposal for the King. Why not make Kew the very hub of a mighty botanical empire, he suggested, gardens that reflected the breadth and magnitude of the King's Empire itself? Why not assemble a herbarium extensive enough to be the centre of all the world's classifications? Why not send collectors out to gather specimens of the entire Empire's flora, and use Kew to grow and acclimatise them all? As Banks saw it, the King of such a powerful and vast Empire should surely have gardens that reflected that status.

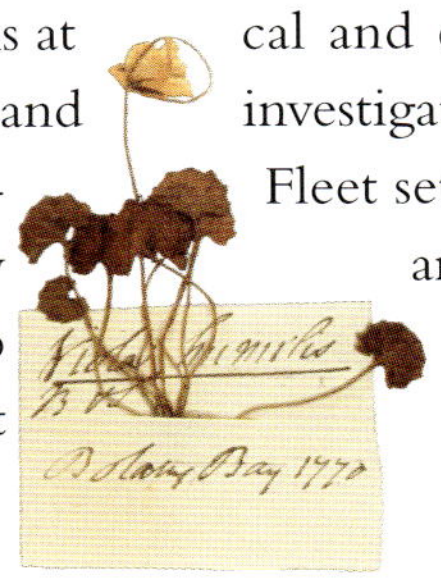

The King approved the funds, and the foundation was laid for a huge network of gatherers to pick the world for Kew, chosen, employed, instructed and despatched by Banks. And before long, too, an opportunity presented itself for Banks to augment both his collections and Kew's with more exciting new items from those florally rich Antipodes. The English government needed a penal colony, and Banks heartily advocated Botany Bay. More than a decade after his frenzied week of collection there, he felt that many of the plants he had seen (and certainly the ones he hadn't) would 'no doubt possess properties which might be useful for physical and economic properties which we were not able to investigate'. The captain and officers of the settlement's First Fleet set sail with instructions on what to collect for him – and how.

The first herbarium of Australia's plants was set to expand exponentially.

The field

Species: Sowerbaea juncea Sm.
Common name: Vanilla Lily
Family: ANTHERICACEAE
Collector: G. Caley
Locality: New South Wales
Date: September 1802

The strange environment of New South Wales – its residents, its weird animals, its even weirder plants (who knew that trees might shed their bark, rather than their leaves?) – filled the journals and letters of the people sent there, their impressions ranging from one extreme to another. Where some saw 'rare and beautiful plants', others saw a place 'so very barren and forbidding that it may with truth be said that *here nature is reversed*'.

However the country looked, its new settlers began their floral gatherings: both herbarium specimens for Banks and whatever they could find that was edible. Saturday was collecting day, every able-bodied person heading out from

Port Jackson (Botany Bay had proved uninhabitable, despite Banks' memories of its fecundity) to pick and pluck any 'vegetables' from the bush – things like wild spinach, and a liquorice-flavoured creeper that earned the name 'sweet tea'.

Yet for all Banks' preoccupation with the colony's plants, he had – as its first governor soon felt obliged to mention – entirely failed to provide it with either an official botanist or gardener: 'it is not therefore in my power to give more than a very superficial account of the produce of this country, which has such a variety of plants that I cannot, with all my ignorance, help being convinced that it merits the attention of the naturalist and the botanist'. Someone with some sort of knowledge should be sent over – someone with a better idea of what to pick or, in an increasingly hungry settlement, what to plant.

The personnel Banks chose didn't get off to a good start. Two were shipwrecked during their voyage; another shot himself on a duck-hunt in his first 18 months in the colony. Then, as the nineteenth century opened, Banks despatched two more men. One, George Caley, he financed himself (the colony had no position for him). Fifteen shillings a week, he said, for which he expected 'to be supplied with new or rare plants for his own Herbarium and with seeds for Kew'. Caley, sailing to Sydney, would collect from there, while Banks' other man, Robert Brown, was assigned to Matthew Flinders' circumnavigation of the continent to collect wherever the expedition went ashore. While Caley gathered up shipment after shipment of specimens, Brown would build up his own herbarium – one that would lead him to the first attempt to describe every plant known to grow in Australia.

The first of the two to arrive in this 'remote country', Caley sailed into Port Jackson in 1800, almost immediately assuring his employer that 'in the course of one year, provided no obstacles fall in my way, I could collect half the specimens [i.e. species] of plants in the colony'.

He began at once, walking, stopping, picking, and then pausing to compare what was in his hand with other Australian plants he'd seen before leaving – in gardens, or in the files of Banks' *hortus siccus* – and plants he'd read about in the journals, books and papers that were starting to make sense of Australia's vegetation. Orchids, boronia, peas, heaths, buttercup-ish *Hibbertia* and lilies (including the delicate scented flowers of the Vanilla Lily) were all parcelled up with pages of notes, descriptions, suggestions of which plant was what, and what altogether new things might be called – and sent to London. 'Your descriptions of plants do you credit,' Banks told him as another bundle was swallowed by his herbarium.

As Caley's first year ended – without his collecting that proportion of the colony's species he'd promised – Brown waited to sail with Flinders, preparing himself for a continent on which some 370 plant species had so far been identified. The possibilities of his assignment were enormous. No naturalist before him had had the opportunity of visiting so many different parts of this landmass. Even Banks, whose collections formed the basis for the entire notion of Australia's botany, had only touched at spots along its east coast and on some of its islands.

Possibly with the lesson of Banks' unpublished collections before him, Brown resolved to complete his investigations as he went along; he wanted his herbarium written up by the time he sailed home. He also resolved to spend the voyage out (Flinders proposed the west coast as the *Investigator*'s first landfall) learning as much as anyone already knew of where he was going. He read Cook's journal. He studied the *hortus siccus* he'd made for himself from samples of Banks' huge collections and the other Australian herbaria taking shape in London, and he'd already identified a new grass among these before his ship left England.

The *Investigator* reached Australia on 8 December 1801, Brown stepping onto its soil for the first time the following day. Describing the moment in his journal, he was nothing but calm: 'the plants grew pretty freely in the loose sand their variety not so great as we expected'. He didn't go ashore the next day, noting instead that he 'remain[e]d on board' and 'describ[e]d a few plants'. December 11, he remarked, was 'more successful in Botanizing'.

However astonishing or exciting it may have been to stand among Australian flora, his journal entries remained brief, direct, unemotional. He was a man of science, facing a continent full of material that needed to be picked, pressed, dried, classified, and taken back to England. Dramatic phrases about 'nature reversed' were not for him: he would simply do his job. In his first two months, he collected upwards of 700 individual specimens (he estimated that nearly 500 came from that first landfall at King George Sound alone), including a beautiful pitcher plant, *Cephalotus follicularis*, with lusciously striped lips, 17 different sorts of banksia, and some tall kangaroo paws.

His herbarium grew steadily, as did the trouble he had preserving it. The problems anyone faced trying to keep plants dry on a ship were daunting. Space was always at a premium, which made it hard to carry much paper to press and dry specimens, let alone finding a spot to store everything. Mice, on the other hand, were plentiful – even with a famous sailing cat like Flinders' Trim aboard – and they liked to eat paper. In addition to which, the *Investigator* was one of the leakiest ships the Admiralty had ever sent out, making conditions even less ideal.

The voyage made its way along Australia's southern edge before plunging into Bass Strait (touching King Island), returning to mainland Port Phillip (not yet settled) and tracing the *Endeavour's* route up to Port Jackson. Kangaroo Island, for example, gave Brown 'a species of eucalypt not before seen'; Doubtful Island Bay gave him a parasitic native 'Christmas tree'. In two places – both of which he found 'too late in the season for botany' – sore legs prevented his going anywhere.

But in Sydney, as the *Investigator* was patched up, he was mobile again and, meeting Caley, they began to discuss what they had found: 'Mr Brown has informed me that this is *Elaeocarpus*,' Caley amended his notes; 'Mr Caley separates *Melaleuca laurina* from *Melaleuca*,' Brown wrote later in his, also noting that Caley had 'not yet begun to examine the trees' but 'seems to attend more to plants resembling British than those peculiar to the country'.

While picking and pressing everything from gum trees and cabbage palms to passion-flowers around Sydney's settlement, Brown hoped Banks might send more paper before Flinders' repaired ship headed north: 'The kind of paper I wish to have is Imperial brown paper. It is fully the size of cartridge, and in many respects is much superior, both for drying and preserving specimens, especially of the rigid shrubs of this country, and, what is no small advantage in our situation, mice do not eat it.'

They sailed again on 21 July 1802, spending eleven months travelling up past Fraser Island and the Great Barrier Reef to the Cape York Peninsula and the Gulf of Carpentaria. It was a dismal journey: Brown was stung by 'a single plant of *Morus incendiarius* ... having incautiously touch'd it' and some of his companions were half-poisoned with '[Arum] roots ... wch several of the people incautiously tasted & tho in very small quantity the effects were in some cases rather alarming. They were a continued flow of Saliva a[n]d heat & pain in the tongue, palate and fauces, so great that the patient was unable to speak for some hours.'

By December, the *Investigator* was leaking so badly she was unseaworthy, and scurvy was rife. Continuing to Arnhem Land, Melville Island and Timor, the crew contracted tropical dysentery as well, limping back into Sydney in early June 1803. There were more than 1,400 new dried specimens in Brown's *hortus siccus*.

When Flinders decided to sail to England in the *Porpoise* to ask for a new vessel, Brown combed his herbarium specimens for the finest to send with him for Banks. He'd made 66 landfalls, but was 'upon the whole disappointed ... The number of species of plants observ'd by us in New Holland, exclusive of the few which belong to the class Cryptogamia, scarcely amounts to 2,000, and of this number not more than 700 or 800 are nondescript. Even of these the far greater part are referable to genera already publish'd ... and a considerable proportion have been seen only in an imperfect state ...

'The *Porpoise*,' he also worried, '... is so much crowded that she can take but a very small part of the collection of specimens, and even this must be put in the hold. She is, moreover, so wet a ship that I am afraid, small as it is, it may suffer very materially in the passage.'

He was right. The ship sank a week out of Sydney and for Brown, 'the loss of the garden and specimens is to my department irreparable, for altho' I possess duplicates of almost all the specimens, yet those sent were by far the best'.

Flinders, meanwhile, having rowed 1,400 kilometres back to Sydney in a six-oared cutter, sailed again for England on the *Cumberland*. Arrested in Mauritius, he was not released until 1810: his expedition was over. Should there be a long delay between Flinders' departure and his return with the promised new ship, Brown had written to Banks, he would busy himself with collecting trips to Tasmania and around the colony.

Which he did.

In May 1805, he left Caley – now collecting eucalypts – and sailed for home in, of all ships, the again-patched *Investigator*. His luggage included another 1,200 specimens, packets of seed and a live wombat (who, when it saw people it liked, would put its forepaws up on their knee and, if invited, sleep in their lap). His herbarium, he informed Banks, was 'provided for, as well as the reduced state of the vessel and her crazy condition would admit ... while within the Tropics the plants were carefully examined & those that most requir[e]d it were chang[e]d into dry paper but such has been the wet state of the ship that they must again be suffering & that I fear considerably'. This time, his fears were unfounded: the specimens were hailed as 'by far the most excellent that ever resulted from any expedition'. Even Banks was moved: 'employed to gather in the Harvest from the Boundless Fields of nature,' he gushed, Flinders' team had 'reaped plentifully'.

Brown unpacked his *hortus siccus*. He had a book to write.

Settling down to work, it was clear he and Solander had had one experience in common. Both men's papers were thick with revisions, reversals, reconsiderations and reclassifications as they had tried to fit this new botany into any system that accommodated Europe's. It was (as one botanist had commented on Joseph Banks' Australian collections) as if one found oneself 'in a new world. He can scarcely meet with any fixed points from whence to draw his analogies ... not only the species themselves are new, but most of the genera, and even natural orders.' As for Brown's plan to work out and write up his specimens as he went, he had had to tell Banks as early as 1803 that although his descriptions 'amount to about 1,600[,] only a few of these ... are finish'd and none of them rewritten'.

In the first years of the nineteenth century, the world's plants were supposed to fit the sequence of families, tribes and so on devised by the Swedish botanist Linnaeus more than 50 years before. Famously based on a 'sexual system' categorising plants by pistils and stamens, some Englishmen had always struggled with it: one vicar claimed that endorsing it encouraged unauthorised sexual unions between people, and that an association with plants may lead to licentious love; the *Encyclopaedia Britannica* said succinctly that 'obscenity [was] the very basis' of it.

Unable to fit his specimens into its spaces, Robert Brown changed the course of modern botany by abandoning Linnaeus for a more 'natural' arrangement based on Jussieu's system, setting a new standard for the organisational structure of large botanical works. First, the ferns, then 19 families of monocotyledons (such as grasses and orchids), and 37 families of dicotyledons (including Australia's famous banksias and waratahs in the Proteaceae family), with the cycads nestling between the two. He also decided, ambitiously, to include not only his own specimens in the book but 'the generic and specific characters of all the plants known to be natives of New Holland', and in 1806 several botanists agreed that all other publications on 'New Holland plants' be suspended in anticipation of his.

Published in 1810, covering 464 genera and about 1,000 species, the *Prodromus Florae Novae Hollandiae et Insulae Van Diemen* was hailed as 'most excellent' and 'precious', with botany's leaders said to be 'astonished' by it. And decades later, it was still applauded: 'We are indebted to Brown's powers of generalisation for a plan of the entire flora, constructed out of fragmentary collections from its different districts, which requires but little correction from our increased knowledge, though necessarily very considerable amplification'.

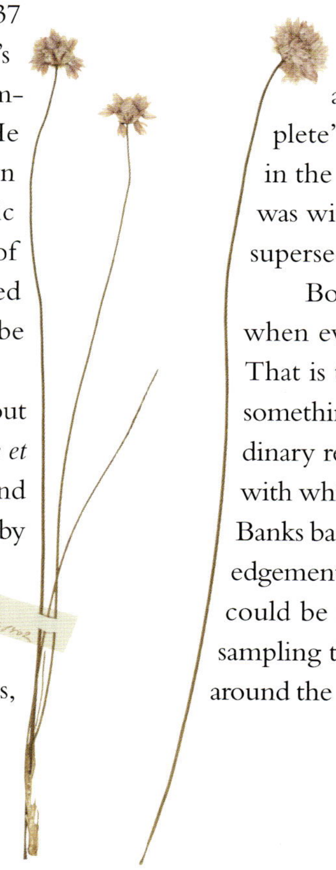

Unfortunately for Brown, the book didn't sell. Botany, as Banks had observed two years earlier, was no longer 'quite as fashionable' as it had been. And the *Prodromus* volume published was only part of the complete flora of Australia that Brown had planned; the rest remained unwritten. He got on with other things.

Still, even incomplete, it had set the formula for a wave of volumes that sought to catalogue the flora of entire regions, even entire countries; it also laid the foundation for Australia's systematic botany. In just over a decade, the number of Australian species that Brown knew had expanded from 370 to 4,200. Even his approach to those thousands of specimen sheets – using a microscope, comparing not just different specimens of the same species, but similar species worldwide – propelled botany's science forward.

Brown never finished that *Prodromus*, and in old age he was criticised for leaving those 'dried plants, or what remains of them,' from Flinders' voyage languishing in cases and 'buried in the recesses of his cabinets'. Years after his death in June 1858, botanists lamented his 'huge collections; the great proportion of which consisted of bundles that had never been opened & were never even dusted'. Yet as his specimens sat, others, collected by different people in different parts of Australia, filtered into botany's knowledge of the colony, the dried gardens of different herbaria expanding again and again.

Quests for completion – for a total set of this, a total exposition of that – are almost always doomed. By the time the next 'complete' description of Australia's plants was attempted in the 1860s and 1870s, its author, George Bentham, was wise enough to know that his volumes would be superseded the moment they were published.

Botany would never reach a conclusion, a moment when everything was known, set down and ruled off. That is the nature and truth of science: there is always something new waiting to be found, something extraordinary ready to challenge the patterns and explanations with which people have codified the world. For men like Banks back in the first decades of the 1800s, the acknowledgement of this simply meant finding more people who could be relied on to travel through these new worlds, sampling their plants and carefully sending them halfway around the globe to the old world's waiting enthusiasts.

Exploration

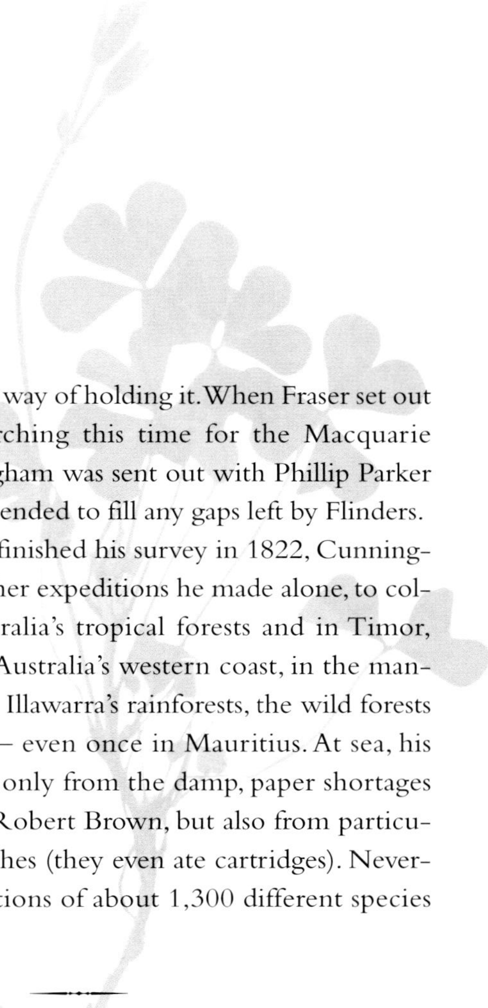

Species: Oxalis perennans Haw.
Common name: A Native Wood Sorrel
Family: OXALIDACEAE
Collector: L. Leichhardt
Locality: Archers Station, Durundur, east of Kilcoy [Queensland]
Date: August 1843

Even when Sydney was less than three decades old, the land on which its Botanic Gardens sat already had a significant British history. Occupying the site of the colony's first farm, just across from its first Government House, the poor soil wasn't the best for gardening, nor had anyone given the Gardens that most necessary facility for true botanical work, a herbarium of dried plants to supplement its living ones. But nonetheless they sprawled on land that ran down to the edge of a magnificent harbour and the very place where the First Fleet had come ashore: the position was superb.

In December 1816, more than a decade after Brown's departure and six years after Caley's, another Banksian botanist sailed into that harbour. Allan Cunningham presented himself to Governor Macquarie, who greeted him with the grand epithet of 'the King's botanist' and sent him out with John Oxley's expedition in search of the source of the Lachlan. Just as maritime expeditions included naturalists, so terrestrial parties were expected to collect, with an eye to anything that might be economically useful, to 'specimens of the most remarkable' plants, and to the seeds 'of any plants not hitherto known'. It would be 'the highest ambition of my life to exert myself in the perform[ance] of the requisite Duties that constitute a Collector,' Cunningham had told Banks when first applying to work for Kew. Making his careful way beside Oxley's slow carts, learning Australia's bush and the way its plants grew, he assembled over 400 herbarium specimens – waratahs, banksias, grevilleas, and more – and 150 boxes of seed: his ambition fulfilling itself.

His botanical companion on the trip was the fledgling Botanic Gardens' first superintendent, Charles Fraser, himself such an enthusiastic collector that he was known to strip off his clothes and carry specimens in them rather than abandon something for want of a way of holding it. When Fraser set out again with Oxley, searching this time for the Macquarie River's source, Cunningham was sent out with Phillip Parker King's coastal survey, intended to fill any gaps left by Flinders.

By the time King finished his survey in 1822, Cunningham had used it, and other expeditions he made alone, to collect in north-east Australia's tropical forests and in Timor, along arid stretches of Australia's western coast, in the mangroves of the north, the Illawarra's rainforests, the wild forests of Van Diemen's Land – even once in Mauritius. At sea, his specimens suffered not only from the damp, paper shortages and mould familiar to Robert Brown, but also from particularly ferocious cockroaches (they even ate cartridges). Nevertheless, he made collections of about 1,300 different species for his employers.

For nine more years Cunningham combed Australia and sometimes beyond – travelling to Norfolk Island, to New Zealand. He gathered and collected wherever he went: grevilleas and eucalypts, pines and acacias. On the day he was to return to England in 1831, a gale forced his ship to wait inside Sydney Harbour. Cunningham went ashore and found a perfect orchid specimen. He'd been hunting it for ten years.

It was Sydney's Botanic Gardens that brought him back, and their potential had clearly played on his mind. The possibility of extensive plantings; the possibility of a good herbarium – and all the scientific knowledge that could spring from this.

At the end of 1831, Charles Fraser had died, his final report stating that some 1,800 herbarium specimens had been sent to Glasgow's botanic gardens, 1,200 to Edinburgh's. There was still no suggestion of creating a *hortus siccus* of Australia's plants in Australia itself.

Cunningham, miserable in the cold damp of his first winter back in England, was recommended as the new superintendent. His health increasingly poor, he declined and instead recommended his brother Richard. But he did sit down and write a long memorandum on how the Gardens might work in the future. The superintendent should join all possible expeditions; the plants already growing should be classified and catalogued; and, finally, it would be highly desirable, he said, to begin a herbarium.

Richard not only took the job, he also took note of the memo, purchasing 'an excellent Botanical Press with patent screws, 22 inches long ... a most useful appendage ... in preparing for an Herbarium'. He also paid heed to his brother's endorsement of exploring.

In April 1835, he wandered off from Sir Thomas Mitchell's expedition as it headed into the dry west of New South Wales. Despite repeated cautions 'about the danger of losing sight of the party,' it was Richard's habit to peel off in search of plants. Unfortunately, this time, he was lost and while Mitchell later found his horse – dead – there was no sign of his botanist. Wandering delirious and disoriented, Richard had blundered into an Aboriginal camp and was clubbed to death.

The position of Sydney's superintendent again open, Allan Cunningham was again asked to apply for the job. He was back in Sydney by March 1837. But his plans for scientific scholarly work, and the possibilities of a herbarium, were a long way from what was really asked of the Gardens: that they provide vegetables for important people. It was, thundered the *Sydney Herald*, a scandal that 'a kitchen garden, under the pretext of being a Botanic Garden, is supported in Sydney at the expense of from £800 to £1000 a year', and Cunningham thought so too. After a very short period of time, he would, as the *Herald* put it, 'no longer consent to remain a mere cultivator of official cabbages and turnips, and ... resigned the management of the *Botanic* garden in disgust'. His replacement, James Anderson, was not a botanist. He was a gardener.

Increasingly frail, Cunningham made optimistic plans to join another survey of Australia's coast. But he wrote to Robert Brown, who after Banks' death in 1820 was hailed as the pillar of British botany, describing himself as 'a poor, decrepit, prematurely old traveller who ... formerly strove to advance, for years, botanic science here from pure love'. As the southern winter of 1839 crept in, he died.

The death of Banks may have also elevated Robert Brown to the position of Australian botany's greatest expert, but it had in no way stemmed the tide of interested collectors, botanists and otherwise adventurous souls heading south. If anything, the numbers increased as enthusiasts in other European countries took up the British passion for the continent's plants.

Less than three years after Cunningham's death, a young Prussian naturalist sailed into Port Jackson. Like Cunningham, he was both botanist and explorer. Like Cunningham, he had grand ideas of all that Sydney's Botanic Gardens might do and be. Like Cunningham, he would see it remain 'a kitchen garden'.

His name was Ludwig Leichhardt, and as he 'sprang ashore' on 14 February 1842, he could hardly describe 'with what joy I greeted every new plant, and how the wealth of novel sights almost turned my head'.

Through a somewhat peripatetic education (including studies with Brown's model, Jussieu, in Paris), Leichhardt had yearned, above all things, to become a discoverer. Among the many disciplines he had studied, he relished botany. Europe, he knew early on, was too small for him: 'You know I have a life full of perils before me in which success and failure alike depend only slightly on human foresight,' he had told his father.

But after years of casting through different courses of study, different thoughts on where to travel, Sydney, it seemed, was about to offer Leichhardt something particular and fixed. Shortly after his arrival, James Anderson, Cunningham's successor at the Gardens, died. Leichhardt was encouraged to apply for the position which, friends assured him, was bound to be his. 'Leave no stone unturned,' one counselled, and Leichhardt didn't. He spoke to people who might be able to help. He wrote long letters outlining his suitability.

'My mind,' he confessed, 'was teeming with the scientific possibilities and I would have accepted the position even on a low salary for the sake of securing a point of support from which I could make myself better known'. But the dead man, as Leichhardt noted, was 'an ordinary gardener, a man without scientific knowledge,' and that – unfortunately, perhaps, not only for the Gardens but for Leichhardt too – was precisely the sort of replacement chosen. Leichhardt may have advocated preserving Australian specimens in Australian herbaria, but the colonial bureaucracy didn't see the need for it. They gave the job to William Naismith Robertson, who'd been principal gardener to New South Wales' powerful Macarthur family for years.

Leichhardt's gaze shifted further afield.

From Sydney, he worked his way up the coast: Newcastle, the Hunter Valley, the Liverpool Plains ('one of those areas,' he felt, 'that still hold out the hope of something new to the botanist'), on to Moreton Bay and inland to places like Durundur, where he found an elegant little oxalis. He learnt new forest trees – could one tell different species of eucalypt by smell, he wondered? Would 'a good nose, with a good nasal memory … distinguish them easily'? He found new waterlilies, new ferns, new 'leguminous plants, some very pretty ones'. He saw mighty bunyas 'lift their majestic heads, like pillars of the blue vault of heaven'.

'You can perhaps form some idea of what hard work it is,' he wrote to an English friend about gathering, preparing and classifying herbarium specimens in Australia's bush:

> The character of the bark is often so indeterminate, particularly amongst closely allied species, that you have often to fell a tree to make sure that it differs from another. If you've a black with you he can climb up and throw branches down; but … it's very hard to get them to do much climbing … If they see an opossum or a beehive right at the top of a tree 180 [feet] high, they will at once try to get it; but, because they don't understand why I want branches and fruit, they hold back obstinately as long as they can …

Unassisted, he had to press and dry plants (often 300 or 400 of them), chop down trees for timber specimens, even skin and boil animals for the skeletons that people requested. He was 'cook, groom, washerwoman and naturalist all together'.

When Naismith Robertson died in 1844 ('he suffers rheumatism and probably of dropsy,' Leichhardt had diagnosed, returning to Sydney and seeing him 'on his deathbed'), the superintendent's position was vacant again, and Leichhardt again applied, unsuccessfully. Half the Committee of the Gardens may have found 'Dr Leichhardt to be a person more eminently well qualified for the situation than we are likely again to meet with' – and certainly 'something far superior to a mere Botanical Collector' – but he was a foreigner. England's pre-eminence in 'the Arts and Sciences,' the more liberal of the Committee felt, was 'certainly not to be attributed to the exclusion of scientific men merely because they happen to be foreigners'. However even Leichhardt, seeing the politics of the situation, knew the job wouldn't be his. And this time, he had other things on his mind. He wanted to mount an overland expedition to Port Essington, at the very top of Australia.

From Moreton Bay, he went north, leading a handful of men out beyond any maps, and out beyond anywhere botanical science had touched. It was like arriving in a new country again, bagging things by the armful. But the journey dragged: it was eight months, ten, then a year. His men weakened; one died. His animals failed; his provisions dwindled. There was 'nothing but the blue skies above me and the bush of New Holland around me'. He reached the worst day:

> Our pack bullocks were overloaded, and it was now imperative upon me to travel as lightly as possible. Thus I parted with my paper for drying plants, with my specimens of wood, with a small collection of rocks … and with all the duplicates of our zoological specimens. Necessity alone, which compelled me to take this step, reconciled me to the loss.

It wasn't enough. Two weeks later, three horses drowned:

> This disastrous event staggered me, and for a moment I turned almost giddy; but there was no help. Unable to increase the load of my bullocks, I was obliged to leave that part of my botanical collection which had been carried by one of the horses. The fruit of many a day's work was consigned to the fire; and tears were in my eyes when I saw one of the most interesting results of my expedition vanish into smoke.

How snappily his herbarium had burnt – that collection he knew was 'almost complete in blossoms, fruit and seed'. More than 3,000 specimens, sputtering and crackling and shrinking to ash. He broke camp, walked away, and five days later gave up even his leather collection case, throwing it into a wallaby soup 'to render the broth more substantial'.

If, another fortnight on, the bullock carrying what little remained of his herbarium left Leichhardt 'almost crying with vexation' when it took itself for a swim, the fact that the horse that next took the specimens got itself bogged in 'a saltwater and Mangrove swamp' was almost too much.

Worst of all, perhaps, was to keep going. Here was a new tree with flat green fruit; there, one with oval blue fruit. All in all, 'many new trees and plants are observed, but I am almost dead. I despair of breaking even a specimen.'

After fourteen-and-a-half months, he reached Port Essington, assumed lost by all but the most persistent optimists in Sydney and still wishing for his abandoned herbarium. He'd make up for it: 'I hope my next expedition will be a famous one for botany,' he said on his return, wishing also 'very much to establish a good well-named herbarium in the Museum in Sydney, that we have some means of ready comparison'. What was left of his specimens he sent to a botanist in Paris: 'send back a whole set of the collection I send,' he asked, 'in order that it may facilitate not only my future studies but those of other botanists who will follow my path'. He would give Australia a *hortus siccus* yet.

Next, he announced, he would skirt the Gulf of Carpentaria 'again at the head of its waters', cross to the north-west coast, and 'travel down to Swan-River [now Perth] parallel to the coast'. From which continental crossing, he would return 'like a bee loaden with rare treasures'. He would concentrate on seeds, he said, the half ream of paper he would limit himself to making it 'scarcely ... possible to collect more than one or two sets of specimens'. For very rare plants he'd stretch to 'perhaps five or six'. From his practice of having everyone in his party collect, he knew that 'those who are not versed in botany will only take the most showy objects'. This left his eyes peeled for those plants 'less conspicuous but perhaps equally interesting'. And every three or four weeks, he planned to pause and sort all the specimens, throwing away those of which there were too many.

He made his preparations, selected his men, wished for camels, found none, and then he went out – twice, as it happened, having to abort the expedition the first time after illness and dissent ran through his party as they entered unknown country. On his second attempt, he rode out in December 1847, stopping a while later with a Mr W. P. Gordon who later recalled the night with pleasure. Leichhardt 'kindly showed me his maps,' Gordon remembered, 'pointing out the route he intended ... and where he hoped to find water in the unexplored parts'.

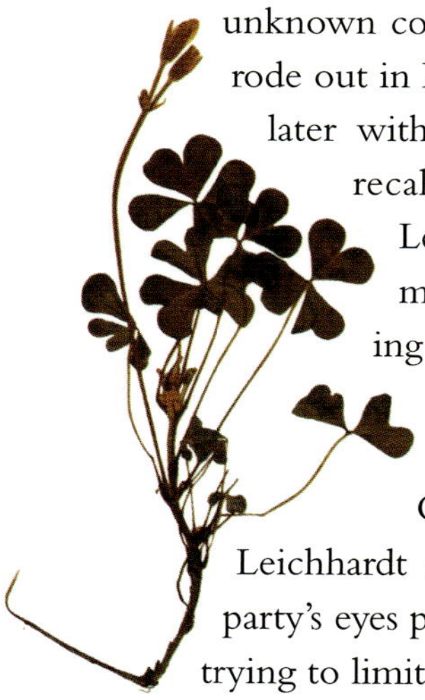

On 3 April 1848, Ludwig Leichhardt crossed over Mt Abundance, his party's eyes peeled for botanical riches and he, trying to limit its load, ready to drop those little heaps of abandoned specimens every three or four weeks. Marking a path.

The party was never seen again.

W. P. Gordon said he knew what had happened to the group and its tidy collections of plants and seeds. He heard it from some blackfellas out on Wullumbilla Station, near Roma in Queensland. They'd seen a white man moving with strange animals (Leichhardt's mules; Leichhardt's bullocks) along the Maranoa River. One night, 'the blacks had mustered in a large body, had surrounded the sleeping camp and speared everyone in it ...

'As for any relics of Leichhardt's equipment,' Gordon concluded, 'all that would survive the first bushfire would be scraps of iron from the pack saddle buckles and such remnants would not show above the grass. None but a black who had actually seen the camp could hope to find the spot a few years later.'

Leichhardt had wanted the job at Sydney's Gardens 'for the sake of securing a point of support from which I could make myself better known'. There was no question of his fame now. Portraits were published; some were defamatory. Searches were organised; they were never successful. Poems were written about him. Ships and suburbs were named for him. Some said Leichhardt had founded Utopia in the continent's centre. Some claimed to know where his last camp was, but always came back empty-handed.

The Gardens in Sydney remained as they were, beds of plants, with no herbarium and no foundation for science – although the specimens Leichhardt had urged his French botanist to hurry and return were sent back to Australia. They ended up in Melbourne where the government had felt itself able to appoint a foreigner as Colonial Botanist: a German, Ferdinand Mueller. Mounted and identified, these plants Leichhardt had so carefully picked and dried and carried did become part of the first, great Australian herbarium – the one established by Mueller in the Victorian capital during the second half of the nineteenth century.

Women

Item: Louisa Atkinson's fern pressings – various species, collected in various localities around New South Wales, c. 1860s.

Still safely under British directorship, Sydney's Gardens continued to send its specimens back to England for identification (assuming the position in 1848, Charles Moore would become the Gardens' longest-serving leader). At the same time, in Melbourne, Mueller began his own long encouragement of the investigation and classification of Australian plants in Australia. And if there was one group of the population he sought to cultivate as submitters of specimens from around the country, it was the ladies.

Determined to collect as many different plants from as many different parts of the continent as possible, and only able to make so many journeys and expeditions himself, he knew the importance of encouraging collectors. The women of Australia, he felt, could be particularly useful: at one point, he advertised in the *West Australian* asking readers to send him the flowers, fruit and bark of the plants amongst which they lived. This might prove a useful occupation, he suggested, for 'many ladies living in these far distant parts of the colony, bereft, to a great extent, of those intellectual resources to which many of them have been accustomed'.

In a world where so many occupations were proscribed and prohibited for young ladies, botany had always been regarded as a suitable source of what was gently called 'rational amusement' – even when based on Linnaeus' naughty sexual system. As one reverend gentleman put it, it was a pastime 'particularly fitted, to attract the attention of the fair sex ... who admire the beauties of nature, and tend to them with womanly care and anxiety'.

Circumstances had delivered a number of botanically curious women to Britain's Australia, starting with the energetic Elizabeth Macarthur who, having sailed into Sydney Cove (with her equally energetic husband, John Macarthur) in 1790, found herself 'anxious to learn some easy science to fill up the vacuum of many a solitary day'. An officer of the marines obliged with lessons in botany and astronomy, and Macarthur's perception of the landscape she had first met as 'all ... novelty ... [and] noticed with a degree of eager curiosity and perturbation' began to change – subsiding, as she put it, into 'calmness'. Soon she was reporting 'a small progress' in her new disciplines. 'No country can exhibit a more copious field of botanical knowledge than this,' she said, learning Linnean ideas and categories and beginning to 'class and order all the common plants'. To Macarthur's eye, 'everything was new ... every bird, every insect, flower,' and she was intrigued to watch and learn the way the nature of her new place turned and behaved.

Forty years later, on the far, western side of the continent, Georgiana Molloy stepped ashore on the coast Robert Brown had pronounced to be 'by far the richest in new and remarkable plants' in Australia. Like most settlers, Molloy was more immediately interested in trying to grow English plants in her new surroundings, but the beauty of the indigenous flowers soon caught her eye. After only a year she was studying her garden so closely as to notice one 'remarkable feature in the Botany of this country ... the numerous kind of leaves with identical flowers'. Yet when she was recommended as someone who might collect for James Mangles, a retired Royal Navy captain in England with botanical aspirations, she demurred, describing herself as one 'who does not enter the lists of a Florist, much less a Botanist'. He sent her a gift – an English *hortus siccus* and a box of English seeds – and she began to consider his proposition. The notion of collecting grabbed at her, and his antipodean collection was underway.

Taking her children out with her (their eyes, closer to the ground, might notice plants she missed, she said), Molloy collected, annotated and despatched her Australian specimens to the other side of the world. Thank you, she wrote, 'for being the cause of my more immediate acquaintance with the nature and variety of these plants'. When her family moved, there was new vegetation to choose from, including something 'I have almost been panting for, a very small, neat white blossom, on a furze looking bush'.

A collecting trip was, she said, 'one of the most delightful states of existence, free from every household care, my husband and children, all I possess on Earth, about me'. When her botanising was occasionally interrupted, she described it as 'all my former pursuits ... necessarily ... thrown aside (by the peremptory demand of my personal attention to my children and domestic drudgery)'. Her husband, she said, had brought her 'a bouquet of beautiful scarlet flowers' once when she was ill – which she dried, 'and which please God I ever get out again I shall send and mark. I was surprised by my illness to receive a Nosegay from a Native who was aware of my floral passion. These are under preparation for you,' she assured Mangles. There was nothing that would not be sent.

But when *A Sketch of the Vegetation of the Swan River Colony* was published in England, drawing on specimens she had shipped to him, she was not acknowledged. And his name, not hers, was honoured in the epithet decided for the beautiful kangaroo paw she had sent him in 1837. (Previously unclassified by science, *Anigozanthos manglesii* would later become the floral emblem for Western Australia.)

Still, as she wrote her last letter to Mangles, after seven years of correspondence and just before she died (at only 38, in 1843), she believed she had 'sent [him] everything worth sending'.

Mueller, at least, seemed better at acknowledging the often arduous work of collecting and preparing specimens that his network of collectors, both male and female, undertook. 'I hope this acknowledgement will encourage you to continue your searches,' he would write to one woman later in the 1880s as he named a species for her. It was a shame, he felt, that there were 'very few Ladies in all Australia, who have any taste for botanic science, in contrast to what is observed in all Europe and North America'. One he did find, and come to rely on, was immortalised in a new genus, *Atkinsonia*.

Miss Louisa Atkinson (Mrs Calvert, when she married one James Snowdon Calvert, who had been in Leichhardt's expedition to Port Essington) did better than most women in terms of official recognition for her botanising. As well as Mueller's new genus, she was honoured by the names of three other taxa. Of all the herbarium specimens she sent to him (there were more than 600), over a hundred were shipped to London to be used by George Bentham in his *Flora Australiensis*, that next attempt to classify the whole country's flora after Robert Brown's abandoned *Prodromus*.

Her fascination with nature was complete. She wrote about it (in articles, which appeared in the *Sydney Morning Herald* from 1860, and in novels); she drew it; and she made long excursions into it on foot and by horse, topping off days of collecting with 'employment for the evening in laying [her plants] in the press'. From one trip to Kurrajong's waterfalls in the Blue Mountains west of Sydney, she came home with 26 different sorts of fern. Another, to the abandoned settlement of Maroota Forest, gave her a entirely new batch of treasures:

> The trees, chiefly eucalypti, were low and stunted, but the scrub, even at this season, was decked with flowers, many of them new to me, and which, owing to the assiduity and observation of one of my friends, on the homeward way yielded specimens, and quickly filled the wallet suspended round my shoulders.

Partly her enthusiasm was driven by a compulsion to observe and sample whatever the bush offered up. But partly too, as Darwin's scandalous notions of evolution began to rattle the world of natural history, she was celebrating a firm and fixed sense of 'nature ... as an outward manifestation of Him – the source of all perfection and beauty'.

Of all the plants she collected to press and preserve, it was the ferns she loved most. 'Scrambling over rocks, fallen trees, leaping the creek, and so on, a mile or so was passed,' she wrote of one journey,

> when a tuft of pendous [sic] green fronds caught my eye. Fern-gatherers will understand the eager nervous fingers which grasped them, as if they were a living thing which would presently elude the extended hand; not so, however. The prize, an asplenium, was one not found by me before, with long attenuated fronds, of a dark green, not divided.

Some she prepared as herbarium specimens (usually for Mueller), some she sketched, and others she pressed and dried and arranged in the pages of her albums, such as the ones her granddaughter (Janet Cosh, herself a naturalist) later gave to the library of Sydney's Herbarium.

This nexus of botany and art was where the work of many women collectors did achieve recognition, although often for their depictions of plants, rather than the rare specimens with which these were inspired. Serious botanists may have despaired of hobby albums, such as Atkinson's, that mounted different species on the same page and paid more attention to arrangement than taxonomy. But the people who created these were also contributing to Australia's botany and its stash of herbarium specimens.

The painters, too, enlarged herbarium repositories. Marian Ellis Rowan, whose depictions of Australian plants beat canvasses by Tom Roberts and Frederick McCubbin to the gold medal at Australia's 1888 Centennial Exhibition, is rarely remembered as one of Mueller's collectors, although she traded the scientific identifications he provided for her images for rare and striking blossoms picked at the farthest points of her travels – places he would never visit. 'My love for the *flora* of Australia, at once so unique and so fascinating, has carried me into other Colonies, Queensland, and some of the remotest parts of the great Continent of Australia,' she said. 'The excitement of seeking and the delight of finding rare or even unknown specimens abundantly compensated me for all the difficulties, fatigue and hardships.' As she worked her way into Queensland's north and beyond, she sketched her flowers, revelling in their fresh, bright colours, before pressing them into the specimens to which Mueller, in Melbourne, would assign names.

It's the way of herbaria that collectors often disappear behind the scholarship of the publishing author, the classifying botanist historically more eminent than whoever found the plant for them in the first place. Even Mueller, who did recognise the efforts of many who contributed to his enormous *hortus siccus*, placed far more emphasis on the scientist who named the plant – which was often him. Bentham's *Flora Australiensis* at least acknowledged 'collections presented to Dr Mueller by friends, chiefly resident in Australia … amongst these numerous amateur contributions, I notice those of … Miss Louisa Atkinson from the Blue Mountains'.

Perhaps the esteem in which the country's early female collectors were held, and the recognition given to their contributions to botany, might have been better if they were men – as has been suggested about Louisa Meredith, another of Australia's finest observers of flora and fauna. Much of the early correspondence of female collectors was lost (a large portion of Mueller's correspondence with women collectors and botanists was recycled as part of the war effort in the 1940s), with many of their stories only recovered later in the twentieth century. But Australian herbaria bristle with their specimen sheets. They are the evidence not only of how, where and which plants used to grow in the Australian colonies' first hundred years, but also of the variety of people who, stopping to collect them, propelled knowledge forward.

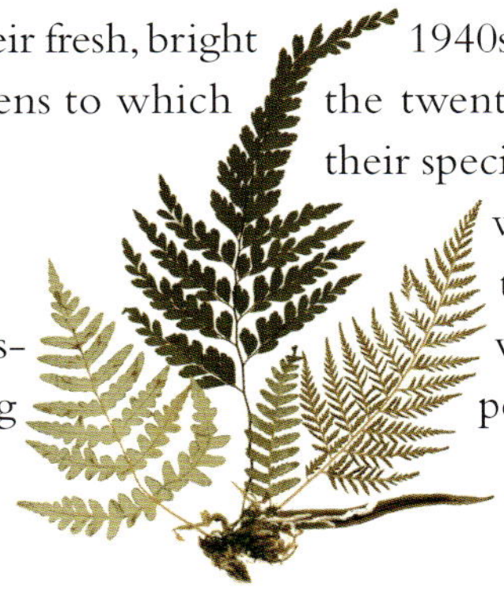

For hire

Species: Heterosiphonia muelleri (Sonder) De Toni
Common name: A Red Seaweed
Collector: W. H. Harvey
Locality: King George Sound, Western Australia
Date: 1854

In colonial Australia's brief botanical history, there had always been those who collected its plants for financial return. And as long as there was an overseas market for Australian herbarium specimens, there were people willing to sail halfway around the world to procure them. In 1837 in Hamburg, Ludwig Preiss tried to sell shares in a collecting tour, promising 'dried plants' (shareholders were requested 'to denote the names of those families, which are of special interest') as well as live plants, seeds, mammals, molluscs and more.

With less backing than he'd hoped (he'd hoped, for instance, to employ a number of assistants), Preiss sailed for west Australia, spending time with Georgiana Molloy and promising to send her duplicates of the specimens he collected. This he failed to do. The British government declined to buy his full collection for £3,000, upon which he offered its finest set to Robert Brown for £200, addressing him solemnly as 'you Father of the Australian Flora'.

For some people, selling specimens was a way of making money in a young settlement. For others – even internationally renowned scholars like William Harvey, Professor of Botany at Dublin – it was a means of financing expeditions.

Leaving Australia (where he found a red seaweed that took Mueller's name) and the South Pacific for South America, Harvey proposed a deal to Kew: 'I have heard you say [Mendoza] is famous for *Cacti* – & I should willingly collect a quantity for *Kew*, [...] – it being understood that I should be allowed to sell the duplicates as remuneration for my trouble, of course after giving you what you want'. Similarly, he was hoping to have an '*Acting Keeper of the Herbarium* appointed' to his staff back in Dublin, 'who will look to my collections as they arrive; & who can distribute the Algae of Australia to subscribers – so as to bring in some *money* for my travels'. Sets of 100 specimens were subsequently sold for £2 5s 0d.

For Amalie Dietrich, employment as a paid collector of curiosities for the Godeffroy Museum in Hamburg certainly meant an end to financial troubles. The only woman offered such a position, she'd learnt her plants from her mother and how to preserve herbarium specimens from the husband who later abandoned her. The reference she presented to the Godeffroy described her as 'a diligent botanist', her collections 'always commendable ... carefully preserved and ordered with taste and understanding'. On top of which, she had an 'unusual talent for her profession, a sharp well-trained eye for everything nature has to offer, a great certainty in identifying the collected material. On her long and mainly difficult journeys,' her referee concluded, 'she has constantly shown great perseverance and bravery' – skills she would no doubt need working on her own on the edges of Australia's settlements, and beyond, for close to a decade. In 1863 she arrived in Queensland to fulfil her contract: the Museum wanted plants, for one thing, but also snakes, skeletons and spiders as well, and everything in between.

If, for the majority of women who compiled or contributed to herbaria, botany was seen in terms of pleasure or virtue rather than professional occupation, Dietrich's luggage alone dispelled any idea of genteel collections. She had, amongst other things, gunpowder, percussion caps and retorts, bottles for live snakes, insect pins, quires of paper for pressing plants, a microscope, two pounds of plaster of Paris, and two boxes of poison. Heading north from Moreton Bay through Gladstone and on to Rockhampton, she sent case after case of specimens back to Germany: ferns, 'amongst which I disappear entirely'; and large orchids that hung 'by almost invisible threads'. These were 'so wonderfully formed, have such beautiful colour, and look at me so mysteriously, that I pick them with a certain awe'.

Through the end of the 1860s and the early 1870s, she went farther north, through Mackay and beyond, and out to the Holborn Islands, before sailing down to Sydney, whose Botanic Gardens (still, in 1872, without a meaningful herbarium) she found 'splendid' in their combination of 'the loveliest plants of tropical and subtropical vegetation'. Of the thousands of items sent to the Godeffroy, the herbarium specimens were her favourites and perhaps the least risky to obtain – although there had been one luscious blue waterlily for which she was said to have risked her life in a swamp. Alongside bagging the world's first taipan, and killing and gutting a number of crocodiles, pressing plants must have been less of a challenge. 'Frau Dietrich continues to work admirably,' her employer wrote in 1867. 'I expect much more from the achievements of this woman.' Although she never published a paper, Dietrich's work as a collector was honoured with medals, with membership of elite scientific societies, and in several species' names (including *Acacia dietrichiana*, published by Mueller). In the late 1870s, she was received at a meeting of Berlin's all-male Anthropological Society, a 'badly dressed woman' in 'worn canvas shoes'. 'Here is Amalie Dietrich,' one of Germany's leading scholars exclaimed as he brought her in, 'who I believe deserves a place of honour amongst us'.

The years Dietrich spent collecting across Queensland overlapped with the career of another German collector-for-hire. Wilhelm Baeuerlen sailed for Australia at the age of 23, although there's no record of either his arrival or the person or event that sparked his lifelong commitment to botanical collection. By 1883 he'd been absorbed into the great web of Mueller's plant-hunters, with Mueller praising him as 'zealous and circumspect'. (Mueller named species in five genera for his zealous collector: *Correa baeuerlenii*, *Pultenaea baeuerlenii*, *Haloragis baeuerlenii*, *Eucalyptus baeuerlenii*, and finally *Acacia baeuerlenii*.)

In 1885, sporting a generous and wiry-looking beard, Baeuerlen went to Papua New Guinea's Fly River: all those 'new, rare and beautiful plants', he told rapt audiences on his return, that 'must be seen in their fresh and living beauty, as dried and preserved specimens give but a poor idea of what those things really are'. At this time, too, he began to send specimens to J. H. Maiden whose herbarium at Sydney's Technological Museum (with a focus on plants that valued their oils, resins, and usefulness in agriculture as much as their purely botanical aspects) was growing in leaps and bounds while the Botanic Gardens remained bereft.

Maiden, too, praised Baeuerlen as a 'painstaking botanical collector' (he was so enthusiastic that he worked on his wedding day), announcing in April 1890 that he was 'exceedingly anxious [Baeuerlen] should go to the northern rivers' of New South Wales to the rainforest known as the 'Big Scrub'.

'Mr Baeuerlen, the botanic collector, has remained on the Richmond River during the past year and has enriched the central and local technological museums with numerous specimens of great value,' Maiden reported at the beginning of 1892. 'I am anxious that he shall remain on the Richmond for a few months longer in order to complete all the material.' As it happened, the area would keep Baeuerlen occupied for long periods spanning from February 1891 all the way to June 1898.

His output was prodigious: Maiden received long letters and even longer specimen notes, often more than one on the same day, in addition to box after box of annotated herbarium specimens and timber samples. And all this, as Baeuerlen liked to point out, was achieved under very discommoding circumstances.

To begin with, there was the problem of accommodation. Initially, he took a room at a pub, with nowhere to work properly, nowhere even to unpack his books. When a house he'd hoped to rent 'out in the country' fell through, he complained again of having to 'work under very unfavourable circumstances, as there is continually a good deal of noise etc going on ... and all manner of inconvenience'.

There were smaller tribulations too. Constant rain meant his herbarium specimens 'require more attention than usual': 'the weather is so warm and wet,' he explained, 'I have to use a good deal more [paper] ... and find it quite necessary to change the papers every day to get on with drying the specimens'. To try to solve the problem, he 'would beg and steal every old newspaper I can lay hold of'.

There was the fact that the resin of trees (in which Maiden was particularly interested) was often hard to acquire: this one with gum 'rather thin, so as to make it difficult to lift with a knife'; that one, 'falling on my head which left a thin brown stain on a white hat and coat I wore'; and another, 'so forcefully astringent' that, when 'some particles dropped in my eye', they caused 'intolerable pain'. Thanks to this incident, he said, 'I may have to stay a few days indoors, though I am glad to say [my eyes] are a good deal better already this morning and may get pretty well again over Sunday. It is also raining today so that there is no bright sunlight which,' he felt, 'is an advantage.'

Then there were problems purchasing timber specimens. 'They all seem to be a lot of sharpers up here,' he commented of the local loggers, 'and think for anything that is done for the government they can just ask what they like. With the sawmills I can do nothing at all ... to get a pine log 4ft long one would think is an impossibility, if one listens to them.'

But all of these difficulties, even in combination, were absolutely nothing compared to the leeches, as Baeuerlen explained to Maiden after two months' work in the region:

> Other localities give you no idea of the leeches compared to what they are here. I always take the precaution you mention [a cautionary tuck of trouser into sock], as that wrinkle is known to me a good few years, but here they come to me from the shrubs, not only from the ground, and I get them on my arms face and neck, I have even had one on my tongue! ... I should think that I am about 10 times more liable to the attacks of leeches than any ordinary bushman, on account of having to proceed slowly as a rule, examining trees and shrubs, and sitting down to sort and lay out specimens between the papers. It requires constant attention to oneself to reduce their attack to a minimum, and in the eagerness of search one is apt to forget all about them until the mischief is done ... I find it quite necessary to drink wine every day in order to keep up my strength, so as to make blood again ...

In 1896, part way through the years of Baeuerlen's visits to the Big Scrub, Maiden left the Museum to assume directorship of Sydney's Botanic Gardens, having endorsed him as '*very good* value', 'most diligent in looking for specified things I send him after'. The samples Baeuerlen sent back were always labelled, described, and enlarged with information contributed by everyone from the druggist and the storeman, to timber-workers and a Mrs Seccombe from Alstonville (an authority on how much sugar was needed to make jam from the local native blackcurrants; Baeuerlen had even included a jar with some herbarium specimens he was sending to Maiden at the Museum). He had often suggested identifications of the plants he picked and pressed; these were often correct.

But without Maiden, his relationship with the Museum soured.

'I cannot understand how, from the sketch I gave you, that the tree ... can be considered to be *below* the Falls,' the new deputy at the Technological Museum chastised him during an 1899 trip to collect eucalypt leaves for distillation. 'If you had gone the way I marked on the plan,' he added, 'you could not have missed it. I will draw it again on the back of this letter.' Eighteen bags of leaves subsequently arrived in Sydney.

When, in the early twentieth century, the Museum staff wanted him to concentrate on collecting pine specimens for their research into its properties, Baeuerlen had had enough: 'I am the Botanical and not merely the Economic Collector, much less merely the drudge for the Chemist, as you, for reasons of your own have been trying to make me.' Under Maiden, he reminded his new superiors, 'I always also had carte blanche for my movements, when I was out collecting'. Now, no one cared for anything he might pick except 'Pines, pines, pines, pines, pines, pines, pines, pines, pines, pines and nothing else but pines'.

Suspended, reprimanded, and then placed on probation, Baeuerlen had resigned by the end of June 1905, suing the Museum for keeping some of his books and notes. He represented himself in front of a bemused judge in 1908. A pension, Baeuerlen informed him, was not enough compensation and, as the *Evening News* reported, he 'continued his narrative in a low, level monotone, dilating about fungus growths and fungus phenomena, until at last His Honour lost his patience. "Oh that is enough," said he, "quite enough. For goodness' sake do cease. I am not a fungus."' And he dismissed the case.

Intending in his retirement to investigate material he had 'discovered and brought to light', Baeuerlen was refused even 'small quantities' of the specimens he'd assembled for the Museum. 'My service, my years, and the public's money is to be wasted,' he concluded, 'and the industrial probabilities and possibilities of our natural resources, especially our forests, are neglected in this scandalous fashion'. Furthermore, he found it scandalous that 'so many of our timbers remain uninvestigated, and some trees several hundred feet in height have not a botanical name yet!'

But whatever became of the material for which he had predicted so much potential, his extensive collections from the Big Scrub did survive and stand as the only full floral record of this remarkable, and now mostly destroyed, rainforest in the north-east of New South Wales. (Previously the country's largest single example of sub-tropical rainforest, covering 75,000 hectares, it's now estimated at less than one per cent of its former extent, mostly cleared for agriculture and other development.) All those specimens, so tedious to dry in all that rain, finally, like Maiden, went to the National Herbarium at Sydney's Royal Botanic Gardens. Incorporated into its vast *hortus siccus* in 1979, they now underpin conservation botany for the area.

If neither Baeuerlen nor his collections could follow Maiden directly to the Gardens, fragments of the Big Scrub could and did make their way there: among the 11,385 letters of Maiden's directorial correspondence in 1916 alone (a year before Baeuerlen's death) came one from a neighbour of the jam-making Mrs Seccombe, at 'Alstonville, near Lismore'. 'Even in this unclassical age,' wrote this Dr Tomlins, 'most people are acquainted with the democratic words of Terence: "I am a man, and I count nothing human alien to me" and to you as a botanist nothing can fail to be of interest. So, in a spell of fine, shockingly-healthy weather' – the likes of which poor Baeuerlen never seemed to encounter – 'I have made a list of the trees, shrubs, bulbs &c, in my garden, thinking that in a leisure time, you may come to peruse it'. Five months later, he offered 'some hundreds of botanical specimens' for Maiden's herbarium, warning that 'unfortunately the rats and mice have ruined some and damaged many'. Accepting them as 'very useful,' Maiden incorporated them into his boxes of sheets at the Gardens.

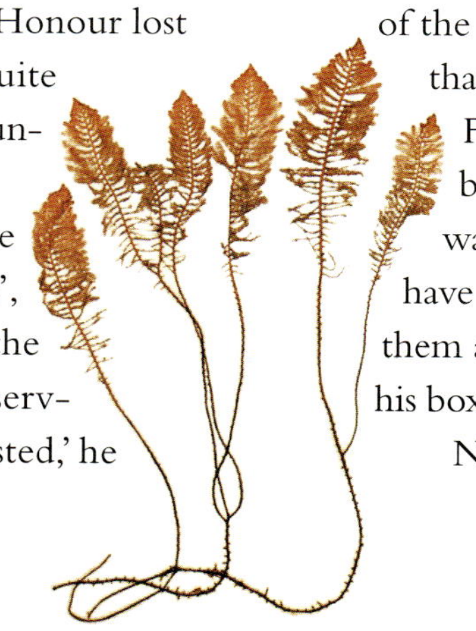

Neither man mentioned leeches.

Science

Species: Rubus rosifolius Sm. var. rosifolius
Common name: Rose-leaf Bramble, Native Raspberry
Family: ROSACEAE
Collector: J. H. Maiden
Locality: George's River Swamp, New South Wales
Date: 25 August 1888

J. H. Maiden's move from Sydney's Technological Museum to its Botanic Gardens in 1896 had a dramatic impact not only on Wilhelm Baeuerlen's collecting life but – more broadly – on the future of the Gardens themselves. Finally, there would be a herbarium. Finally, the Gardens would have their own catalogue, their own *hortus siccus*, of Australian plants. Finally, the Gardens would become the scientific institution that people such as Cunningham and Leichhardt had imagined.

Charles Moore had enjoyed 48 years in the top job (his appointment as director, by the government, had put an end to the power of the Committee that had twice rejected Leichhardt), his career spanning more than half of the Gardens' existence to date. As he retired, the energetic young Mr Maiden took over.

Thirty-five years old, he'd been in Australia for 15 years, having arrived as a botanical enthusiast and quickly taken a position in Sydney's young and faltering Technological Museum. This institution he had seen through disaster (the destruction of most of its collections by fire) and other dismal circumstances (it was housed for years in half a tin shed on the edge of Sydney's parkland Domain) before he left it – enlarged and enriched, and with a herbarium bulging with sheets – in a spanking new building in Ultimo.

The Gardens, in comparison, Maiden found poor: ramshackle buildings (years in the tin shed made him determined never to suffer such accommodation again); disappointing archives (a bundle of letters including some from Banks and Mueller had only been saved from one of Moore's clean-ups by the attention of a man employed to write plant labels); even the number of gas lamps and public latrines was inadequate. As for a herbarium, it was virtually nonexistent.

He was, Maiden confided to Kew, appalled by the meagre 'thousand or two named species ... [and] no exotics for my predecessor *would have no exotics in the herbarium*. There were not even pieces of the plants under cultivation in this Garden, some of which are no longer in cultivation in Australia & some of which are of highly scientific interest.'

Moore, assessing the Gardens' facilities on his arrival in 1848, was known to have pronounced them 'utterly destitute' of a herbarium – without 'a single specimen'. In Maiden's view, nothing much had changed in the 48 years that followed. Because Moore had sent most of his specimens to England for classification, Maiden, for all his searches, could turn up only a dozen portfolios of plants, each holding 150 pages.

To make matters worse, some of the very few collections made by white Australia's pioneering botanists and remaining in Sydney – a few of Charles Fraser's and Allan Cunningham's – had been damaged by storage 'in a damp room from 1870–4 and were burnt'. In 1852, Sir Thomas Mitchell had offered a sizeable set of sub-tropical specimens collected in northern Australia during his fourth and final expedition, which he modestly described as the 'proceeds of some personal trouble', observing that 'scientific arrangement [was] the first step toward the cultivation and domestication of the indigenous shrubs of their country, a duty one owes to his Creator and to himself'. These plants were found in an old seed store.

Had he not brought large numbers of specimens with him from the Technological Museum, said Maiden, 'I do not know what I should have done. Every moment I could spare I have been in the field. Every moment I could snatch from my other duties I have spent in the herbarium and this very arduous, though very pleasant duty has been so time-absorbing that it has seriously diminished the time ... for my other duties.' As his first director's report made clear, aside from taking care of the very garden beds themselves, and the library (similarly parlous), 'the care of the Herbarium has been my greatest solicitude'.

It was beyond belief to discover, then, that Moore had been offered a set of Robert Brown's Australian specimens 'and that the offer had not even been acknowledged'.

As Maiden saw it, a botanic garden could not 'properly perform its functions without the support of a rich herbarium'. The scientific reputation he wanted to build for his institution depended on such support, and he planned a large reputation – and a large *hortus siccus* – indeed.

The only solution, then, was to collect, both from the field and from other institutions. Just as he had bolstered the number of South Australian specimens in the Technological Museum's herbarium years before with an appeal to that colony's botanic gardens, he was almost immediately in touch with a number of herbaria including Melbourne's (which Mueller's commitment to collection and exchange during his 43 years there had made, Maiden assured its director, 'the mecca of Australian botanists'), and with individual collectors around the country and beyond. To Kew, he said simply: 'will you help me?'

Apart from which, he dusted off his huge vascula – the metal boxes in which fresh specimens were carried – and took himself out-of-doors. With the whole continent needing to be sampled for his purposes (he planned a 'National Herbarium' from the start), he had two methods of choosing where to go. First, as he had demonstrated in sending Baeuerlen north to the Big Scrub as logging and development picked up pace, he knew the importance of getting to 'specific localities [that] are becoming botanically effaced'.

'Over 20 years ago,' he explained, 'seeing that Mosman [on Sydney's north shore] was beginning to attract population, I assiduously collected there. Brown collected there! When the Illawarra [railway] line was opened, for months, at weekly intervals, I made it my business to get out at different stations as far as George's River, and preserved specimens' – on one day, a delicate rose-leaf bramble – 'from localities which are now densely built over ... who can foretell the progress of these localities during the next half century!'

His other method was, as he had done with Brown, to 'get on the tracks of good men whenever we can and emulate their good deeds'. And so he went to Bourke, to Wagga and to Cooma, all in his first year as director; then from Port Macquarie to Walcha, to Jenolan Caves, to the Illawarra and New South Wales' south coast, and to different parts of South Australia and Victoria, retracing the expeditions of early collectors including Caley, Brown and Cunningham.

He went up Mount Kosciuszko, enduring 'two days' blizzard of wind and snow'. He even sailed to Lord Howe Island, where landing required a two-hour, two-mile row in a whaleboat, and where he ended up marooned when rough weather caused his steamer to break her anchors, after which she made for Sydney. 'Collecting,' Maiden said of the weather, 'was seriously interfered with ... [but] fortunately I had brought my vascula and presses ashore in the whaleboat'.

All of which enlarged 'the Herbarium [by] several thousands of specimens'. As the nineteenth century ended, it could be reported that it now held 'not less than 15,000' specimens, and that the smart new Gardens' building, with three rooms specifically allocated to the growing *hortus siccus*, was ready for its official opening. It was a 'worthy headquarters' for the 'scientific botany' now enjoyed by New South Wales, Maiden said.

Specific (often honorary) curators were appointed to the different groups of plants – the algae, the fungi, the mosses, and a lovely collection of 20,000 lichens that had just been purchased – while Maiden himself, in Mueller's wake, assumed the title of 'the oldest Australian worker on *Eucalyptus*'. And, as Federation opened the bright new twentieth century, Maiden's institution lacked 'very few described New South Wales Plants'. Only three years on, Ernst Betche, Maiden's chief assistant in the Herbarium's overhaul and replenishment, could quantify that claim to report that its papers contained, pressed and dried, 97 per cent of described New South Wales species, and some 30,000 exotic specimens, in addition to material from the rest of the country.

There was more to come, Maiden divined: 'all the Australian states, except perhaps Tasmania and Victoria, are botanically so imperfectly explored, that in some cases some 10 per cent and more will be added in the future'. He wanted more, often feeling like Oliver Twist, he said, holding up his bowl hopefully.

But beyond his commitment to amassing collections, and to science, perhaps the greatest thing Maiden brought to the Gardens' Herbarium was his sense of history.

For all his quantities of leaves and buds, notes and names, Maiden was missing the 'original' collections of the country's botany, the specimens that had sparked Europe's interest in the continent and its plants, the specimens collected by Joseph Banks and Daniel Solander during 1770. To see those, one had to go to London, to the British Museum's botany department (in Maiden's day – now part of the Natural History Museum). One had to go to its long rows of high wooden cupboards, their jambs softened with slices of scarlet, purple and blue velvet ribbon, their doors latched with tiny metal keys. And there they were: this banksia, collected by Banks and Solander, which Brown had then carried as part of his herbarium on the *Investigator*; that eucalypt, collected by Banks and Solander, its huge leaves fading towards the buff colour of its papers.

Standing in London in 1900, his first visit to a place that no longer felt like 'home', Maiden thought that it was time for some of these specimens to use 'the return half of their ticket' back to Australia. Turning to the department's keeper, he 'begged for a few duplicates'. 'I will give you a few with pleasure when we distribute them,' the keeper replied.

Five years later, in March 1905, a large parcel – 'a set of 586 species ... of Australian plants collected ... in 1770' – was delivered to Sydney. Working his way through them, Maiden confessed to feeling 'quite small as I contemplated the remains of an insect which had spun a cocoon nearly a century before I was born'. He urged his readers 'to imagine Cook saying: "Now Mr Banks, I am going to call those the Glasshouse Mountains, unless you can suggest a better name." To which Banks,' said Maiden, 'would answer, "No, I cannot, Captain, but I would give a hundred pounds for a day's botanising amongst them."'

He could even imagine the contemplative Solander, 'sitting on a rock or on a fallen tree, busy with a lens examining the few plants that he collected with his own hands, and scrutinising the bunches and bundles ... brought in by the emissaries ...'

Now, after 135 years, Maiden was welcoming these broken-off fragments of plant (nothing neatly cut, he noticed) back to their own part of the world. Some scorched by bushfires that had burned before 1770, some munched by long-dead insects, they were, he wrote, 'absolutely the oldest articles ... ever returned to this continent'. They were as priceless to his collection then as so many of his own collections are now, in another century again. 'Dried specimens,' he explained later, flourishing his precious 1770 examples at one government royal commission, 'are [as] permanent as written records' of vegetation long since changed, or disappeared.

Thanks to all this, Maiden was now confident that his National Herbarium of New South Wales stood as 'one of the best herbaria in the world'. Of his own term at its helm, he felt it fair to add, 'I have transferred the Sydney Botanic Garden [sic] from a mere horticultural establishment, as I found it, to a *Botanical* establishment in addition, and Scientific men throughout the world now recognise Sydney as one of the principal botanical centres of the world'.

In his subsequent 20 years as the Gardens' director, Maiden seized every opportunity to improve his Herbarium, often thanks to unsolicited offers such as Dr Tomlins' from the Big Scrub. Even internationally destructive events such as World War I provided additions: one soldier alone donated 300 specimens from Palestine, while others contributed 'from Palestine, Turkey, the Dardanelles and France'. A single parcel from Kew was opened to reveal 58 acacias and 23 eucalypts collected between 1816 and 1839 by Allan Cunningham and Charles Fraser, among others.

Maiden was also occupied by several enormous scholarly projects: there was work on acacias and on the country's earliest botanical collectors, a *Forest Flora* for New South Wales, and a complete *Critical Revision of the Genus* Eucalyptus. Eucalypts fascinated him; he once described them as 'wild, untamed things, like the native bears and opossums which frequent their branches'. And in his work with eucalypts he had such assistance from one of his staff – Margaret Flockton – that he talked of crediting her as co-author of the entire project.

Here was change too: Flockton was neither a casual collector, nor a lady whose work on plants was seen as a hobby or a quiet pastime. She was employed at the Gardens as an illustrator, although it was for her skills and observations as a botanist that Maiden particularly valued her. Nor was this movement of women into the more professional side of science specific to Sydney, to Flockton or to botany. It was a trend mirrored around the country as women began to graduate from science degrees at Australia's universities and began to make their own contributions to research across a score of disciplines.

With the *Forest Flora* finished, Maiden reached retirement, officially leaving his office on 21 April 1925. 'I pride myself,' he said simply, 'that I handed over an admirable staff, and the gardens and parks and herbaria in splendid order'. He had even gone to the trouble of laying in a good stock of specimen boxes, 18-and-a-half inches long, 12-and-three-eighths inches wide, and either three or six inches deep, 'designed,' he said, 'to comfortably take a piece of folded cartridge paper of the standard size ... [and with] a tin label holder' for identification. He'd come up with the design himself, years before, inspired by the mechanism of a church pew. He left a stockpile of thousands.

During his short retirement (he died only seven months later, on 16 November) he continued corresponding, working on eucalypts, and writing articles as he had throughout his professional life. No one would have predicted that,

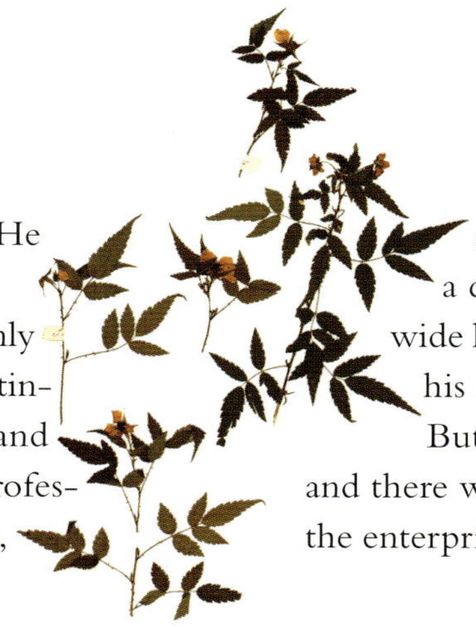

after the celebration of Maiden's achievements, a combination of economic depression and worldwide lack of interest in taxonomy would see a decline in his Herbarium's reputation before a decade passed. But as the 1930s turned, expectations turned again, and there were always new enthusiasts, their commitment to the enterprise as great as Maiden's, ready to take up the work.

The hobby

Species: Dendrobium gracilicaule
Common name: A Dendrobium Orchid
Family: ORCHIDACEAE
Collector: Rev. H. M. R. Rupp
Locality: Bulahdelah, New South Wales
Date: September 1923

What makes an ardent collector? What makes them incapable of passing through a field of flowers without stopping to pick and classify them? What sparks their close attention to one fern's frond compared to another? What shape does their life take when every possible day of it has to contain part of the hunt, the quest, for the next addition to their herbarium?

'The taste for collecting,' said one French thinker, 'is like a game played with utter passion'. That commitment in plant collectors' lives, not to mention the knowledge they spawn through the networks their hobby creates, sits in the stories you can attach to almost anyone recorded on a herbarium sheet. There are years of pursuit, massive correspondences between people who share a complete fascination with a species, occasional discoveries, occasional disappointments. All this is wrapped in their dedication to finding even just one more thing that their collection lacks.

And propping up years of professional botanists' work there's often a raft of collections made by someone properly designated as 'amateur': the person who lacks the appropriate degree; the person whose collecting is pleasurable, not professional. These are the people whose passion for the game and contribution to its knowledge are extraordinary in scope.

In the last days of J. H. Maiden's life, a tall man with a kind face, a warm humour and the clothing of an Anglican priest, arrived on his doorstep from Paterson, near Newcastle in New South Wales. It was one of the few meetings between Maiden and the Reverend Herman Montague Rucker Rupp – a man with whom he'd enjoyed botanical correspondence for 20 years. On this occasion, Rupp may have found Maiden frail, but he was 'full of genial enthusiasm and kindly encouragement', not to mention 'still hard at work on the Eucalyptus'. 'Whatever little contributions I have made towards the botany of New South Wales,' Rupp wrote in his own memoirs another two decades on, 'are really due to him'. In the volume of Rupp's own work and achievements, however, sits one example of just how much a hobbyist, an amateur, may contribute to their field.

If there was one thing that he loved – in the temporal world, at least – it was orchids. Exotic beauties, they're from the largest family of flowering plants, boasting some 30,000 species. Cultivated in the Far East from the eleventh century, they swept into Europe several hundred years later. In 1731, an English botanist received a dried specimen of *Bletia purpurea*, a purple orchid whose 'tuber,' he noticed, 'appeared to have life in it'. He potted it, nurtured it, and thus found himself with the first tropical orchid to flower in England. This sparked a mania almost equivalent to the notorious enthusiasm for tulips that Holland had endured in the previous century, and set off such a frenzy of orchid-collecting that many species found themselves in danger of extinction.

Orchids caught early at Montague Rupp's imagination. As a young boy in Victoria's Koroit in the 1870s, he'd read 'a thrilling serial in the *Boys' Own Paper* about Orchid-hunting in Borneo' and had driven through the bush with his father, revelling in the wildflowers and learning 'to love wild orchids, "spiders" and "doubletails" and the like'.

By the time he was a student at Geelong Grammar, he was well and truly smitten and had a reputation for being 'a bit dotty on wildflowers'. Already orchids were his 'special favourites', and he was often up at four in the morning for one of the school's customary Saturday rambles. Tumbling home hours later with 'about 25 species of orchid collected', he'd borrow the headmaster's copy of 'von Mueller's *Key to the System of Victorian Plants* to work out the identifications of the flowers we gathered'. Which was, he conceded, 'rather heavy going for schoolboys'. (The headmaster was John Bracebridge Wilson, Fellow of the Linnean Society and an expert on Port Phillip algae.)

Leaving Geelong for the University of Melbourne, his weekend rambles continued and by 1896 he'd compiled a complete catalogue of his collections in the outer suburbs near the university. Thanks to a letter of introduction from his headmaster, he even spent an afternoon over lemonade and biscuits with an elderly Ferdinand Mueller. (Was there anything Rupp could do, Mueller asked later, to encourage children to collect; perhaps if there was the attraction of some 'small pay'?)

Although his professional goal was to follow his father into the Anglican Church, Rupp hoped also to graduate in science – Botany I was among the first subjects he took – but he lost a year of study to illness and left it too late to cover the prerequisite of Chemistry. Taking orders in 1899, he conferred the status of 'hobby' on his interest in botany and the status of 'amateur' on himself. 'I am merely an amateur botanist,' he would say, 'who has concentrated on the study of our Australian native orchids, and has written much – perhaps too much – about them'.

This is a little of what Reverend Rupp's 'merely amateur' hobby ran to.

In terms of herbaria, his life fell into two parts. In the first, spanning 60-odd years from his childhood, he collected, compiled and curated a herbarium for himself, also sending regular contributions to Maiden's National Herbarium in Sydney. During the second – the last 17 years of his life – he worked as honorary curator of that Herbarium's entire orchid collection.

Through successive parishes in the dioceses of Grafton and Armidale, Wangaratta, Tasmania and Newcastle, he sought the rarities as well as the suite of plants common to each area. '[It's] a good parish,' he commented ahead of one of his moves, 'and you know what those valleys will mean to an orchid man'. But while botanically rich, many of these more remote areas were fairly devoid of other botanical enthusiasts, which was how Rupp had come to open his correspondence with J. H. Maiden in 1903.

It was to Maiden, too, that he owed thanks for a more accessible way of organising his constantly expanding herbarium. Brown paper folios, each holding a series of sheets, could be easily stored in 'nice strong boxes', the specimens well-protected by 'naphthalene flakes or crushed moth balls'. ('Some folks object to the odour,' said Rupp, but 'personally I don't mind it'.) With the aid of an index and a system of tags, he could find anything 'in a minute'.

He explained this in detail to the University of Melbourne as he discussed donating more than 3,000 herbarium specimens to his old faculty – 'the non-orchidaceous plants,' he said, as it seemed 'hardly likely that I shall in future give much time to the general flora: specialising is the order of the day, and what I am doing in leisure hours in connection with Orchids appears to be considered of some value'.

This was 1926, and if there was one advantage to a vicar's life, it was those constant relocations to new parishes with their new botanical hunting grounds. His collections were enormous and although they were 'a treasury of precious memories' from 'all sorts of outlandish places, nice and nasty', hiving some of them off made practical sense. His herbarium, he admitted, 'has got beyond my management'.

It was a lovely gift, said Melbourne's Professor of Botany, as at the time the university had 'practically no herbarium'. (It now holds more than 100,000 specimens.)

Once he'd made the decision to specialise, Rupp was unstoppable. He'd always preferred writing to 'orchidy' folks, he confessed, and loved the parcels of specimens that constantly arrived at his door. Some had orchids tucked into beds of 'a peculiar spongy lichen of the granite' to keep them fresh; others had sphagnum moss. Some were packed into damp maiden-hair fern; some (embarrassingly, from recognised orchid experts) came without so much as damp blotting paper to keep them perky.

With his vast correspondence, and all the papers he wrote, both academic and popular, he wanted to entice people, to draw them into the fascinating world he knew orchids offered. 'Are [New South Wales' orchids] worth growing?' he once asked *Sydney Morning Herald* readers. 'Of course they are.' There were those that would 'bear comparison with high-priced exotic beauties'; those 'whose dainty loveliness and exquisite perfumes are a joy to the possessor'. There wasn't one, he said, 'not interesting to the student of nature, professional or amateur'. They held him in thrall.

Only once or twice, later in life, did he admit to finding his 'brain … addled with all the orchids stuffed into it,' or to being 'a bit orchid-weary, although not for all the world would I have folks stop sending them'. One week alone saw 74 specimens come by post – all from one collector.

Rupp discovered new species of orchids and of other plants, including a wattle (which honours him as *Acacia ruppii*) and a tea-tree, and as he looked back over his collecting life, he regretted not knowing so much about 'orchids and their ways', nor having much faith in his own ability, early on, to puzzle out the character of a plant. This lack of confidence, he suspected, 'caused me to pass over many variations from type, some of which probably indicated "new" species'. A more knowledgeable person, he felt – like the one he became – would probably have found double the number of orchids in particular locations.

But there were moments of great glory. Of the many new things he turned up, he liked recounting the story of the 'tiny Helmet orchid' in 1924, which puzzled everyone from Rupp himself to the country's most eminent botanists until they realised it was something that had been 'discovered and named by Allan Cunningham in 1833, and thereafter lost sight of for 91 years!' A dried plant, forgotten in a herbarium, suddenly came back to life when Rupp noticed it.

And in 1931, the orchid world bristled with what Rupp felt might 'fairly be termed a sensational discovery'. A Mr Slater of Bulahdelah (the Anglican posting Rupp had held during 1923–24 and described as 'a botanist's seventh heaven') was 'clearing away the rubbish at the base' of some orchids wanted by a couple of enthusiasts when he noticed a 'queer specimen of plantlife which puzzled him completely'. Suspecting it 'of at all events some association with orchids,' he picked it as well, and it was forwarded to Rupp.

Yes, Rupp confirmed excitedly, it was 'an orchid which apparently germinates, grows and comes into flower beneath the surface of the soil' – a subterranean species, like one found only three years earlier in Western Australia. Rupp called it *Cryptanthemis slateri* in honour of its discoverer, and worried that the subsequent lust to find it might endanger populations of the original orchid species Slater had been excavating, as excited plant-hunters scrabbled round underneath them for this new wonder: 'I tremble for [its] welfare.'

The second phase of his orchidaceous life coincided with his retirement from the ministry in 1939. He was 66, and Mrs Rupp might have hoped he'd take a pause, move a little more slowly. But within months he'd been offered – and accepted – an honorary position as curator of orchids at the National Herbarium of New South Wales. For an annual honorarium of £26, the Herbarium's director hoped the Reverend might overhaul and revise the institution's entire Orchidaceae.

No one, as Rupp discovered, had so much as consulted the specimens 'for years … I love it'. So in the great tradition of volunteer work that keeps large scientific establishments moving and growing – and had contributed so many specimens to the Herbarium in the first place – he began 17 years of study.

The only significant interruption to the task came early on, when Rupp was asked to pack and ship all the specimens to Glen Innes, hundreds of miles north. It was World War II: Japanese strikes were expected on Sydney and the Herbarium's collections sat so close to the harbour that everything had to be moved. In an extraordinary effort, all the precious Australian type specimens were relocated in under a fortnight.

When he wanted to get away from thinking about the war, Rupp took himself orchid-hunting in Sydney's northern suburbs, lamenting that while his fellow hunters found themselves knee-deep in potentially new species in floristically fascinating places around Asia and the Pacific, thanks to the war, they couldn't collect anything.

Occasionally the imbalance between the work of botany and how it was rewarded must have niggled too. When a parcel of 'fine' and 'very valuable' orchids arrived on his desk, Rupp commented that 'the National Herbarium ought to pay collectors for their trouble; but while we have the present type of Australian Governments, which regard herbaria as collections of dead weeds, it never will'. He himself must have given them 'over 1000 specimens,' he calculated, 'but I've never had a bean'. Not that it mattered, he said: 'I take pride in supplying material for the national collection'.

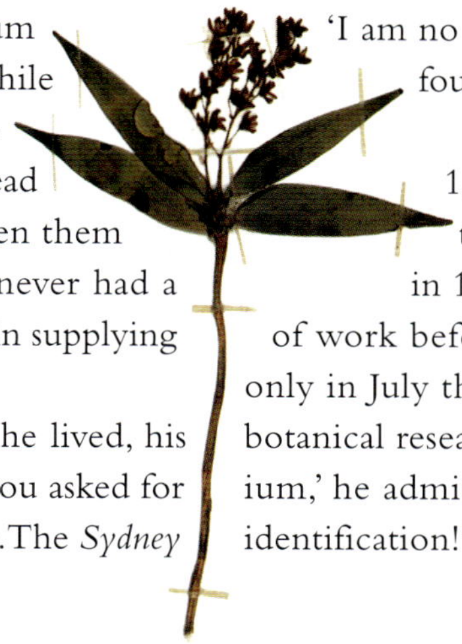

If you asked for him on the street where he lived, his neighbours referred you to 'the orchid man'. If you asked for him at the Herbarium, they used the same epithet. The *Sydney Morning Herald* saw him as 'a pioneer in a land where pioneering was thought to be finished'. But the Reverend Montague Rupp had the same measure of modesty as he had of warmth. 'I am no pioneer of orchidology,' he replied simply. 'Its foundations were laid before I was born.'

Donating his personal orchid *hortus siccus* – 1,500 specimens of 470 different species – to the National Herbarium of New South Wales in 1946, he would still contribute another decade of work before his death in September 1956. And it was only in July that year that he felt he couldn't 'continue my botanical researches any longer. I can't go in to the Herbarium,' he admitted, 'but they still send specimens to me for identification!'

Herbaria

If a botanic garden's beds of living, growing plants form a museum through which anyone can wander, its herbarium is the museum that remains off-limits, behind closed doors. Just as the stories behind specimen sheets can drop out of sight, so the specimens themselves are, for the most part, unavailable to a general stroller, a casual observer. And so the herbarium feels not only like a cumulative garden, but a secret one as well, questions and answers tucked into its leaves.

But here's the clever thing about those rows and rows of crisp, flat plants: they're quivering with the life, the vitality, of constant change, constant movement, constant addition and revision.

The herbarium's people file more and more into its drawers – its botanists, staff collectors and students, and the enthusiasts who donate their discoveries unbidden – while the ancient practice of gathering now intersects with the most modern technologies. Construction of an Australia-wide virtual herbarium, which dreams of making all the country's institutional specimens available online, is underway, restricted only by a want of funding. And classifying plants by morphology – literally examining what they look like – is now complemented in many cases by classification at a molecular level – examining fragments of their DNA, which in turn reveal some surprising relationships between floral families.

Who'd have thought, for example, that there would be so many similarities between some pretty little shrubs (the genus *Tetratheca*) and some of the hugest rainforest trees (*Elaeocarpus*)? Perhaps the two families should merge, their boxes, their sheets, combined. Perhaps it's time for all the herbarium's drawers to be lifted and shifted into a new pattern, just as Robert Brown revised everything from Linnaeus' system to his version of Jussieu's in 1810, as Maiden rearranged his boxes from Jussieu's plan to a 'modern' German one in 1917, and as the entire collection was reorganised from that German method to a new Swedish system, considerate of plants' evolutionary histories, in 1982.

Of course herbaria are in a constant state of gentle shuffle, accommodating the new genera that still turn up. Like someone arriving late to dinner, another chair is pushed into the table, and everyone moves around, squeezes in. The Wollemi Pine's arrival, for instance, meant the creation of a whole new red herbarium drawer for the genus, the other members of its family making room for the newcomer.

Even now, it often seems easy to find new species. On a bush block in central New South Wales, a botanist picked something he thought might be a new *Hibbertia* – his eye caught by the bright yellow of its little buttercup flower – and sent it off to the *Hibbertia* expert in South Australia. It looked promising, South Australia said, but could more specimens be sent for further investigation? Making a special trip back up to the bush, he scooped up more yellow, and on opening this parcel, South Australia calmly informed him that he'd found not one new species of *Hibbertia*, but two. Just like that.

Perhaps most surprisingly, new species turn up inside the herbaria themselves. Among the wealth of algal and fungal specimens sitting unclassified in herbaria collections (often the least studied of all the plant groups, there are perhaps ten times as many of them than those more frequently studied land-dominating flowering plants), botanists are confident they will find thousands of new things. Like one red alga, mistakenly collected as a moss, filed – unnamed and unclassified – and forgotten. So discarded was this specimen that it fell out of its folder and sat there, until something believed to be new and exciting was brought in for comparison with everything else the herbarium held. Yes, this specimen was new; it was a whole new genus. But it was the same as that alga that had slipped from its folder, meaning the first collection of this 'new' species had been made, unknowingly, 25 years before its classification took place.

As for that small packet of violets picked by Joseph Banks near Botany Bay in 1770, its identity changed while it sat in its herbarium drawer. Those violets were found – 234 years after their collection – to be not the well-known native Australian violet, *Viola hederacea*, but a species classified and named as recently as 2003. When Robyn Stacey chose to photograph this specimen from Sydney's Herbarium, flowers that had spent a century labelled *Viola hederacea* were taken from their drawer and announced as this newly recognised *Viola banksii*.

There's a practice in Zen Buddhism that places a single item on a shelf to allow it a different importance, a different meaning, a different beauty set apart from everything else. There's always new life in finding new ways of looking at things – the dried components of a *hortus siccus* no different to anything else. The proudly painstaking portraits of plants produced for illustrated floral works from the seventeeth century on moved beyond the scientific or referential purpose for which they were intended to be valued as pieces of art. Even some of the first herbarium specimens themselves, assembled in Leiden and elsewhere in Europe back another hundred years, had their sheets bound together as books, transforming them into volumes with a different aesthetic, a different story.

Plants and stories have always intertwined. Christianity is busy with fig leaves picked as clothing; the apple deep in paradise (which botanists hypothesise might actually have been an apricot, a fig, maybe even a quince); the olive branch carried across the water by Noah's dove. Some of Borneo's Dayak cultures begin creation with a single sapling falling to the ground and spreading its roots; others have men and women whittled from a *garing* tree. In Norway, the Scandinavian Cosmic Tree promises ascent from the heavy world of matter to an airy place of spirit as you climb from its earthbound roots to branches that touch the heavens, while everywhere flowers are loaded with meanings and symbolisms. Rosemary for remembrance; lotus for enlightenment; and for every woman in this world, says Chinese tradition, a flower blooms in the next.

Beyond uses and purposes, plants give people a focus and a fascination – splendour as well as service – which has always sat well with those other human preoccupations: the quest for beauty, the quest for knowledge, and that overarching compulsion to collect.

Among this, a herbarium is a place of the past, the present, and the future. While each preserved plant is frozen by its moment of collection – one person, on one day, at a certain moment in time – it also remains a potential source of more and more information. A specimen collected 100 years ago, perhaps now the only evidence that this plant grew in that place, can still be mined by the newest scientific methods, contributing to genetics or pharmaceuticals, to taxonomy or land management. Today's scientists draw on yesterday's specimens, just as today's specimens will be used by the scientists of tomorrow.

Selecting from the past the specimens she will catch in her lens – their lushness, their brilliance – Robyn Stacey creates a new story, a new collection. She opens the door to this secret garden, and the rich beauty of its fragments spills onto her pages.

The new world

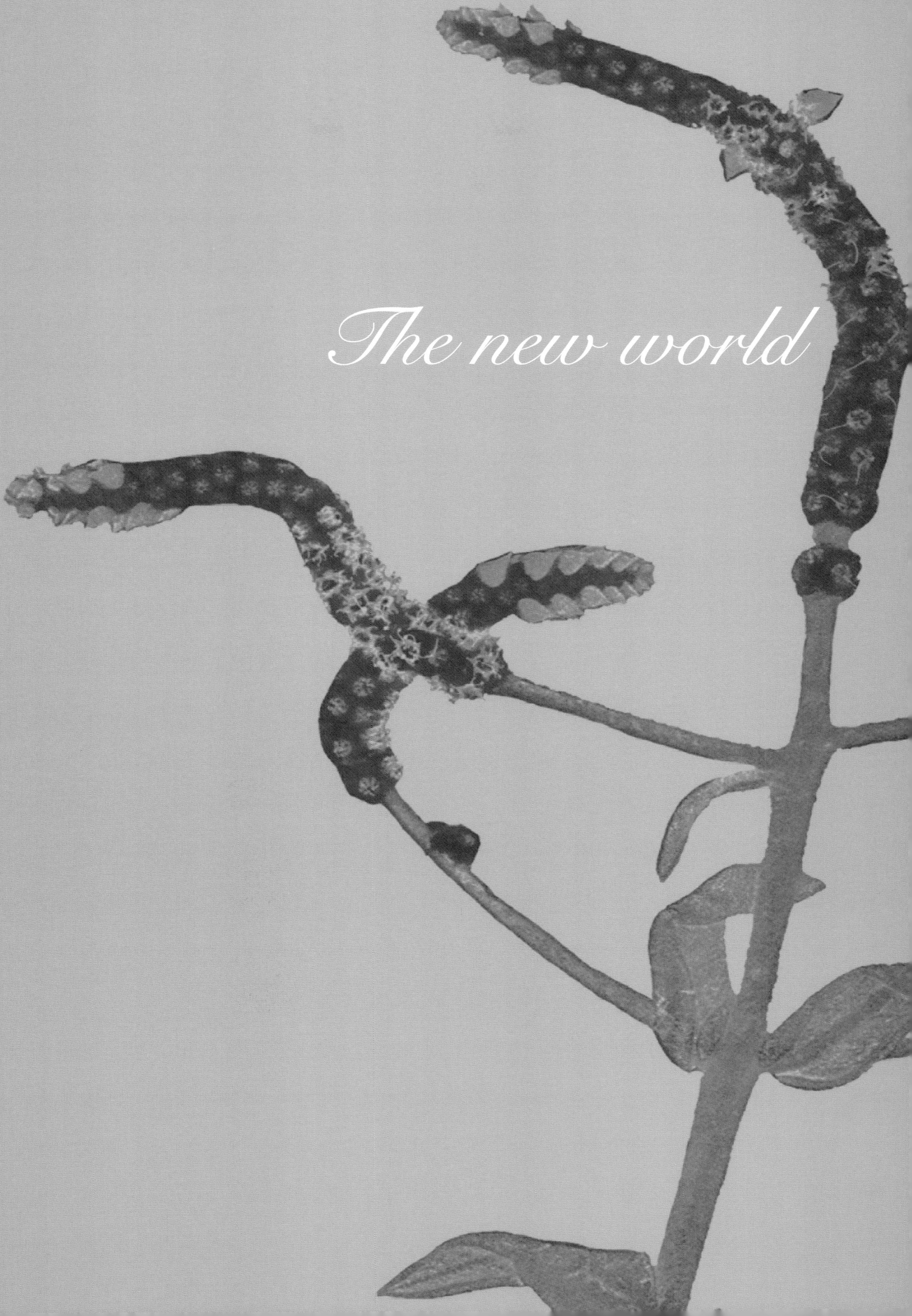

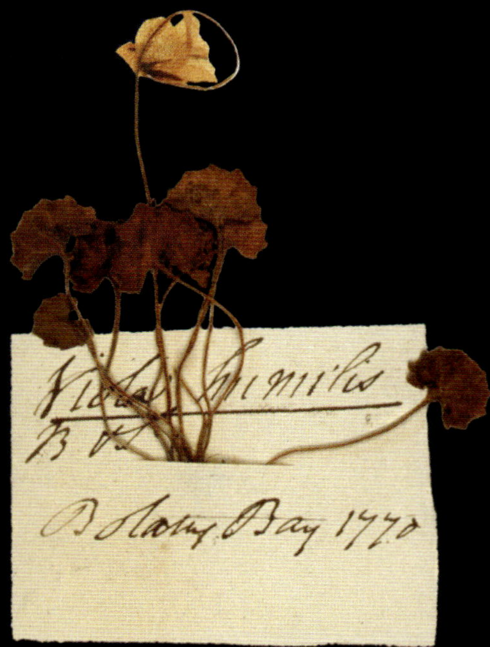

A Native Violet
Collected by Joseph Banks and Daniel Solander, 1770

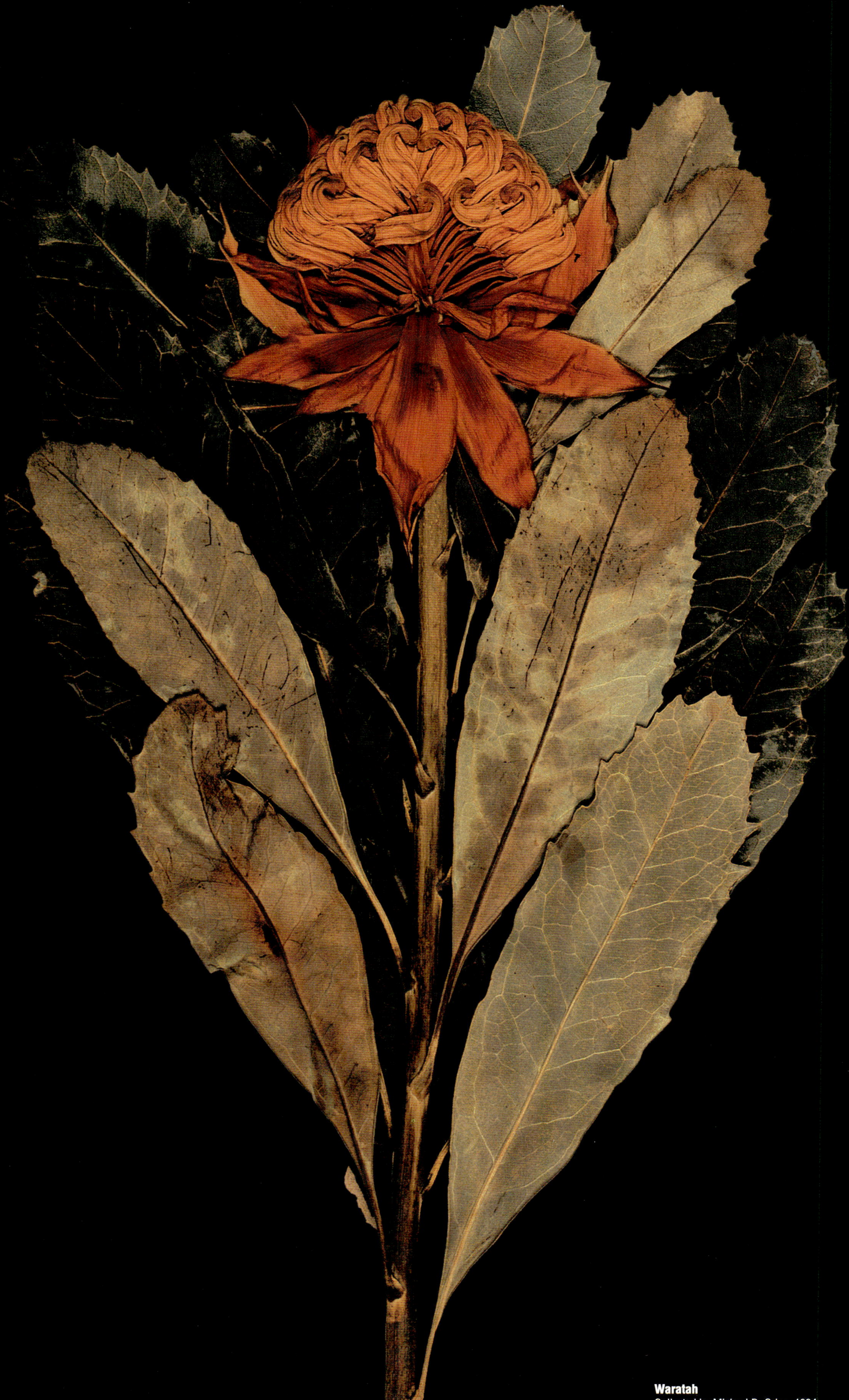

Waratah
Collected by Michael D. Crisp, 1984

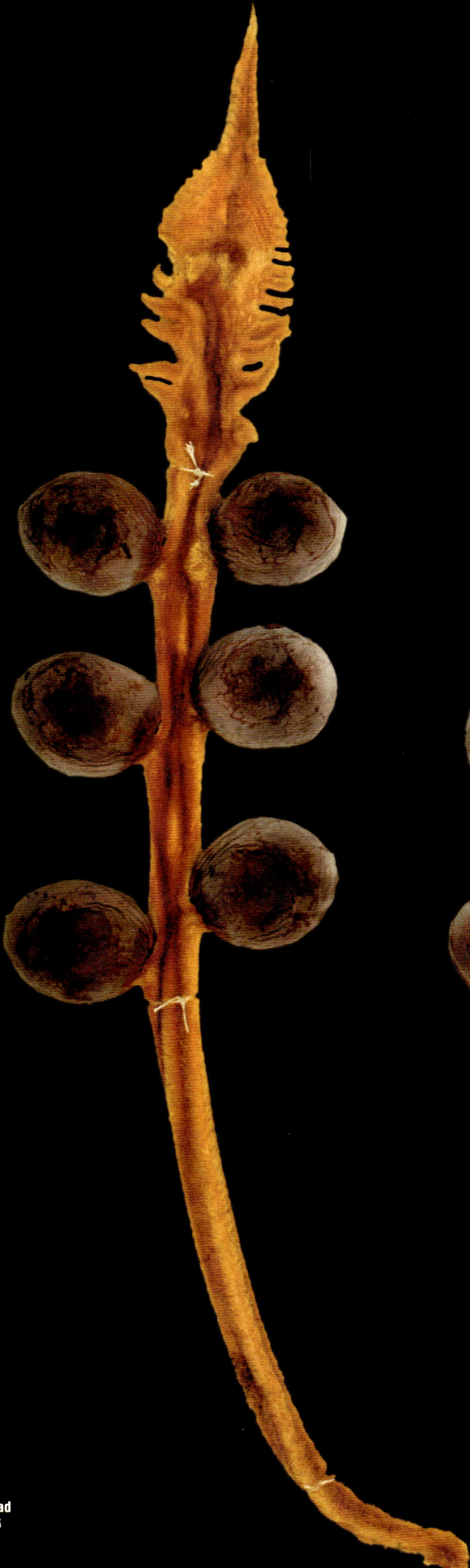

A Northern Australian Cycad
Collected by D. T. Liddle, 1996

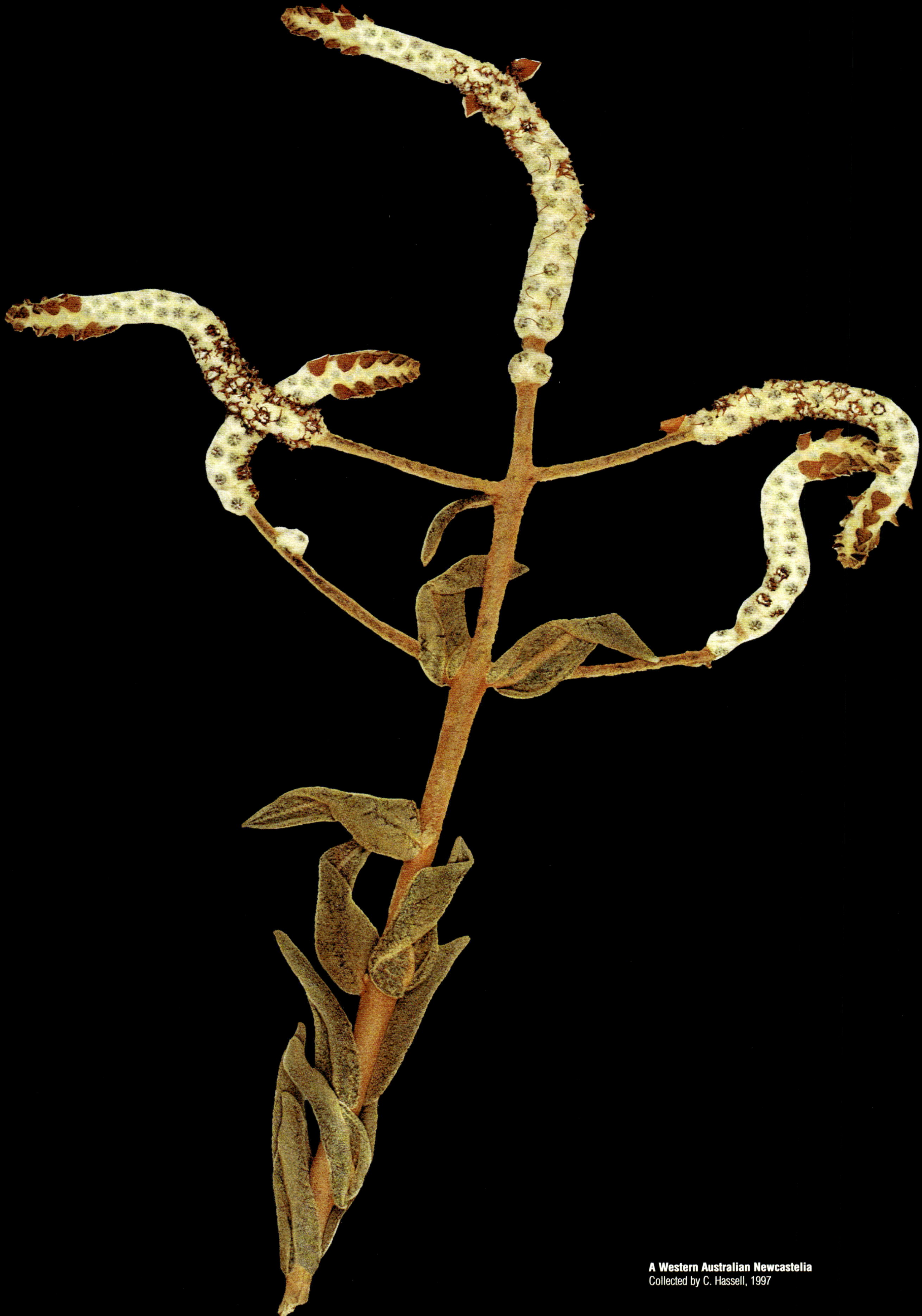

A Western Australian Newcastelia
Collected by C. Hassell, 1997

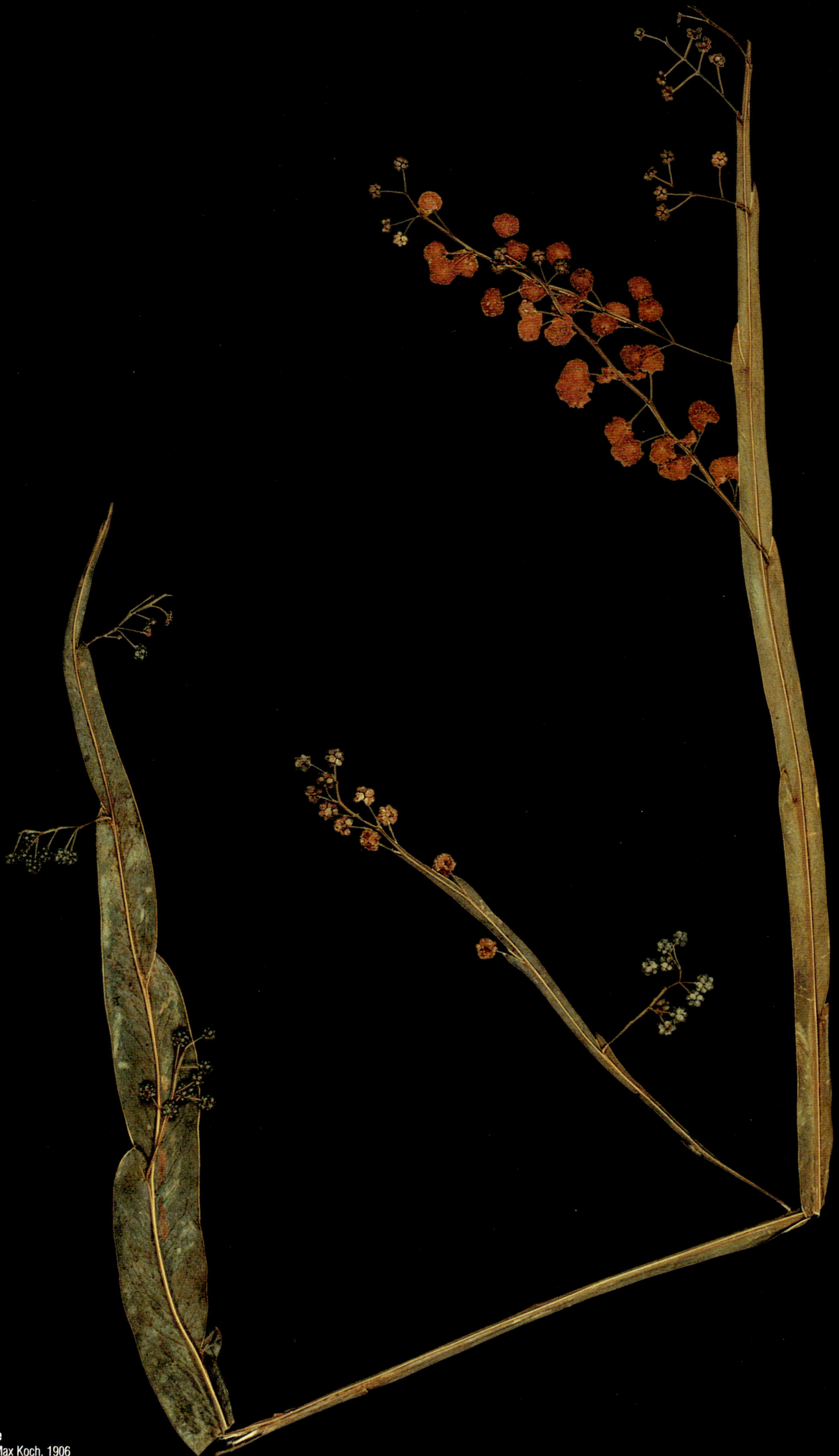

Grass Wattle
Collected by Max Koch, 1906

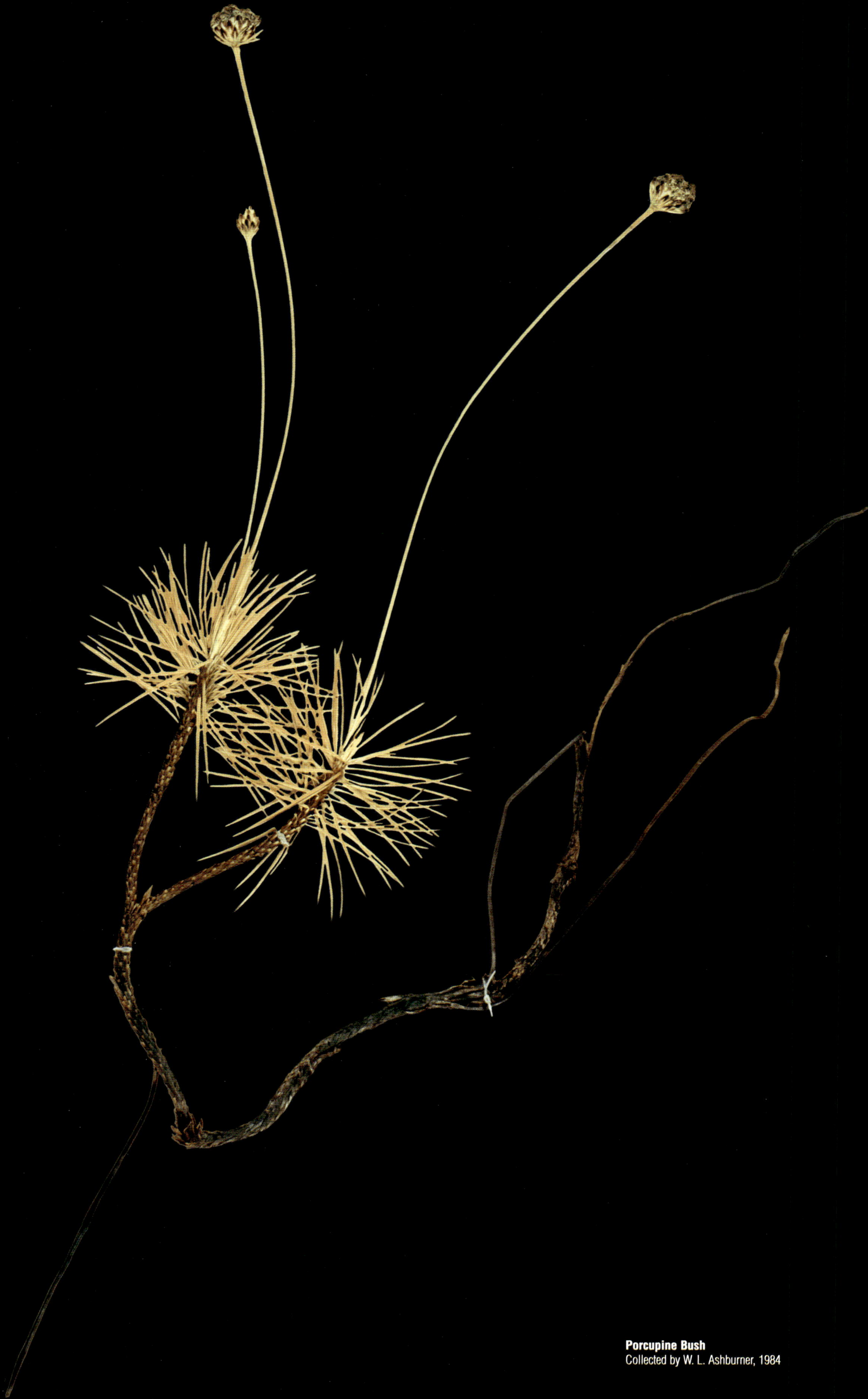

Porcupine Bush
Collected by W. L. Ashburner, 1984

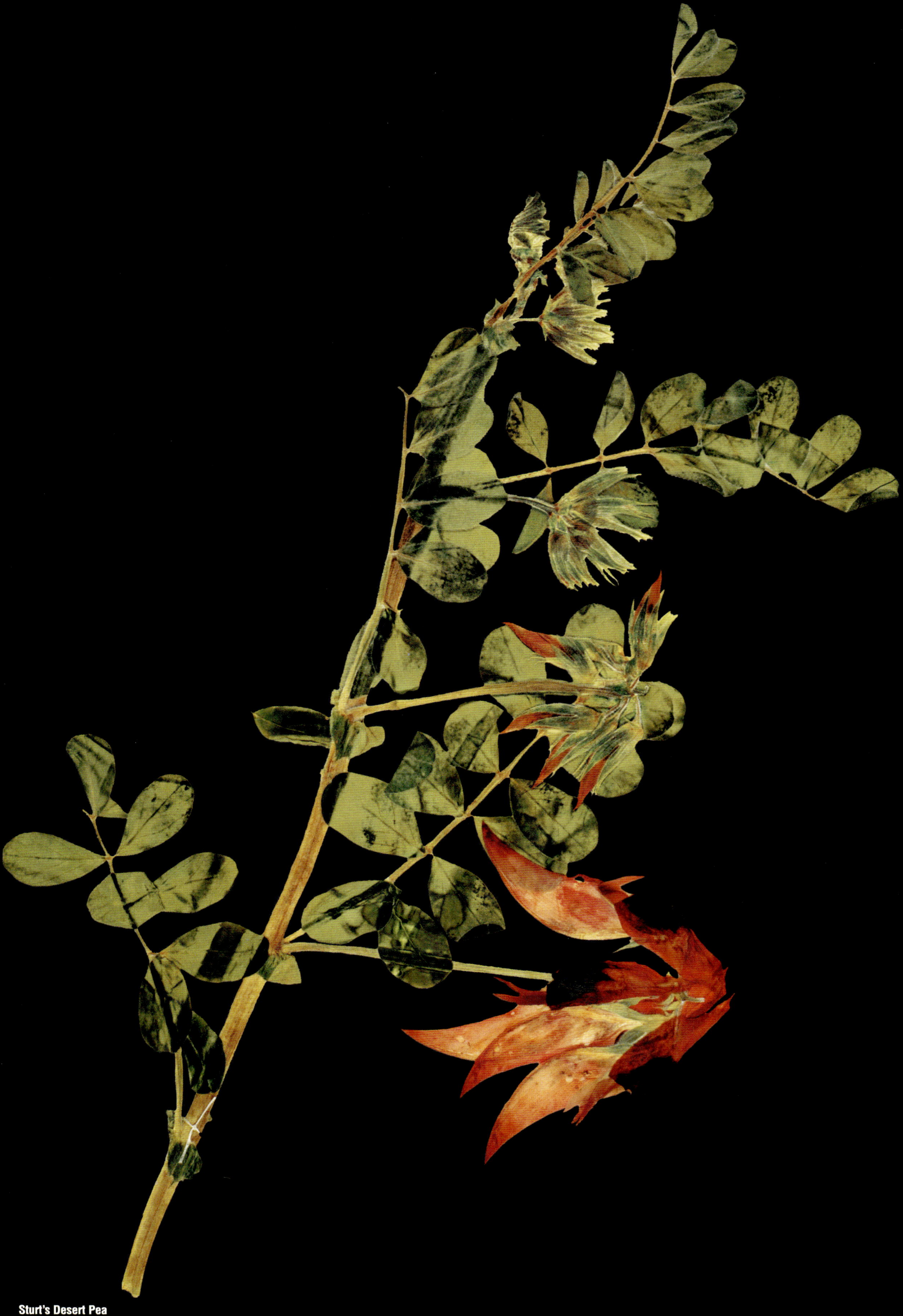

Sturt's Desert Pea
Collected by Ian Southwell, 1969

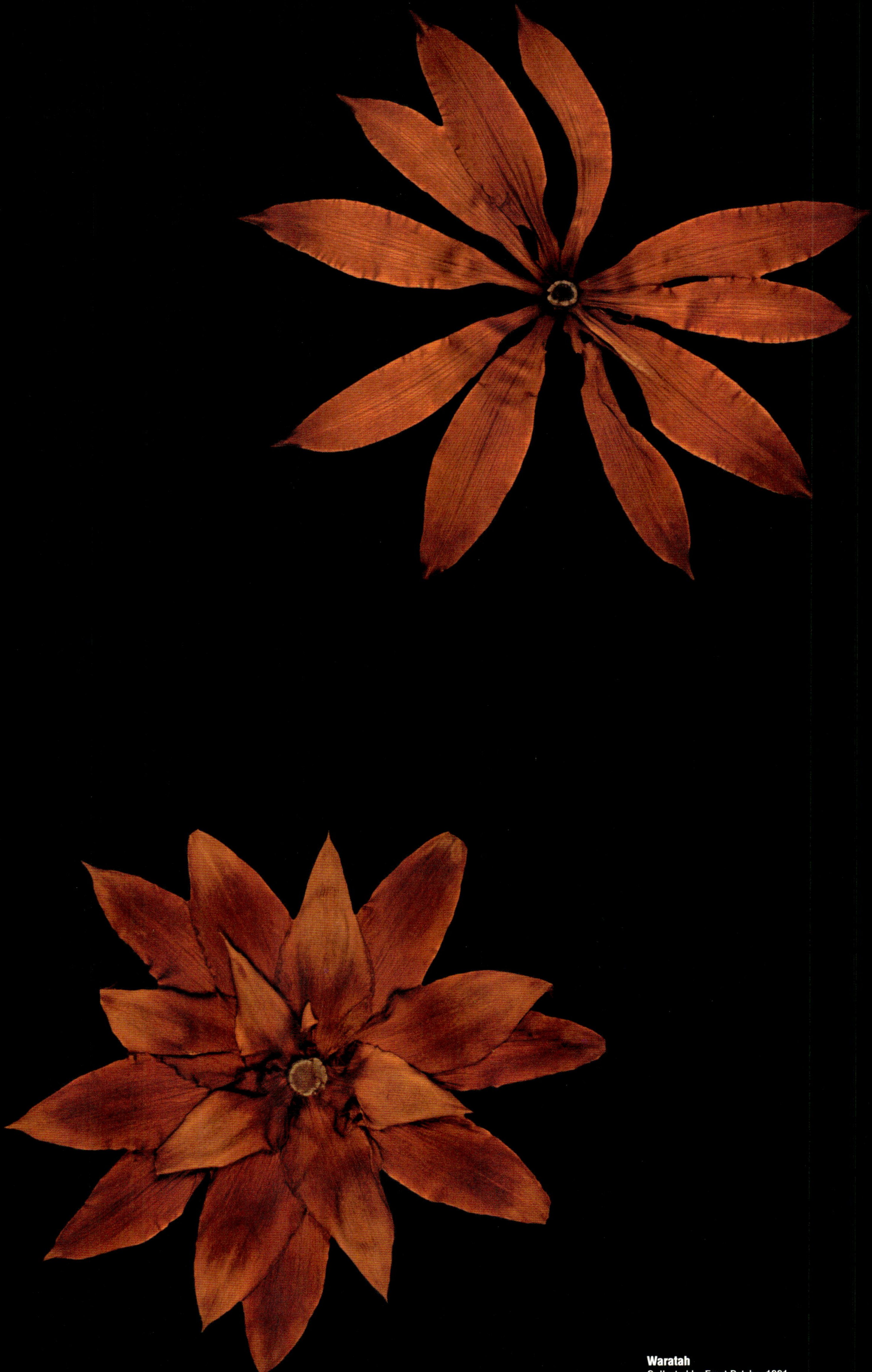

Waratah
Collected by Ernst Betche, 1881

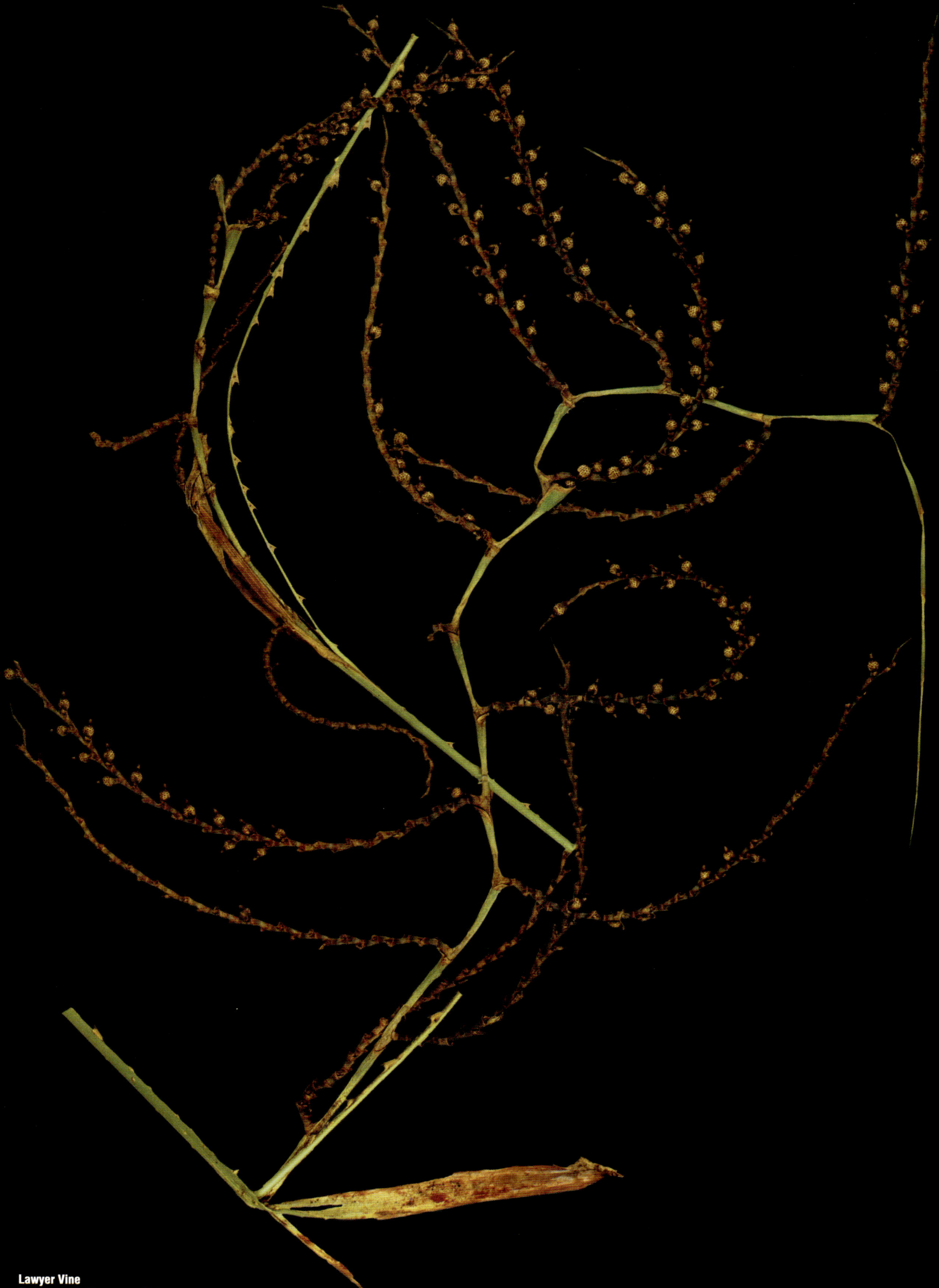

Lawyer Vine
Collected by B. Wannon and R. Jago, 1997

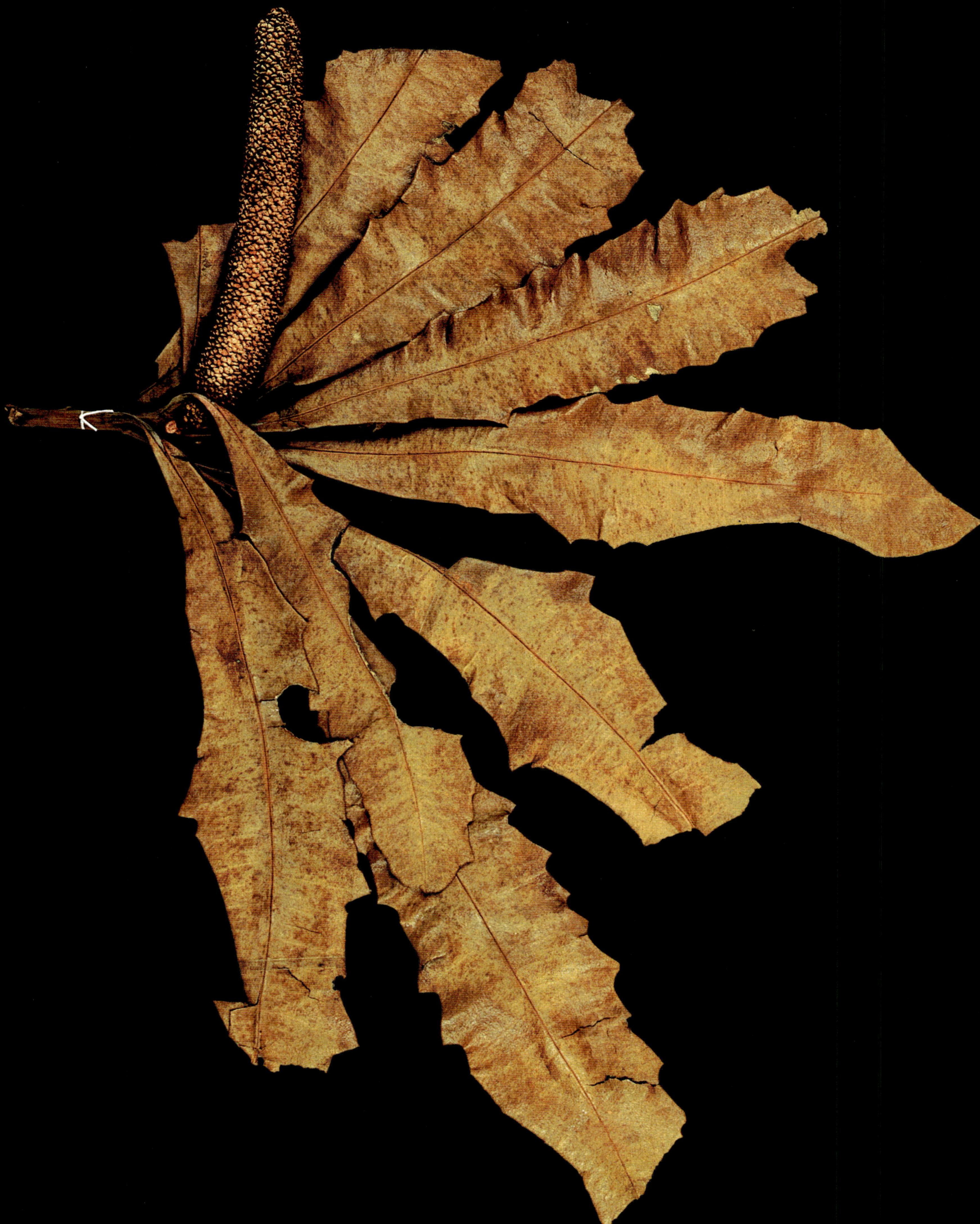

Tropical Banksia
Collected by Joseph Banks and Daniel Solander, 1770

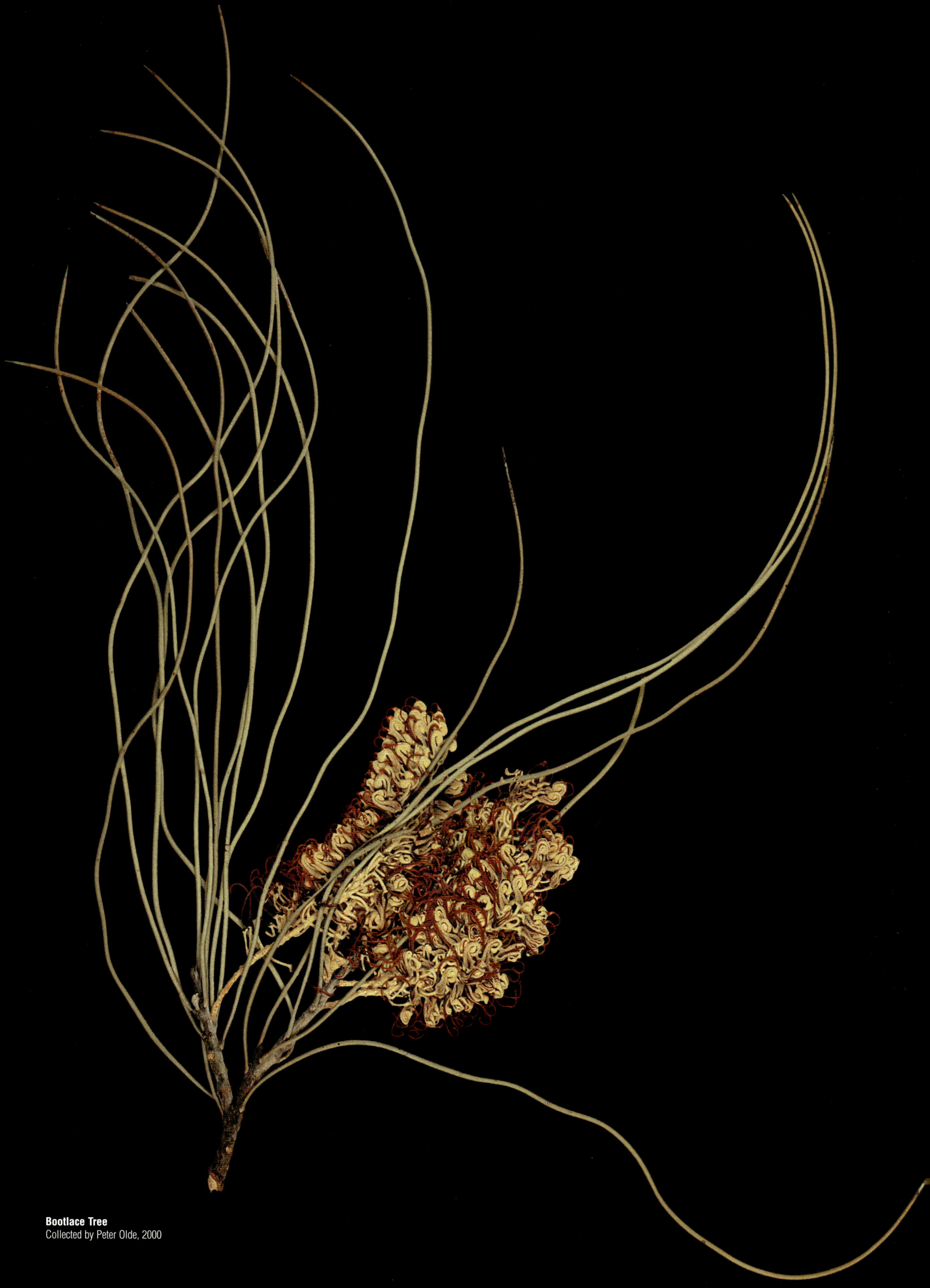

Bootlace Tree
Collected by Peter Olde, 2000

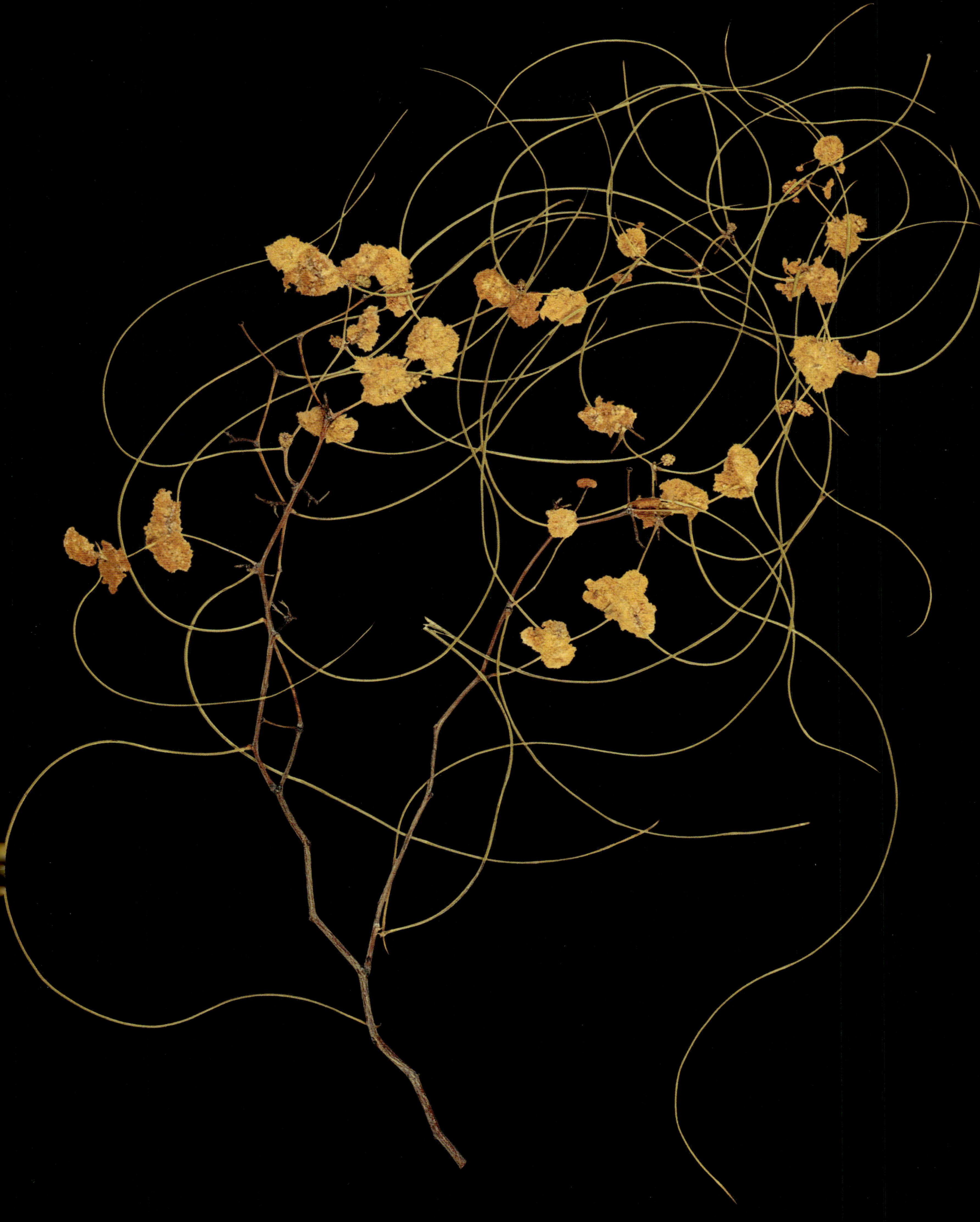

A Western Australian Wattle
Collected by Michael D. Crisp, 1980

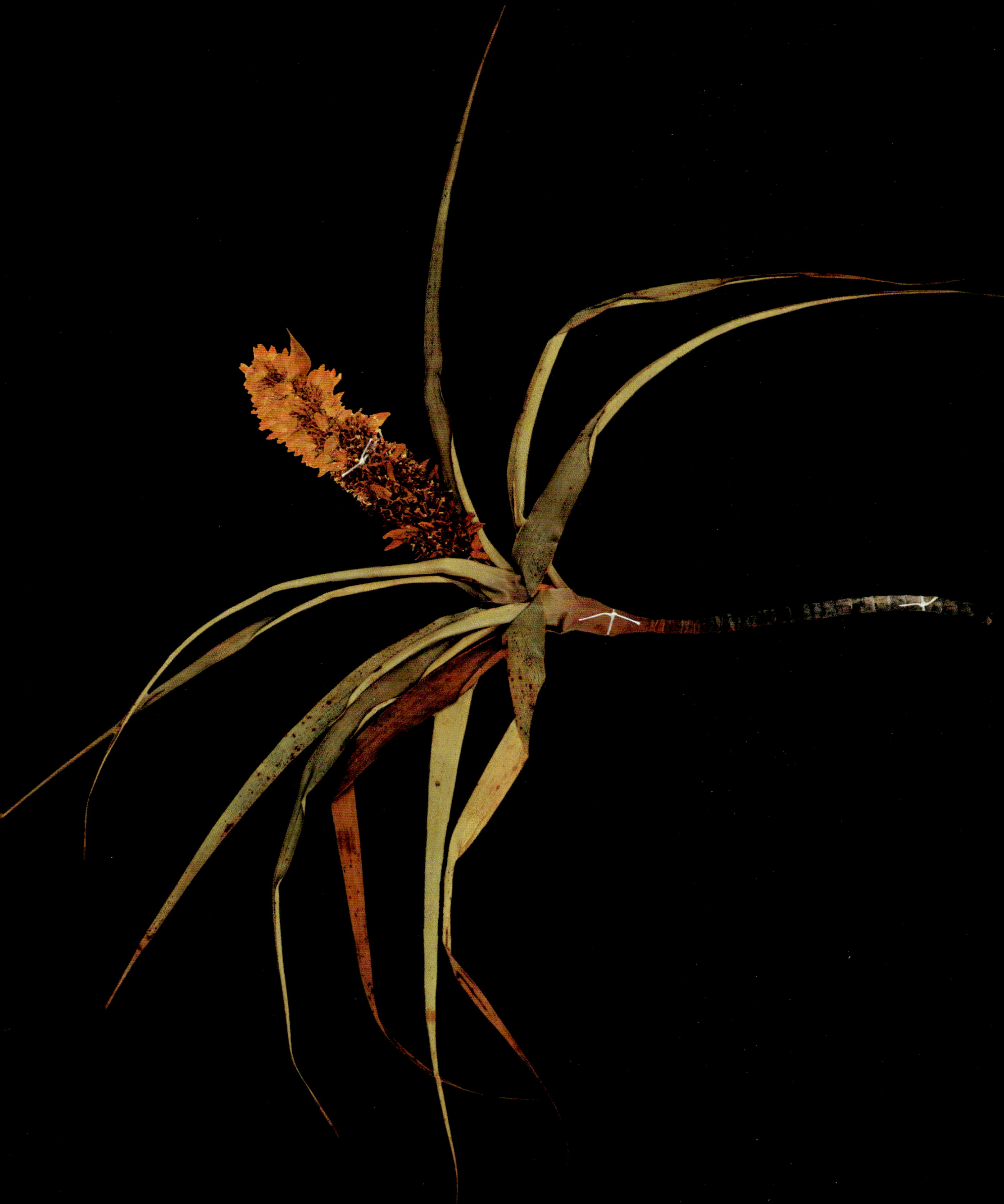

A Tasmanian Richea
Collected by S. Pedersen and colleagues, 1999

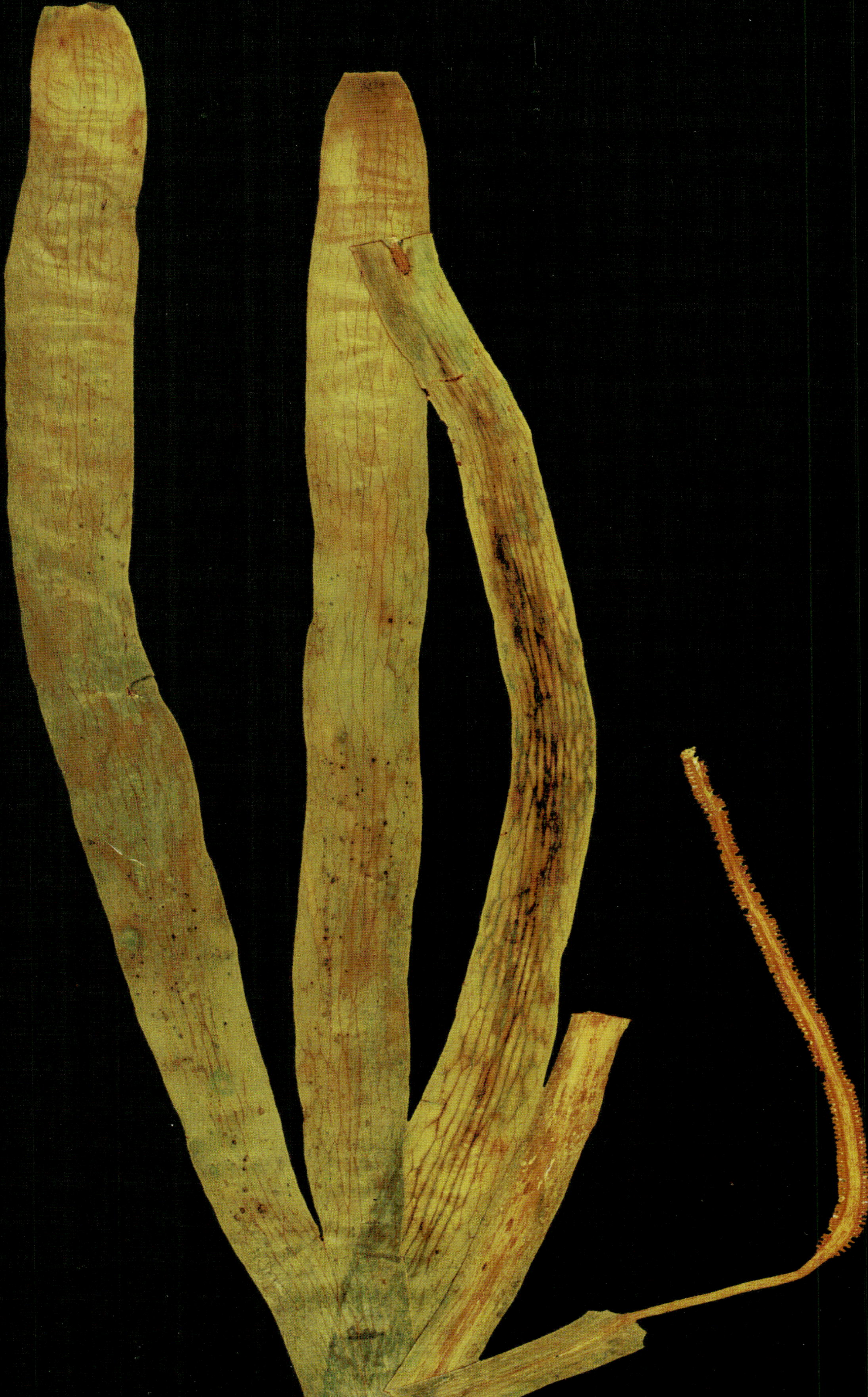

A Belvisia Fern
Collected by the Reverend Copland King, 1891–1918

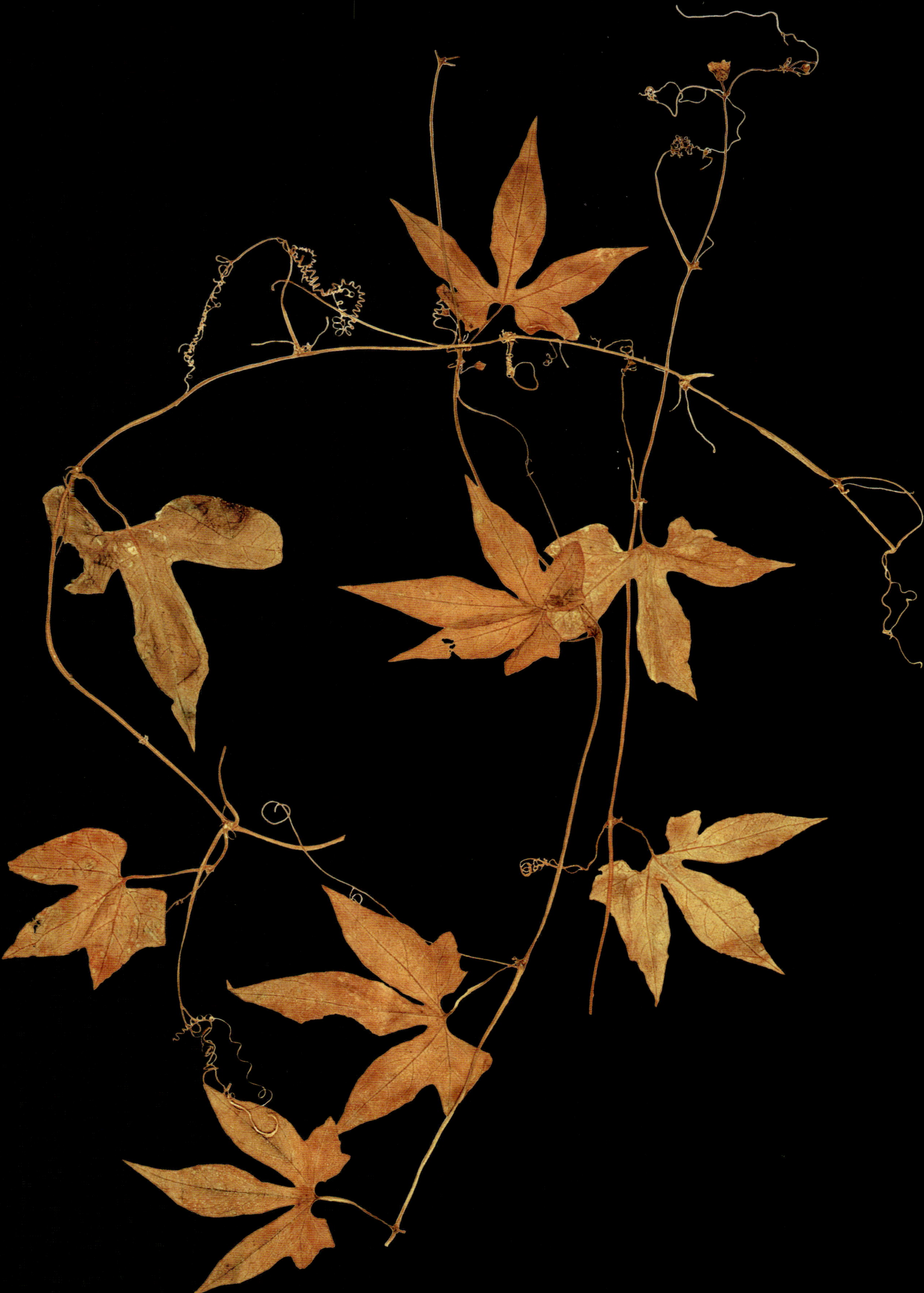

A Palmate-leaved Diplocyclos
Collected by Joseph Banks and Daniel Solander, 1770

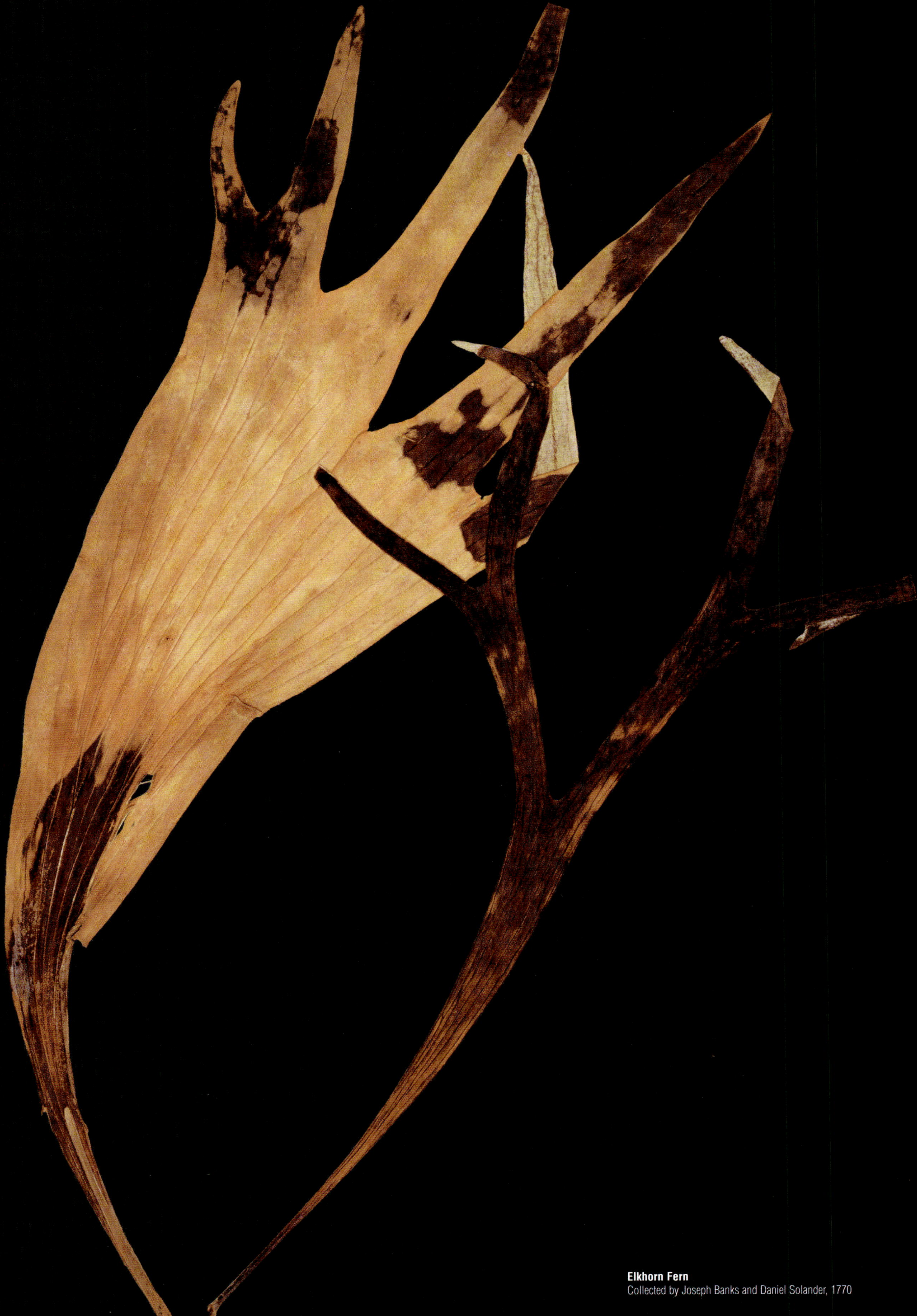

Elkhorn Fern
Collected by Joseph Banks and Daniel Solander, 1770

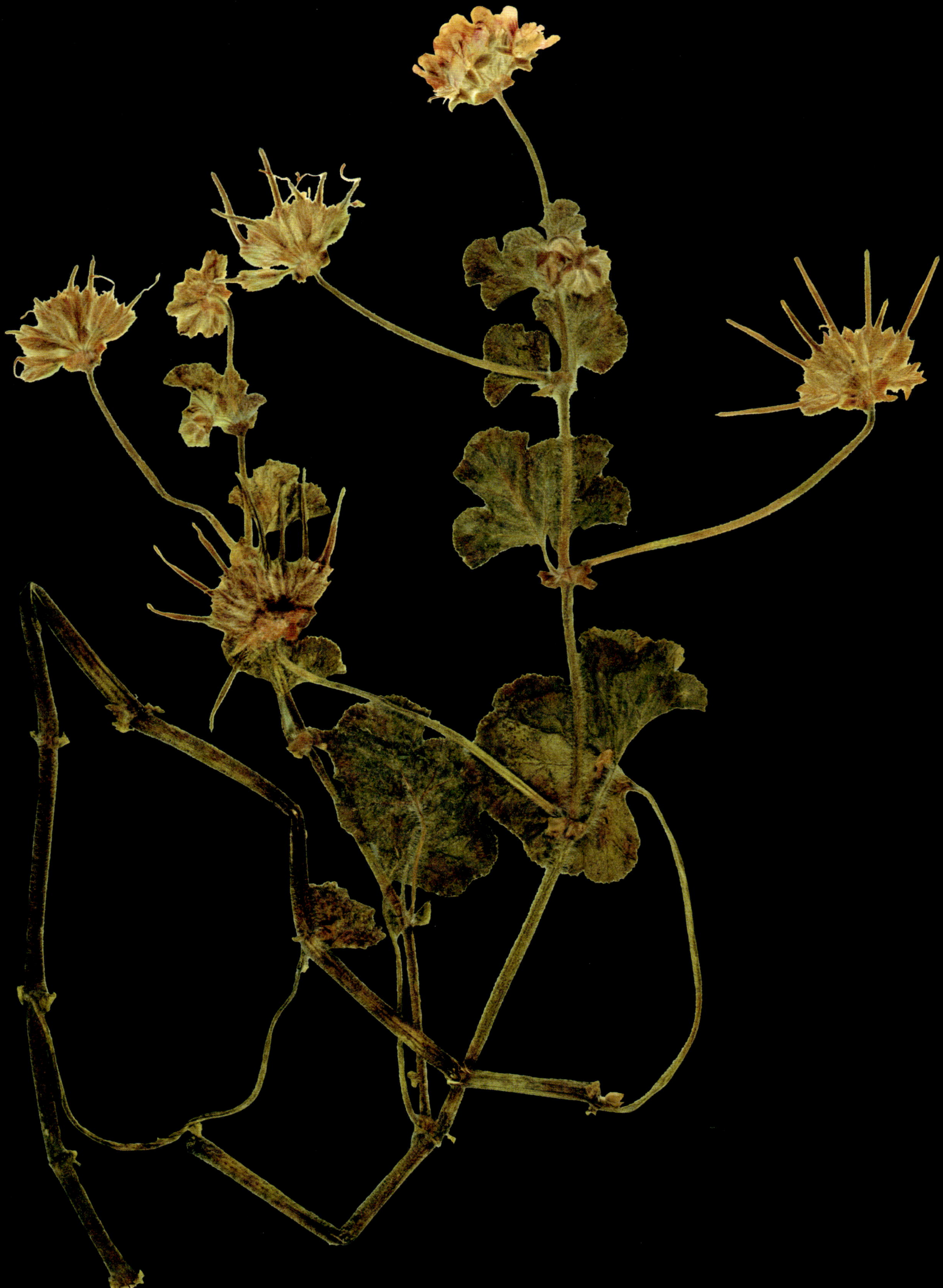

Rose-scented Pelargonium
Collected by Barbara G. Briggs, 1960

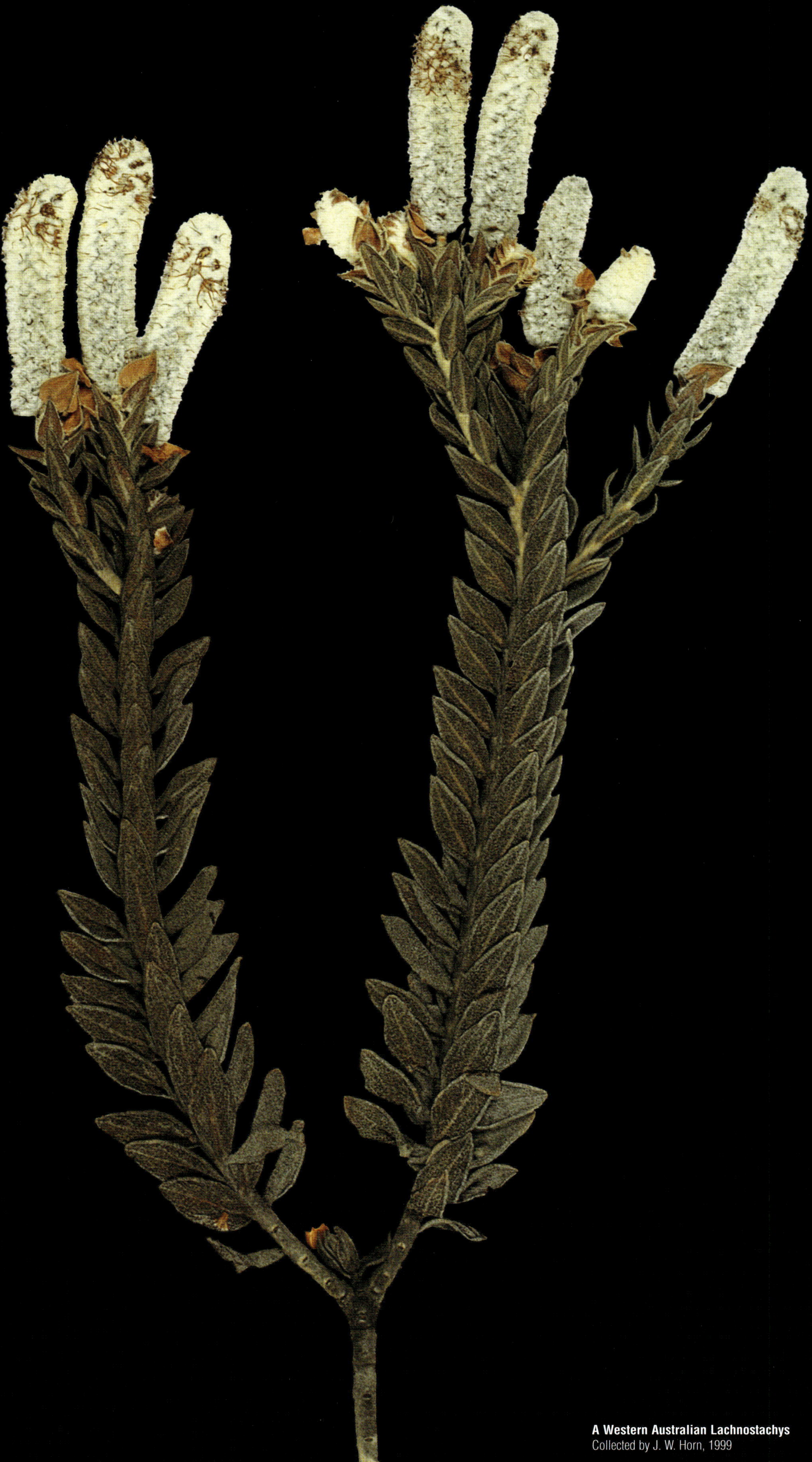

A Western Australian Lachnostachys
Collected by J. W. Horn, 1999

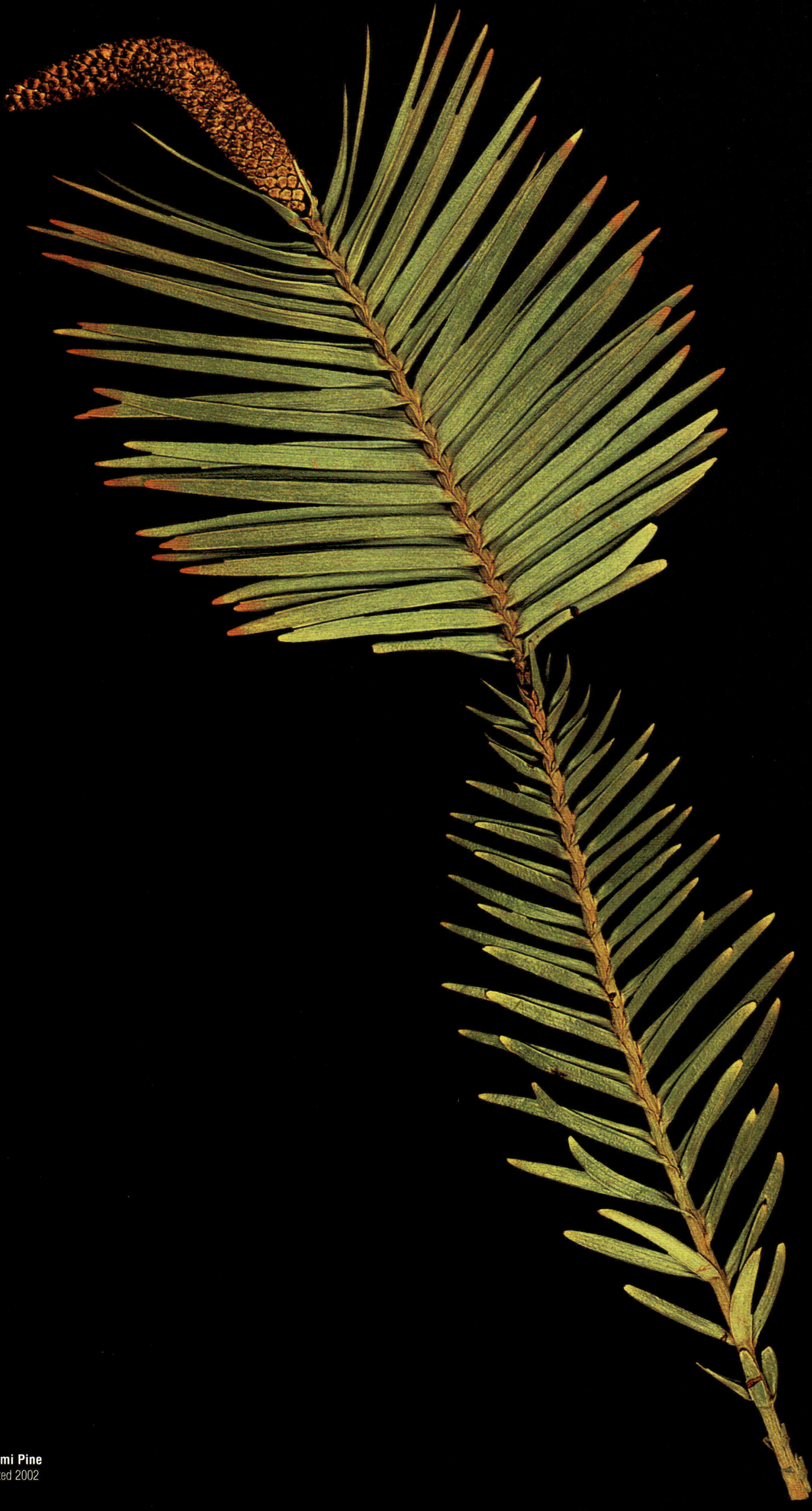

Wollemi Pine
Collected 2002

Scientific fascination

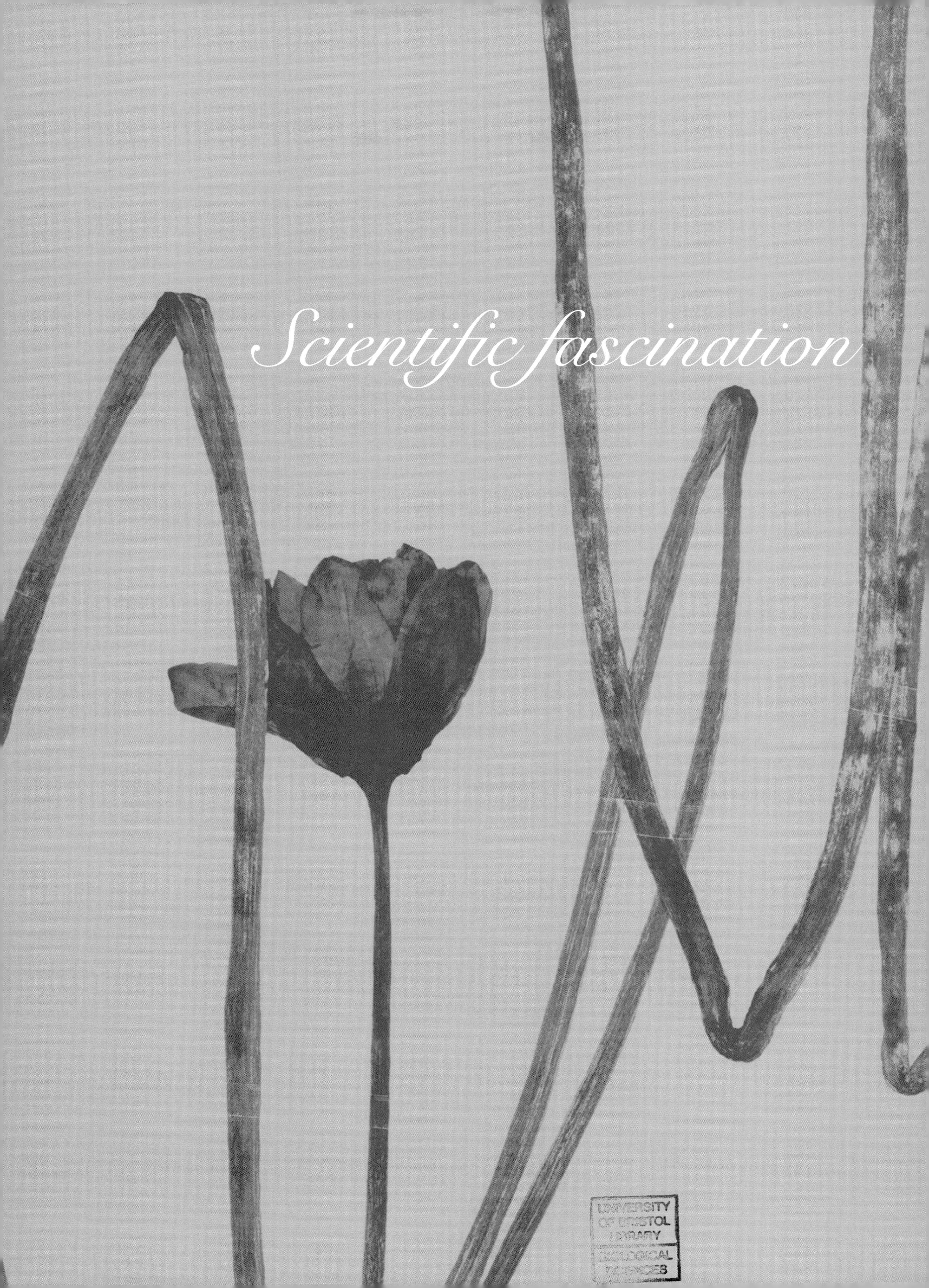

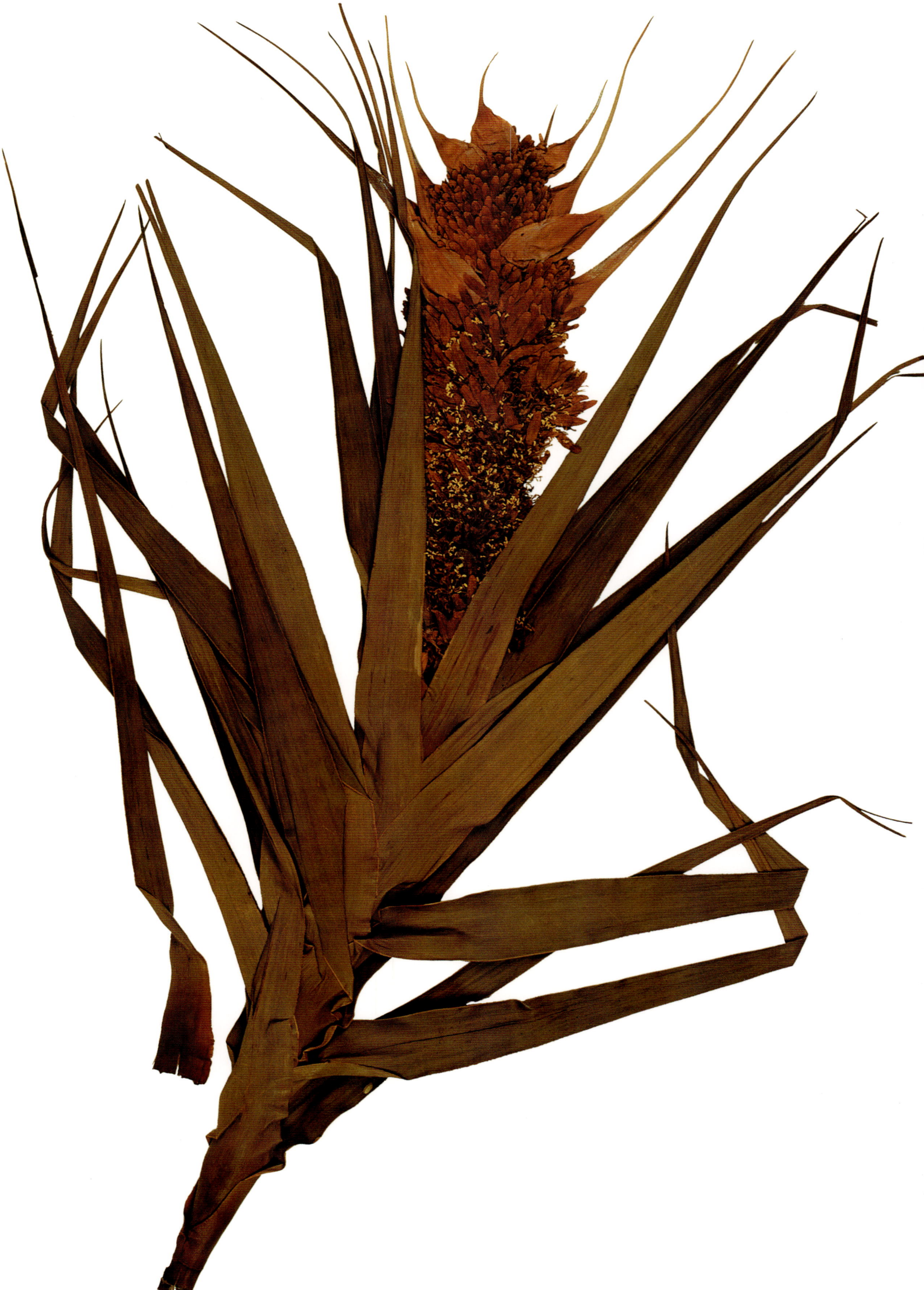

A Tasmanian Richea
Collected by W. H. Archer, mid-nineteenth century

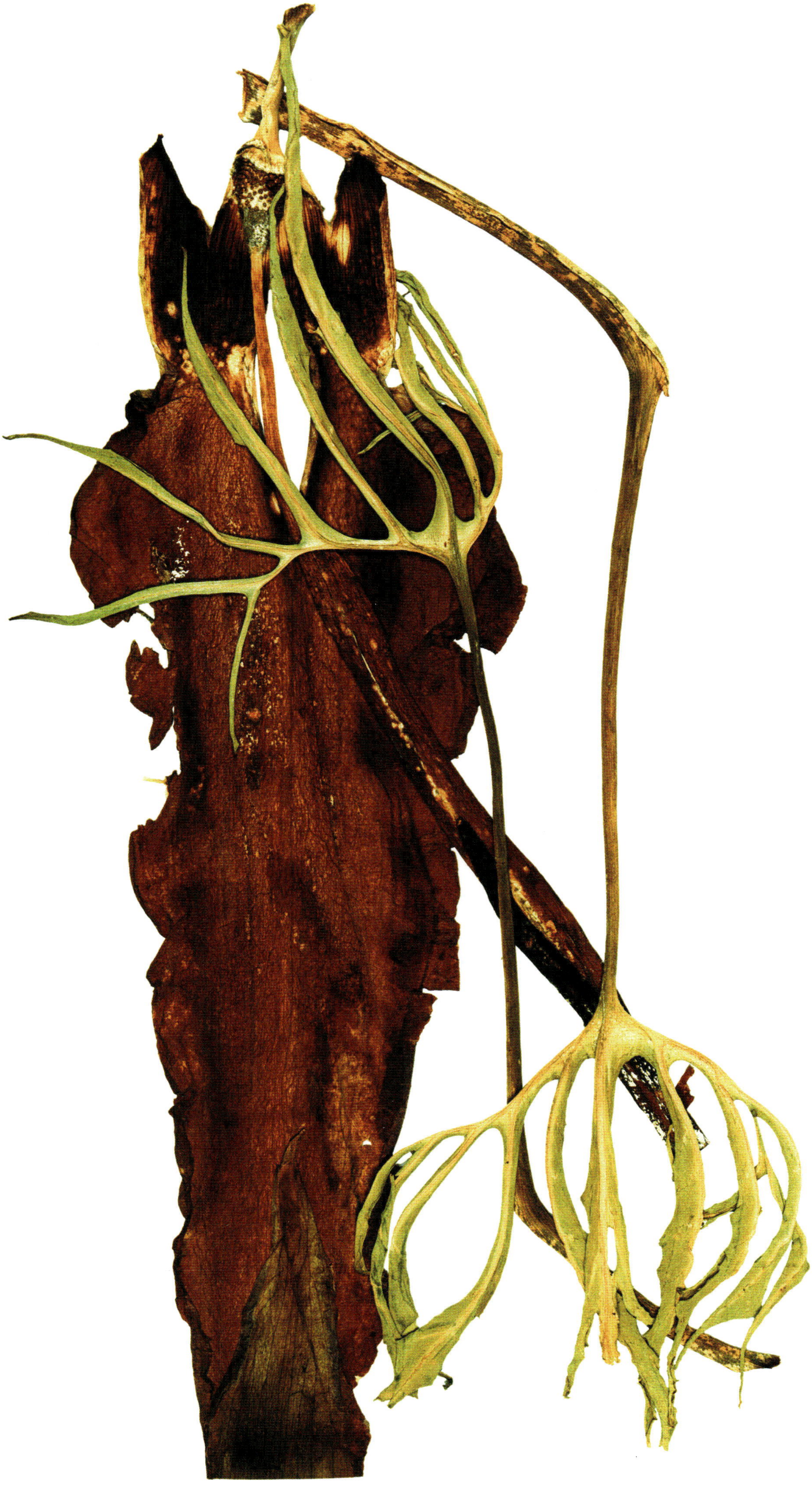

Dragon Arum
Collected by E. G. Knapman, 1946

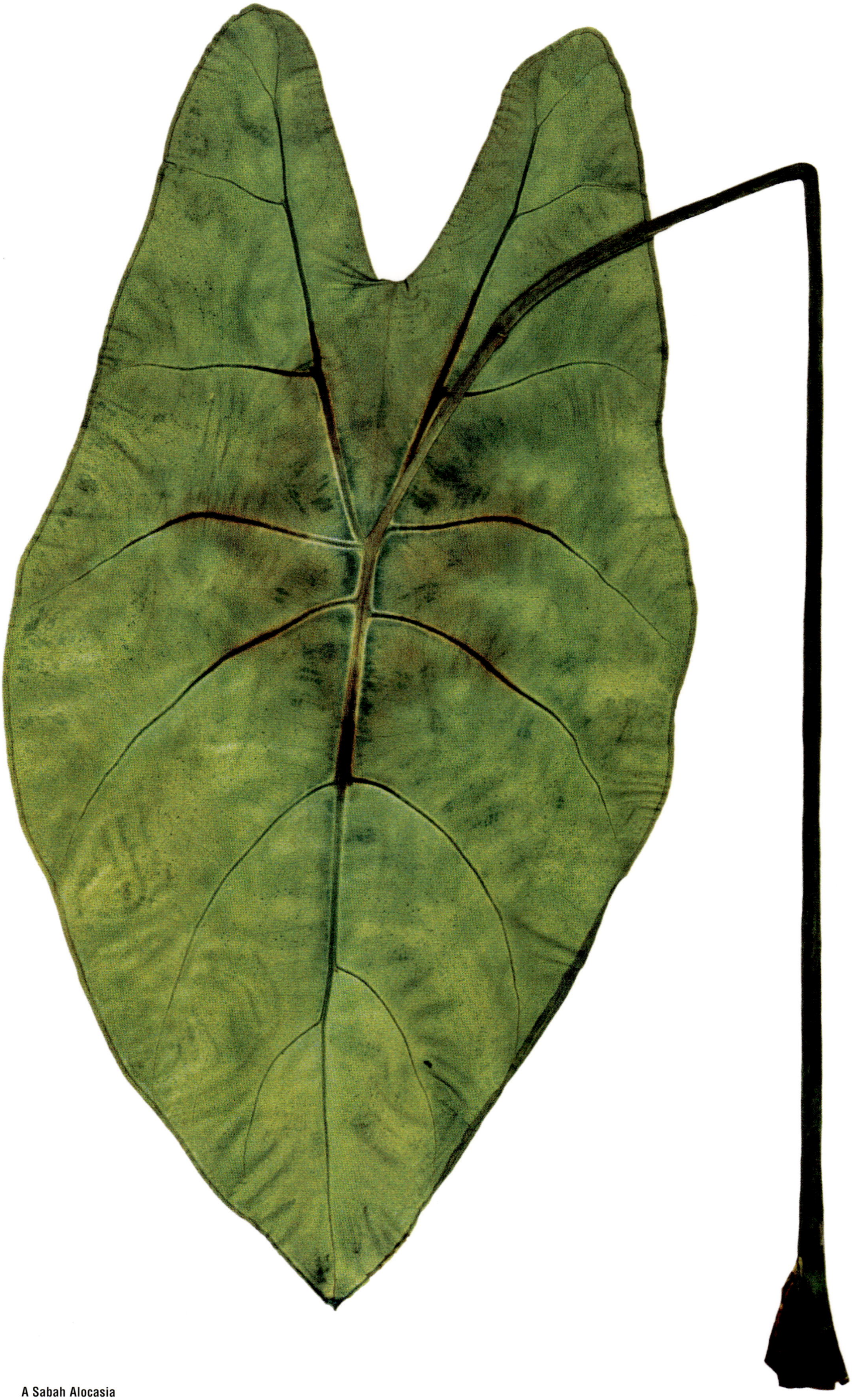

A Sabah Alocasia
Grown from material collected by Alistair Hay, 1996

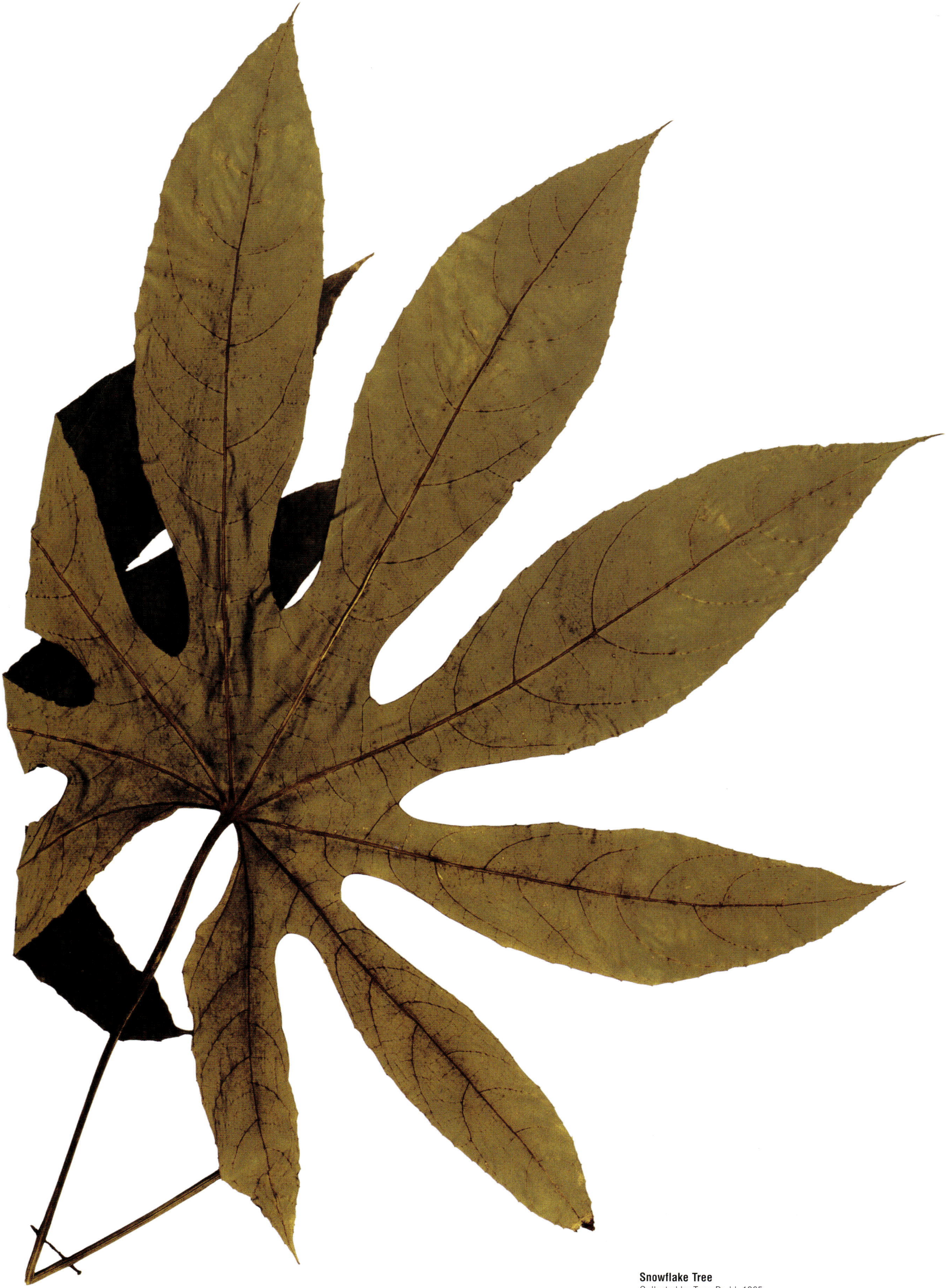

Snowflake Tree
Collected by Tony Rodd, 1965

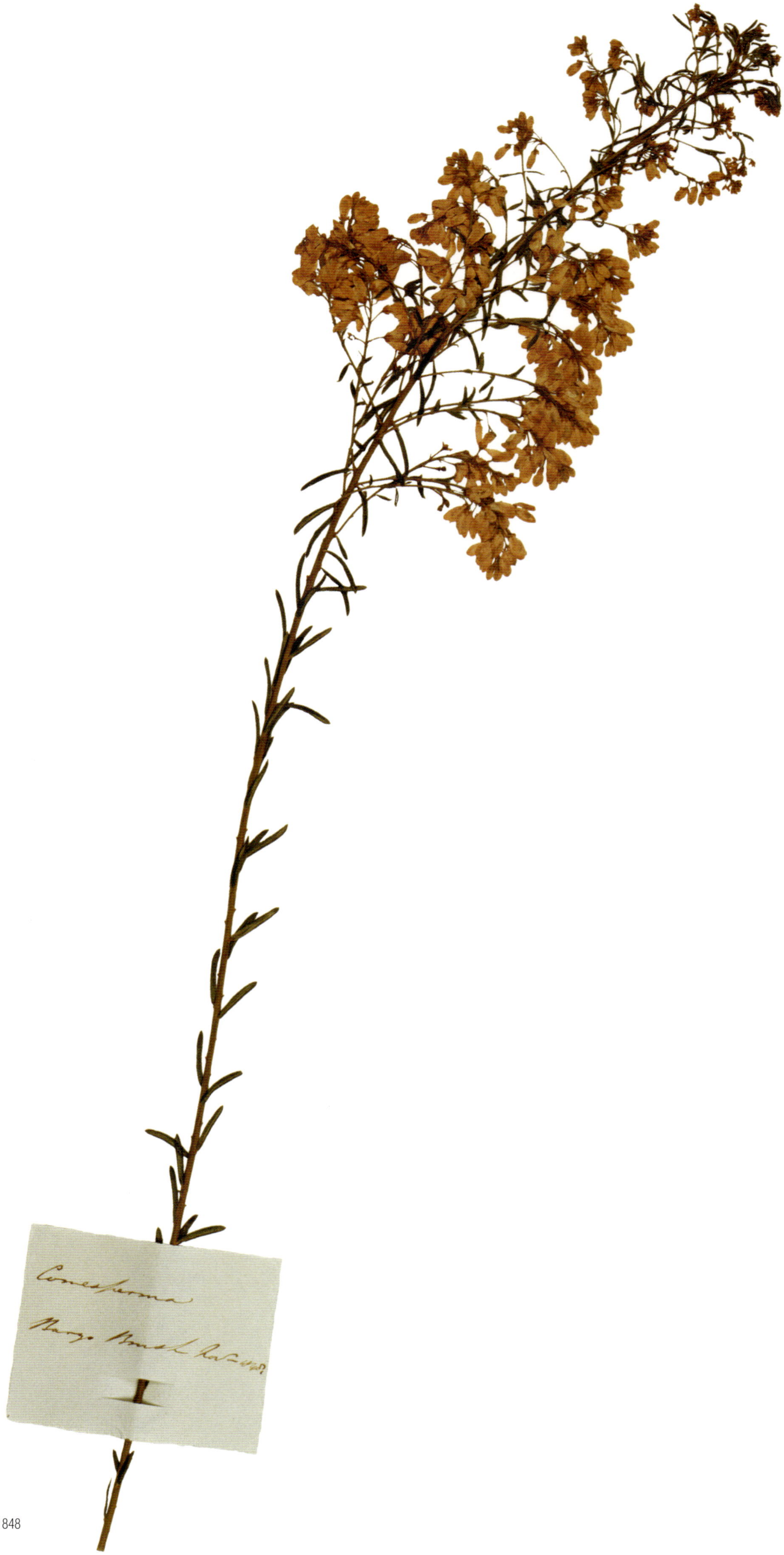

Pink Matchheads
Collected by William Macarthur, 1848

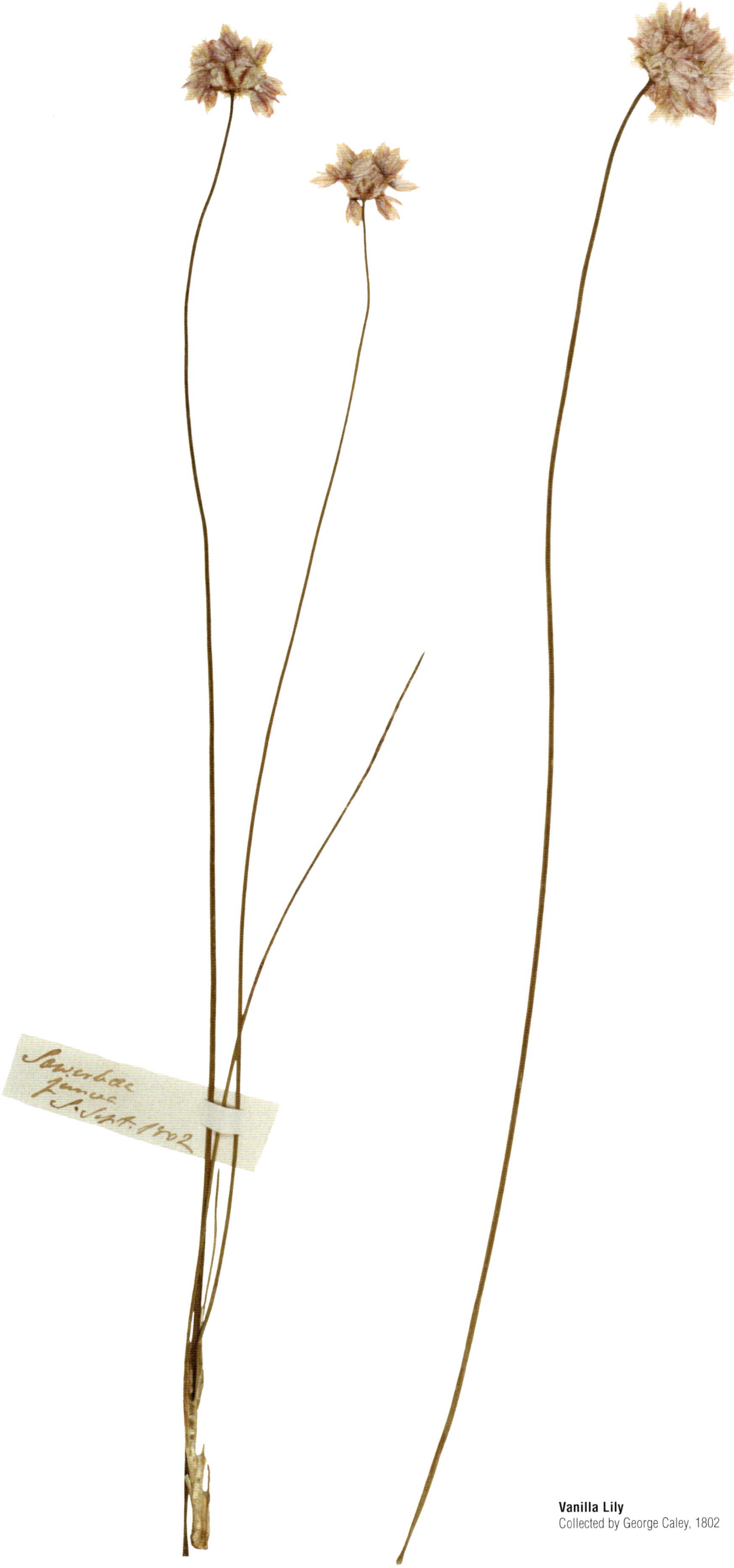

Vanilla Lily
Collected by George Caley, 1802

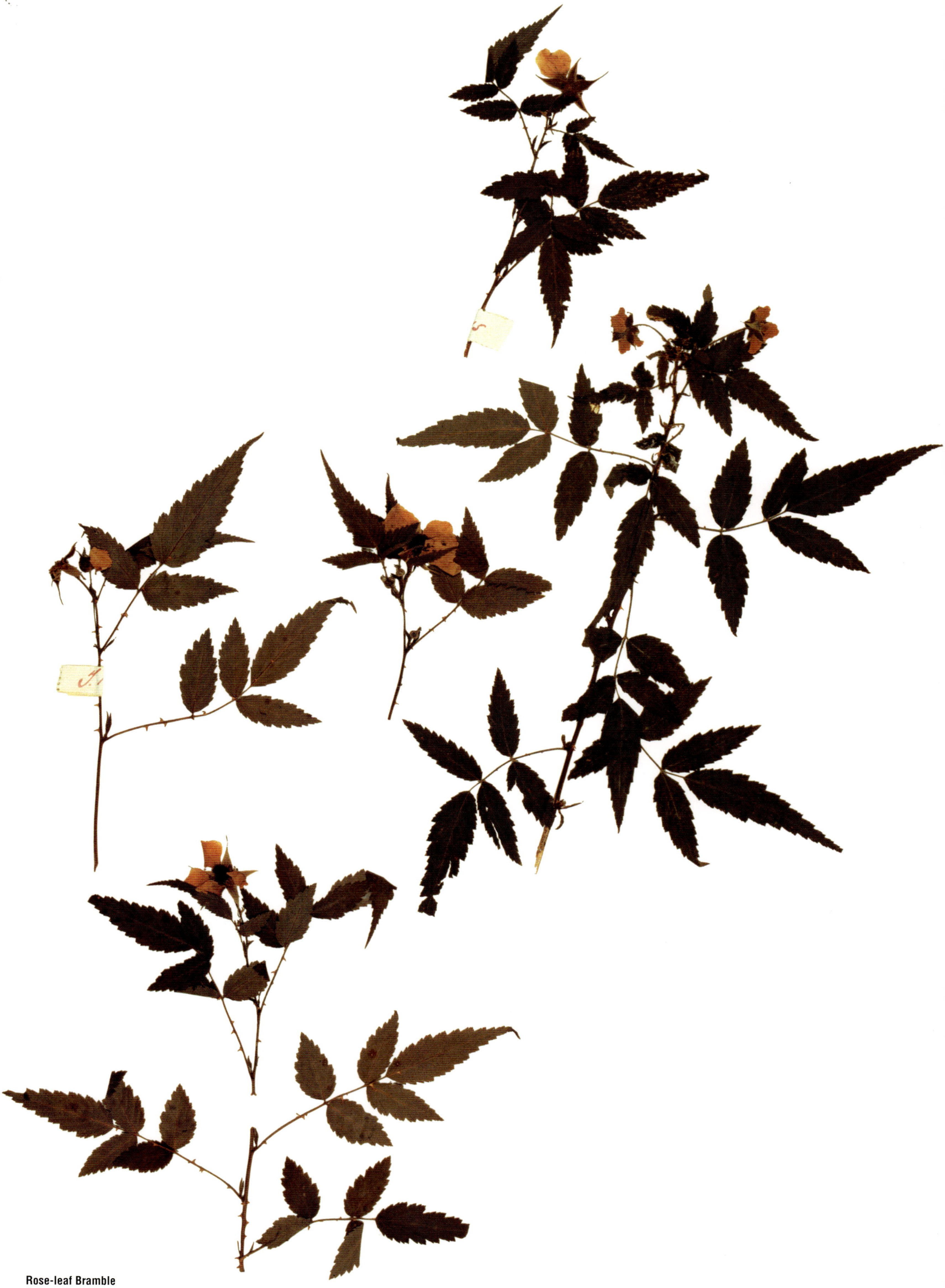

Rose-leaf Bramble
Collected by J. H. Maiden, 1888

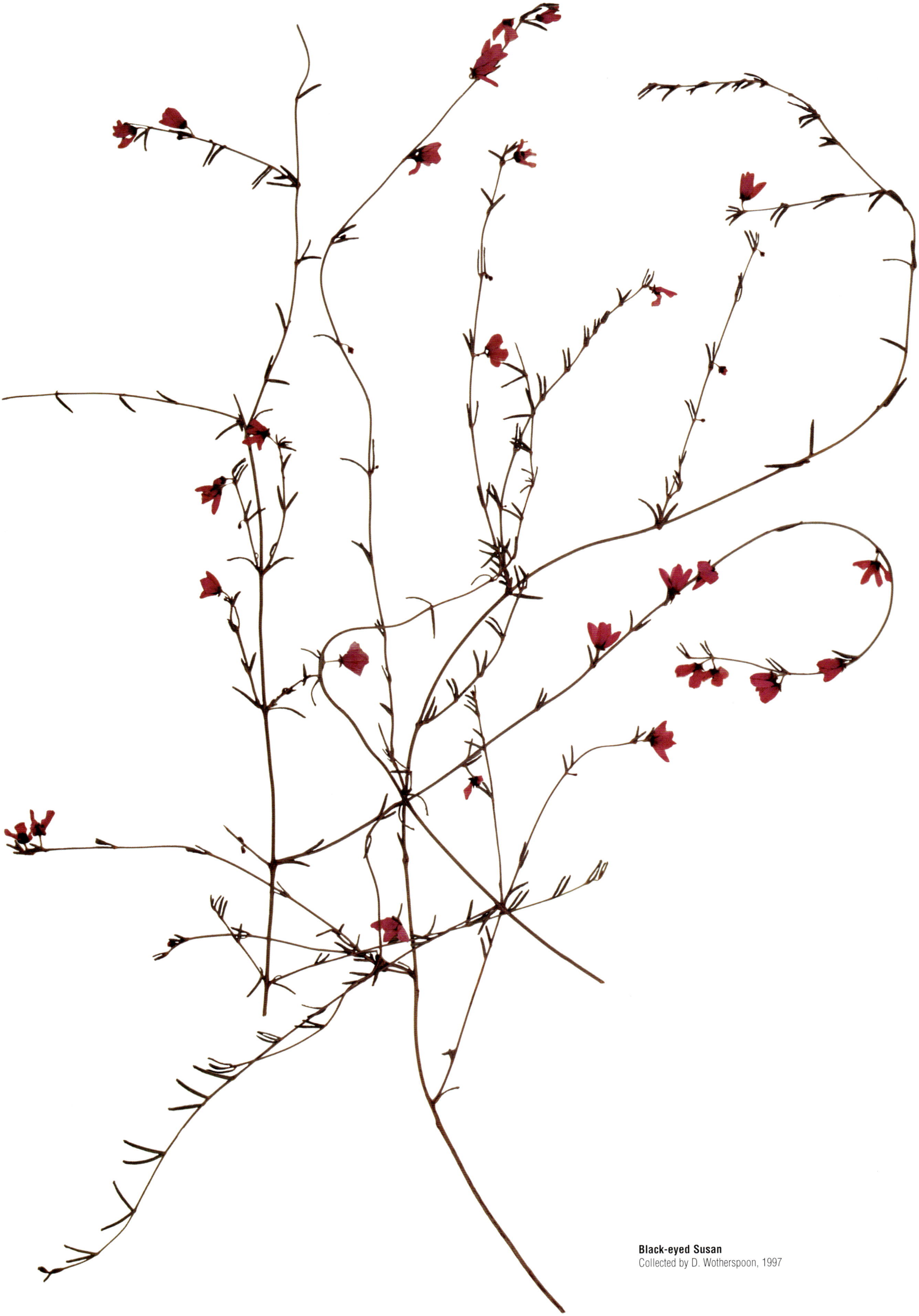

Black-eyed Susan
Collected by D. Wotherspoon, 1997

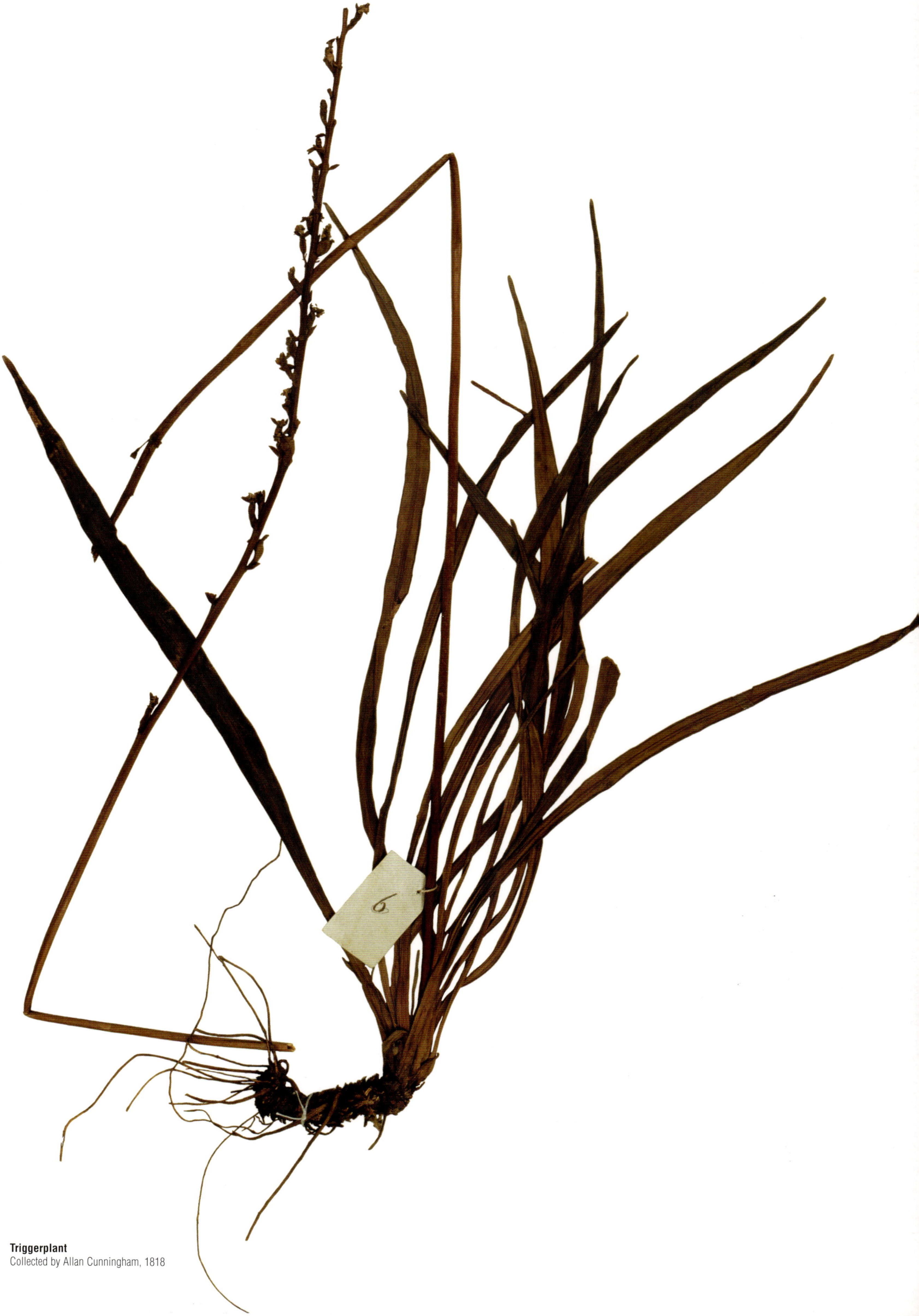

Triggerplant
Collected by Allan Cunningham, 1818

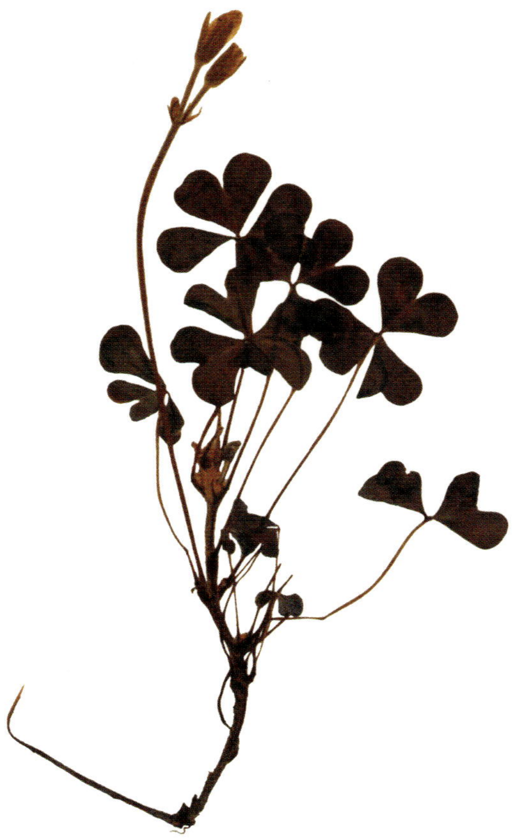

A Native Wood Sorrel
Collected by Ludwig Leichhardt, 1843

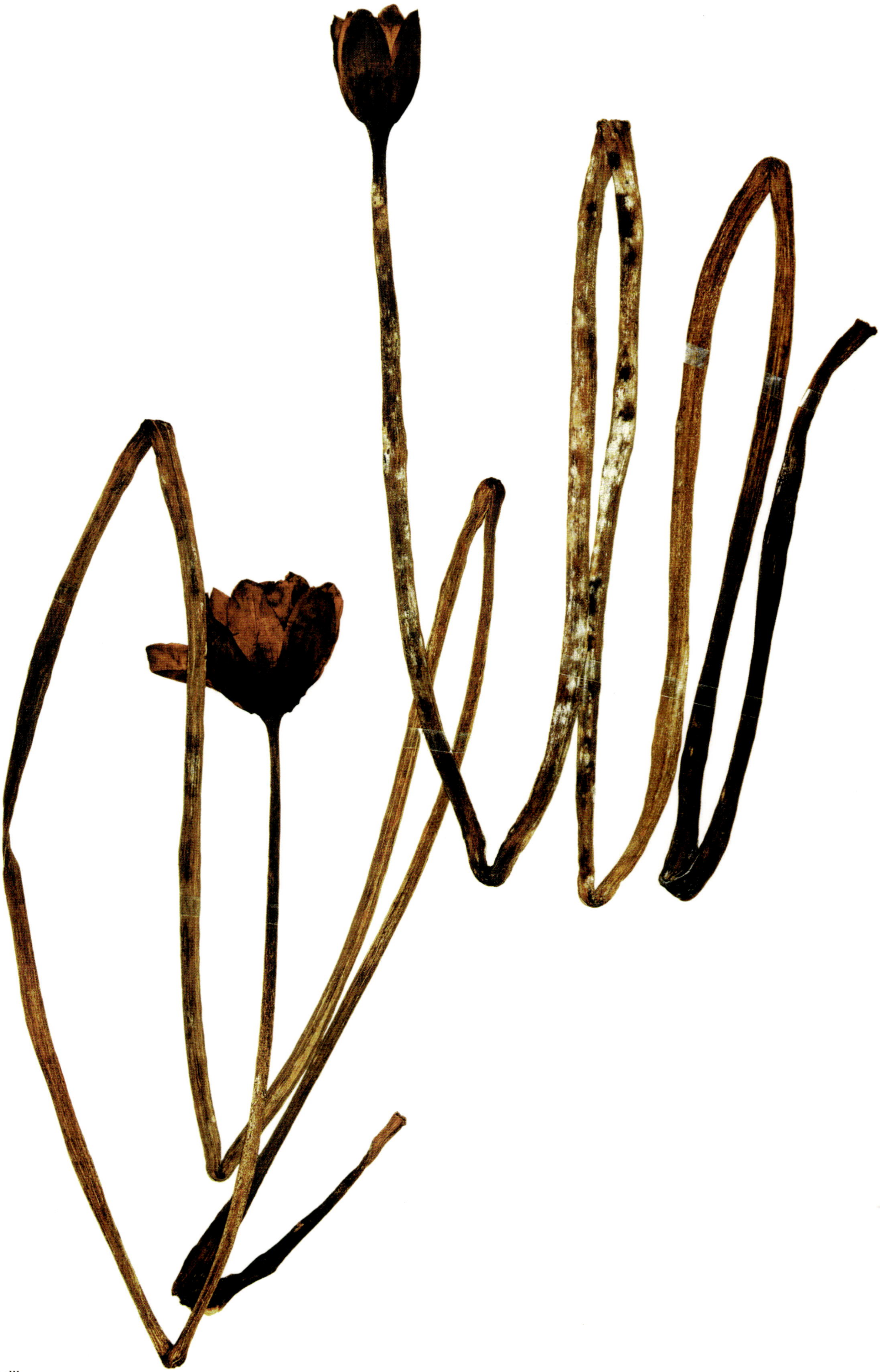

A Native Waterlily
Collected by John Clarkson, 1982

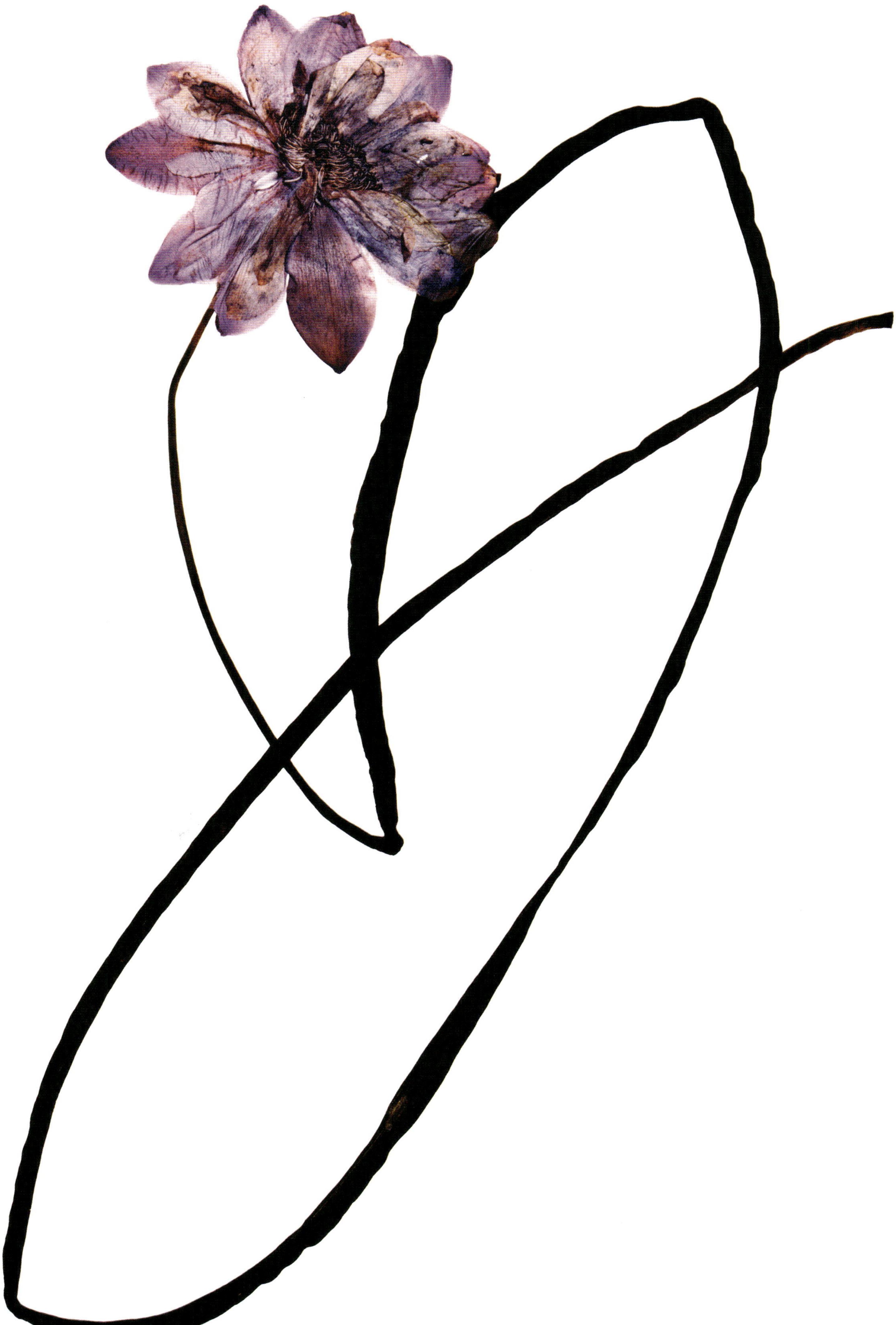

A Native Waterlily
Collected by Barry Conn and John DeCampo, 1983

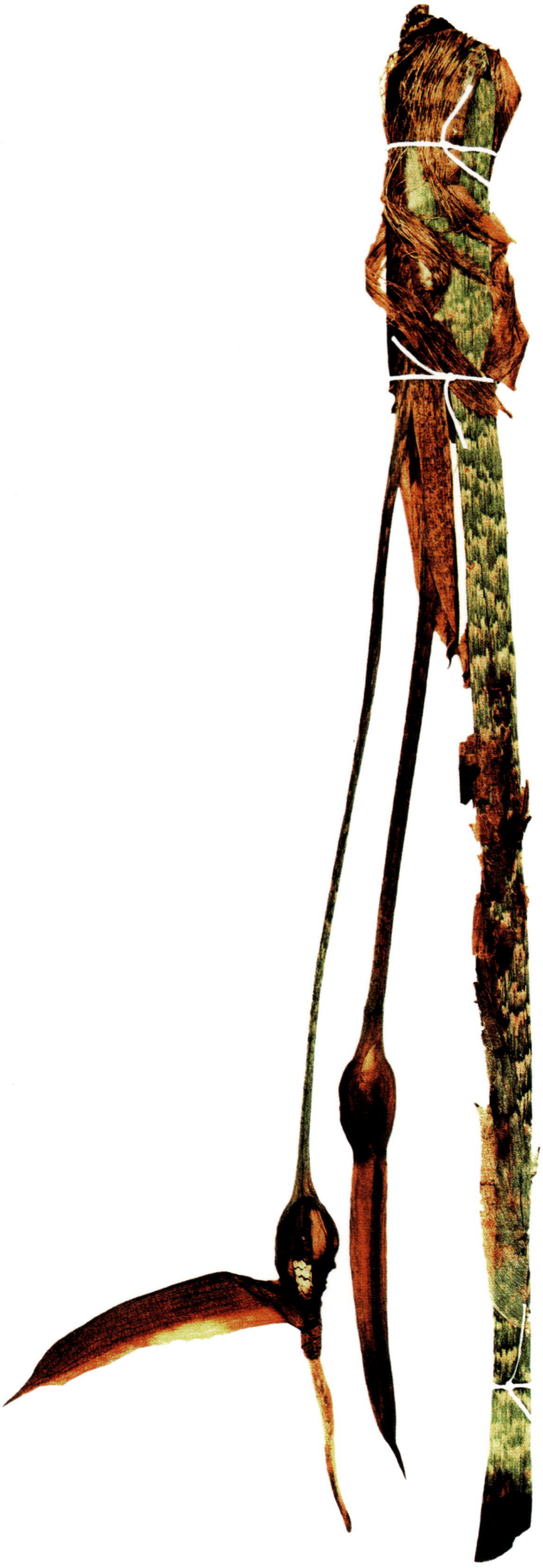

An Asian Alocasia
Collected by Clare Herscovitch, 1996

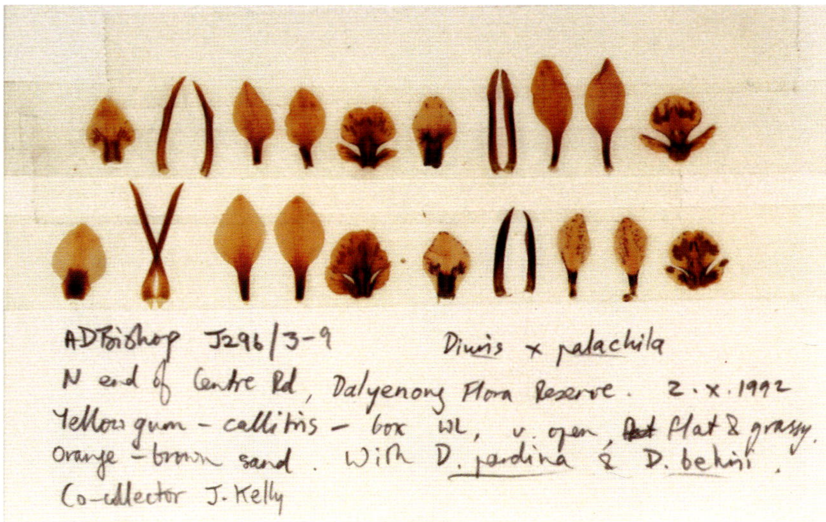

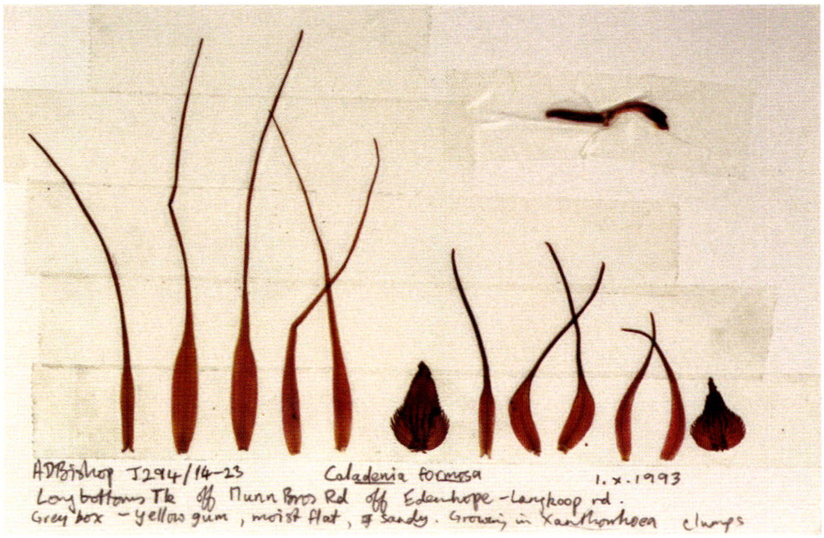

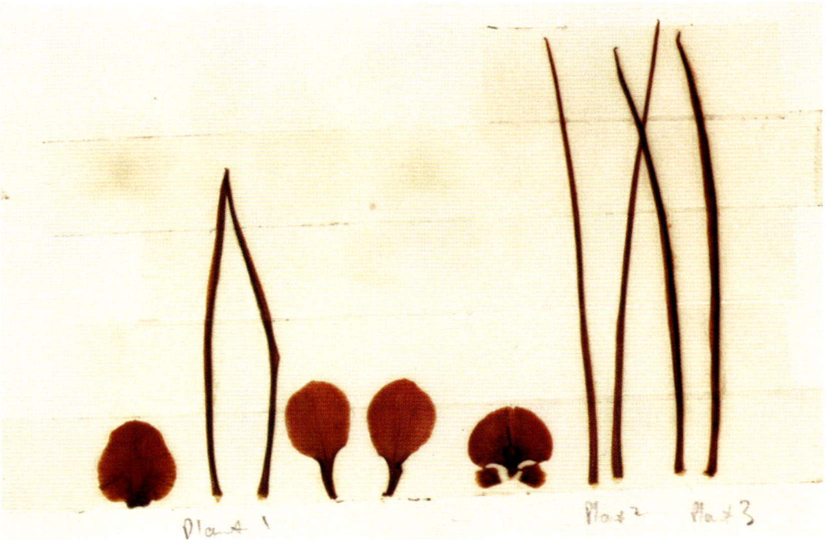

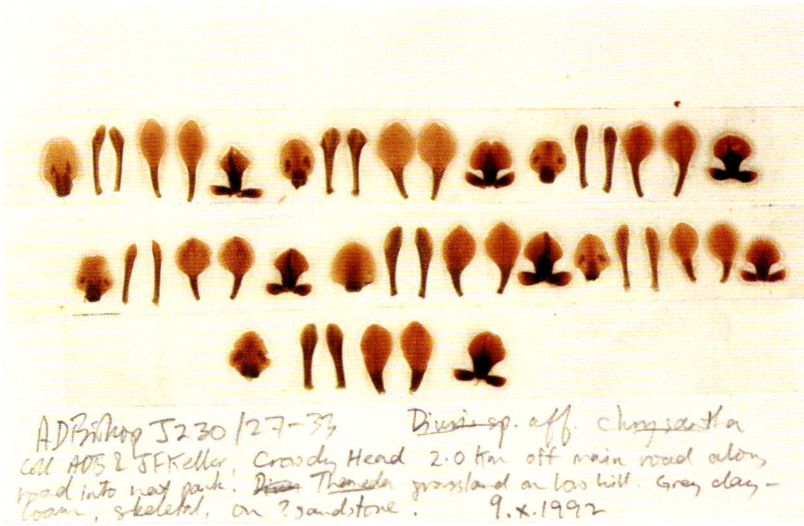

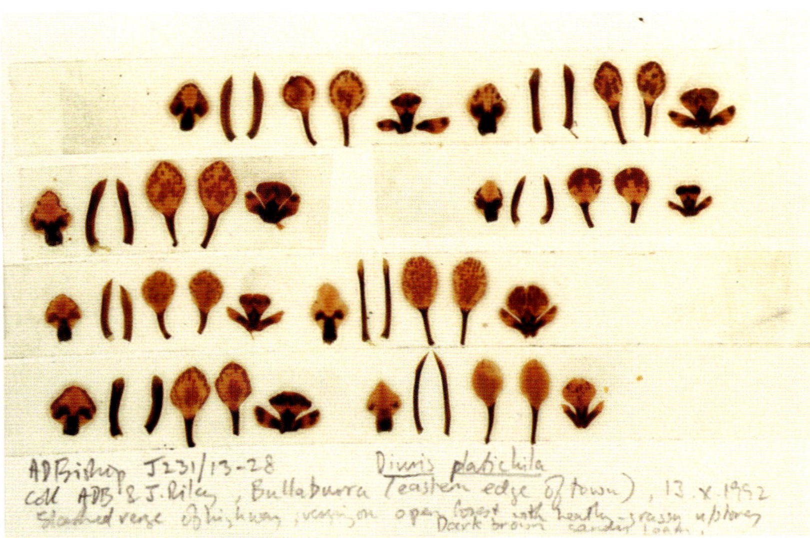

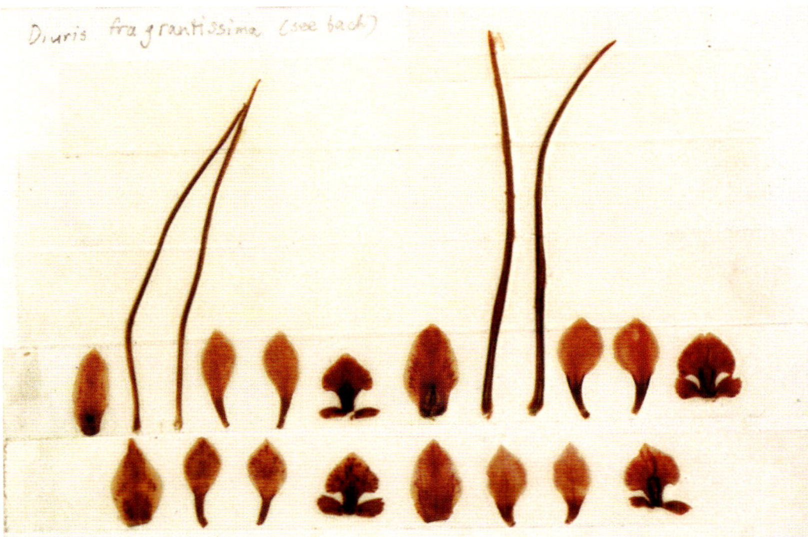

Tony Bishop's Orchid Cards

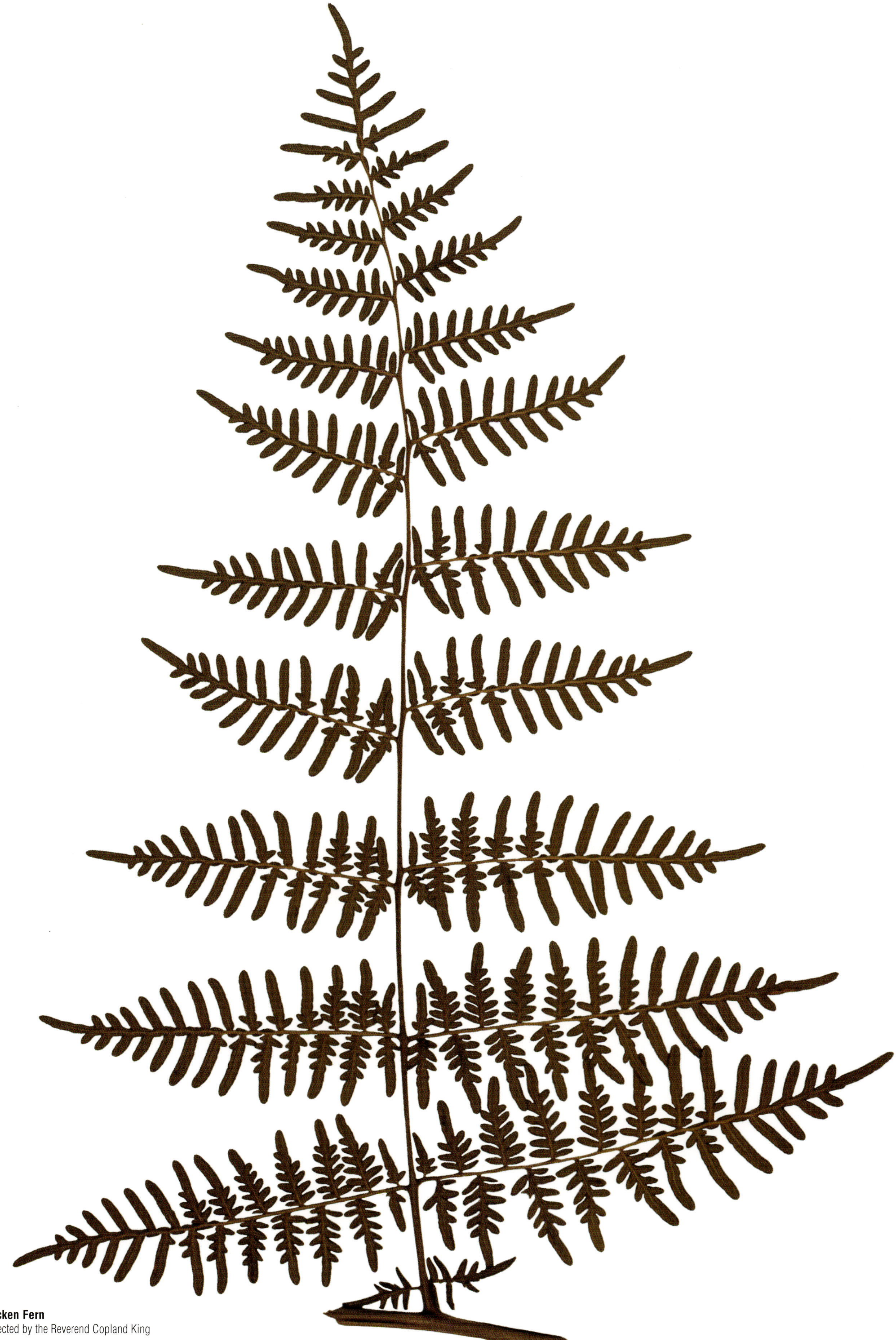

Bracken Fern
Collected by the Reverend Copland King

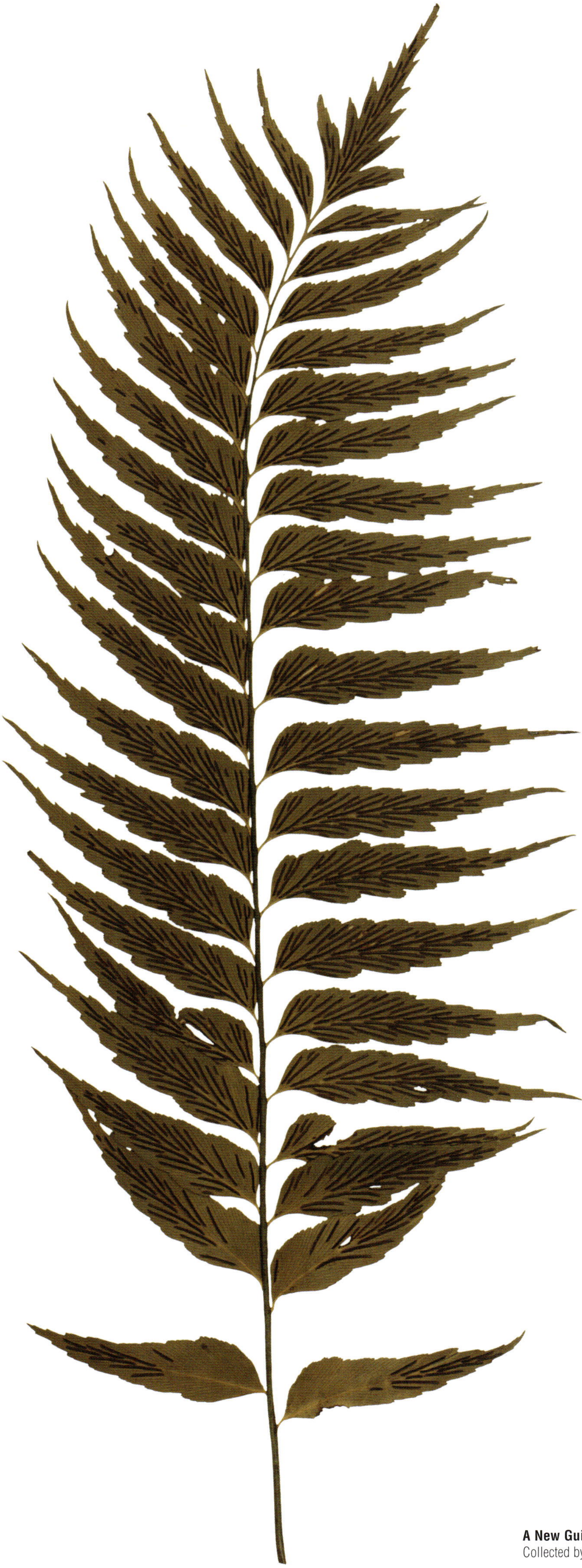

A New Guinea Fern
Collected by the Reverend Copland King

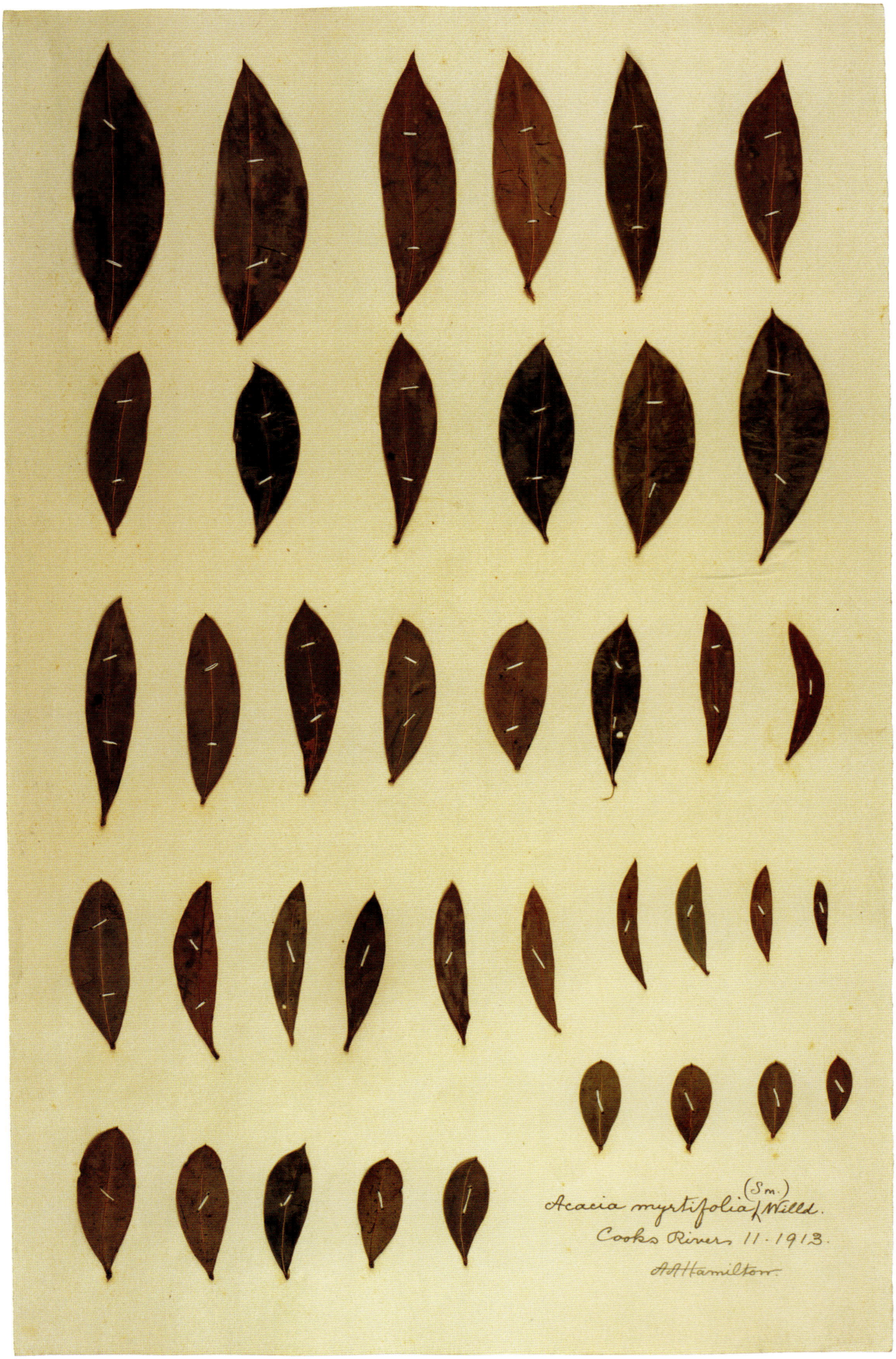

Red-stemmed Wattle
Collected by A. A. Hamilton, 1913

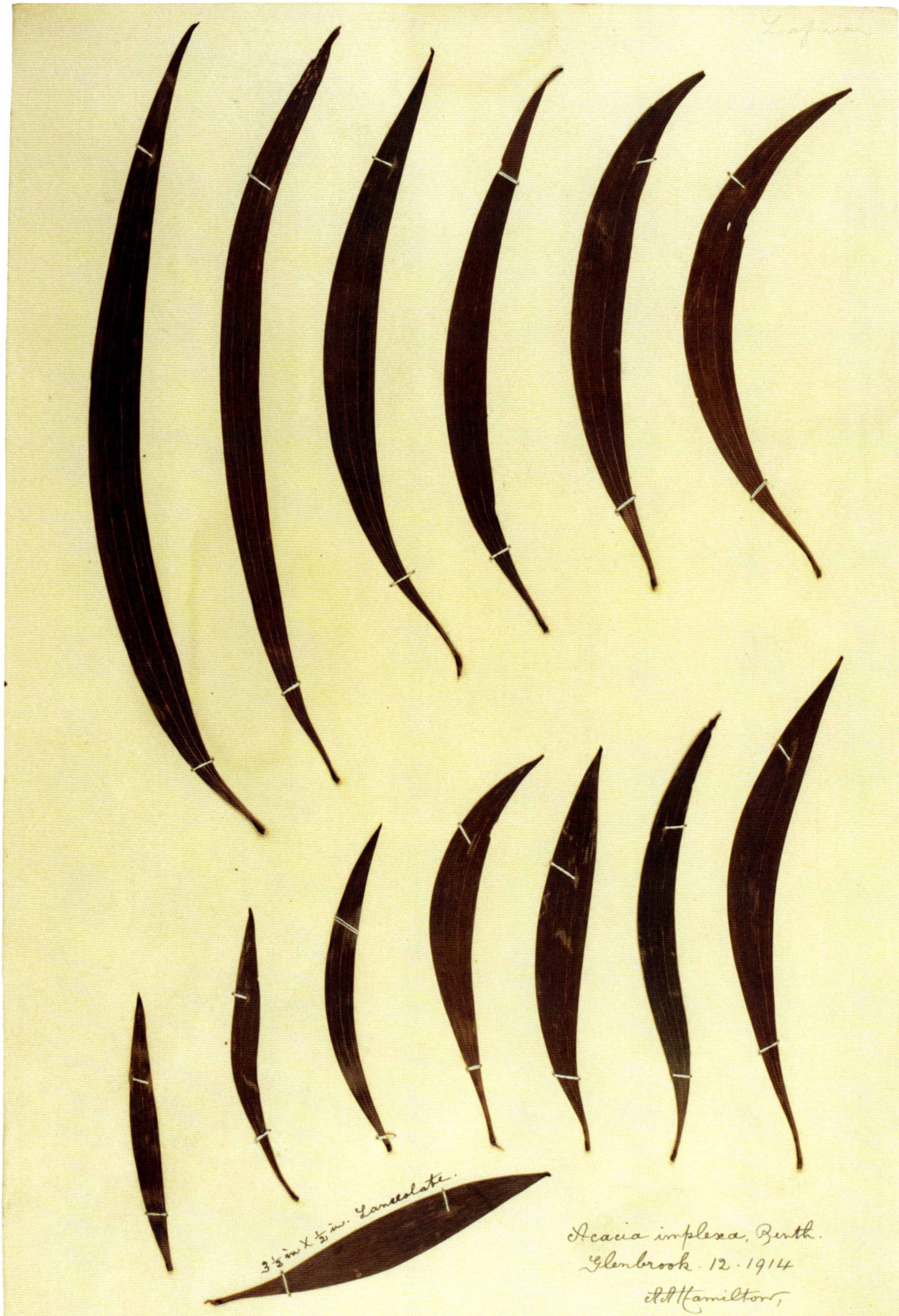

Hickory Wattle
Collected by A. A. Hamilton, 1914

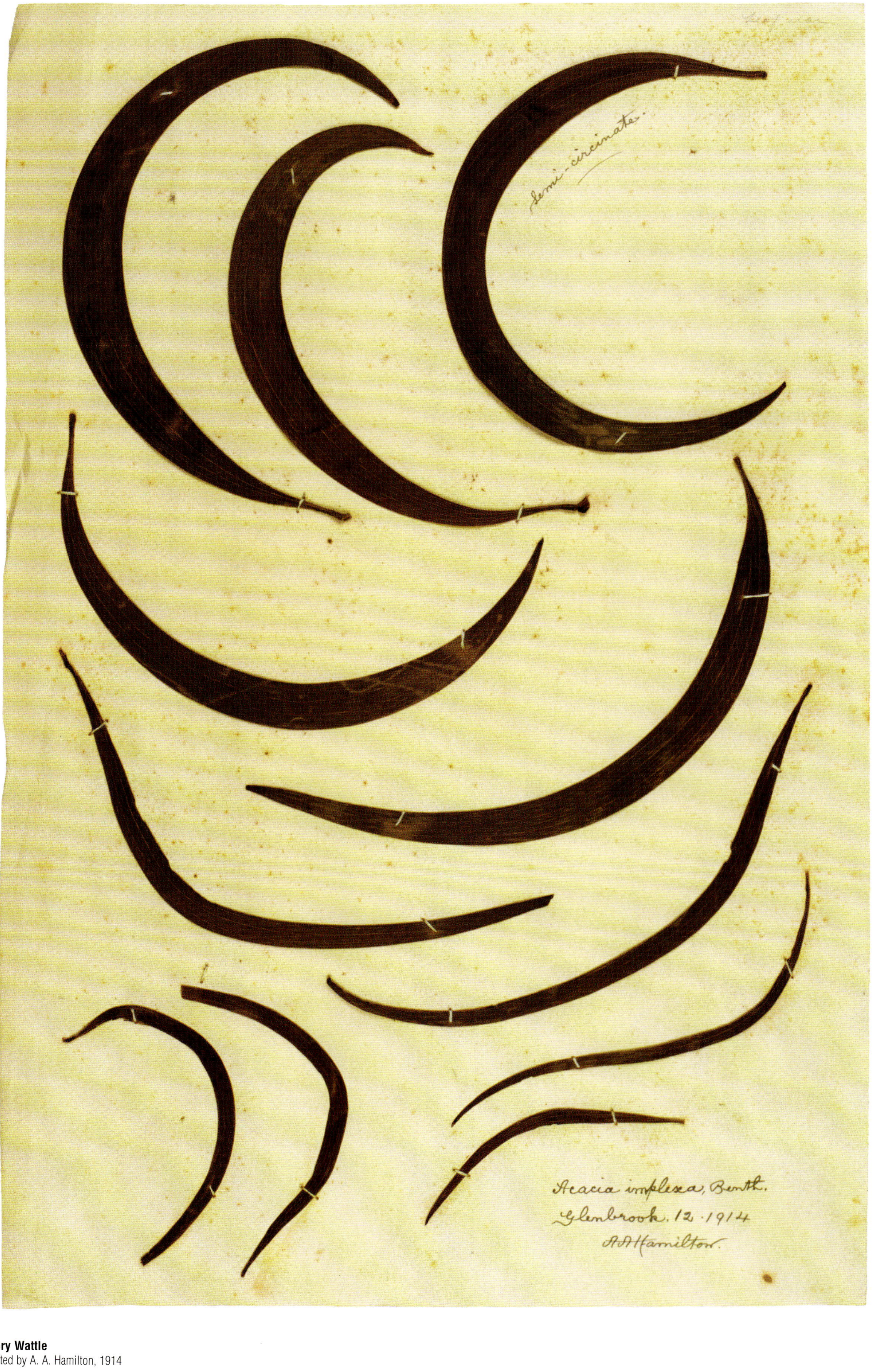

Hickory Wattle
Collected by A. A. Hamilton, 1914

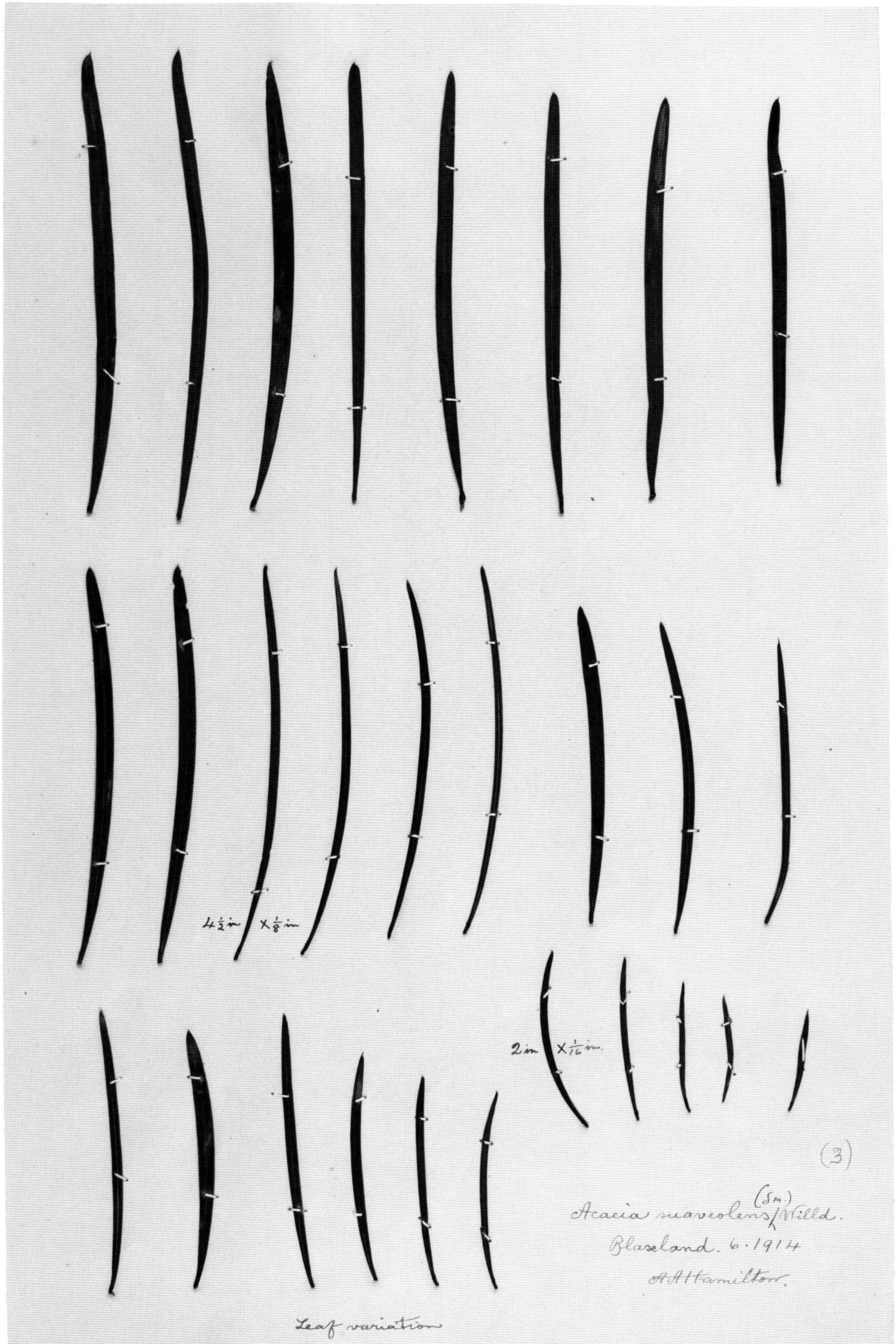

Sweet Wattle
Collected by A. A. Hamilton, 1913

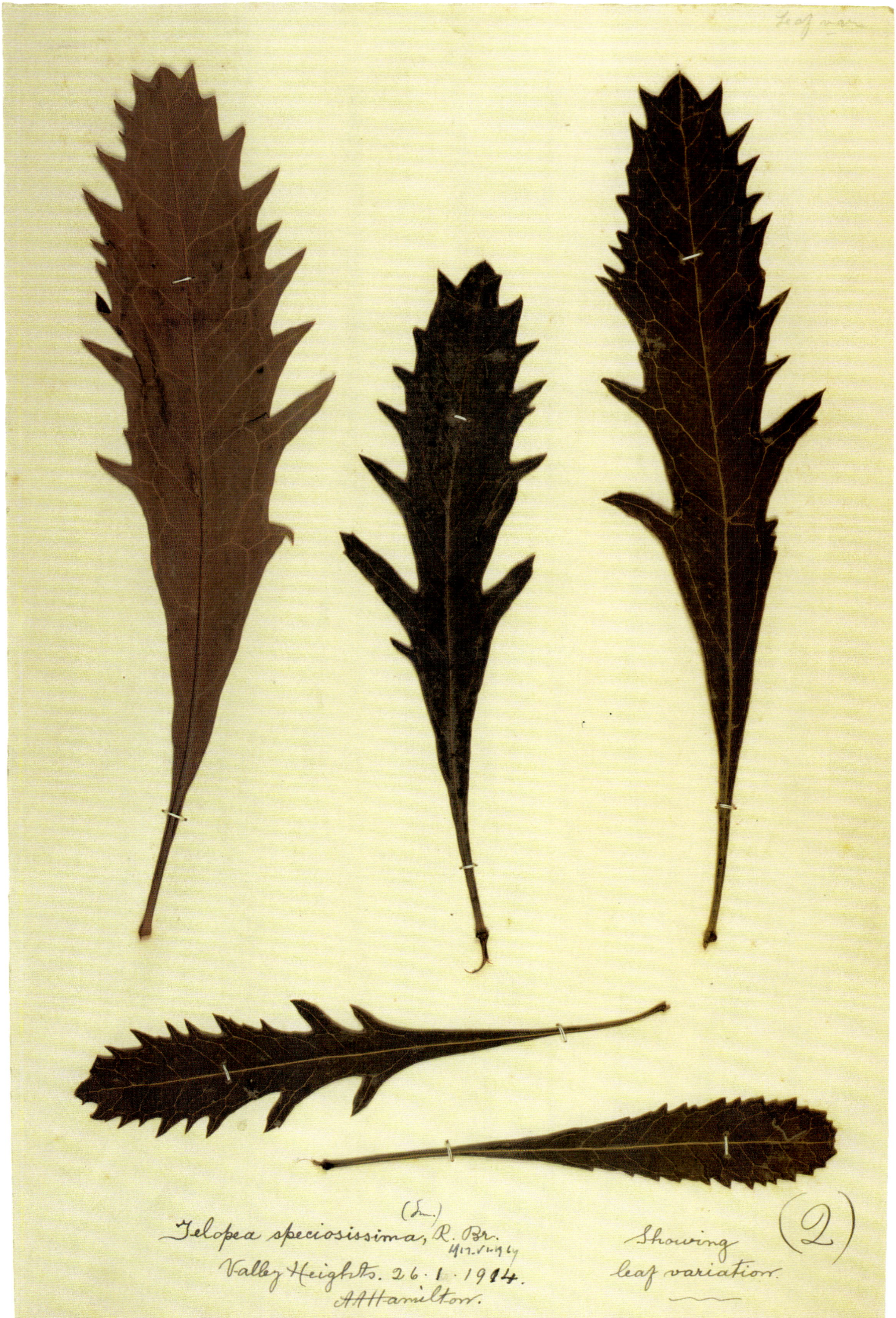

Waratah
Collected by A. A. Hamilton, 1914

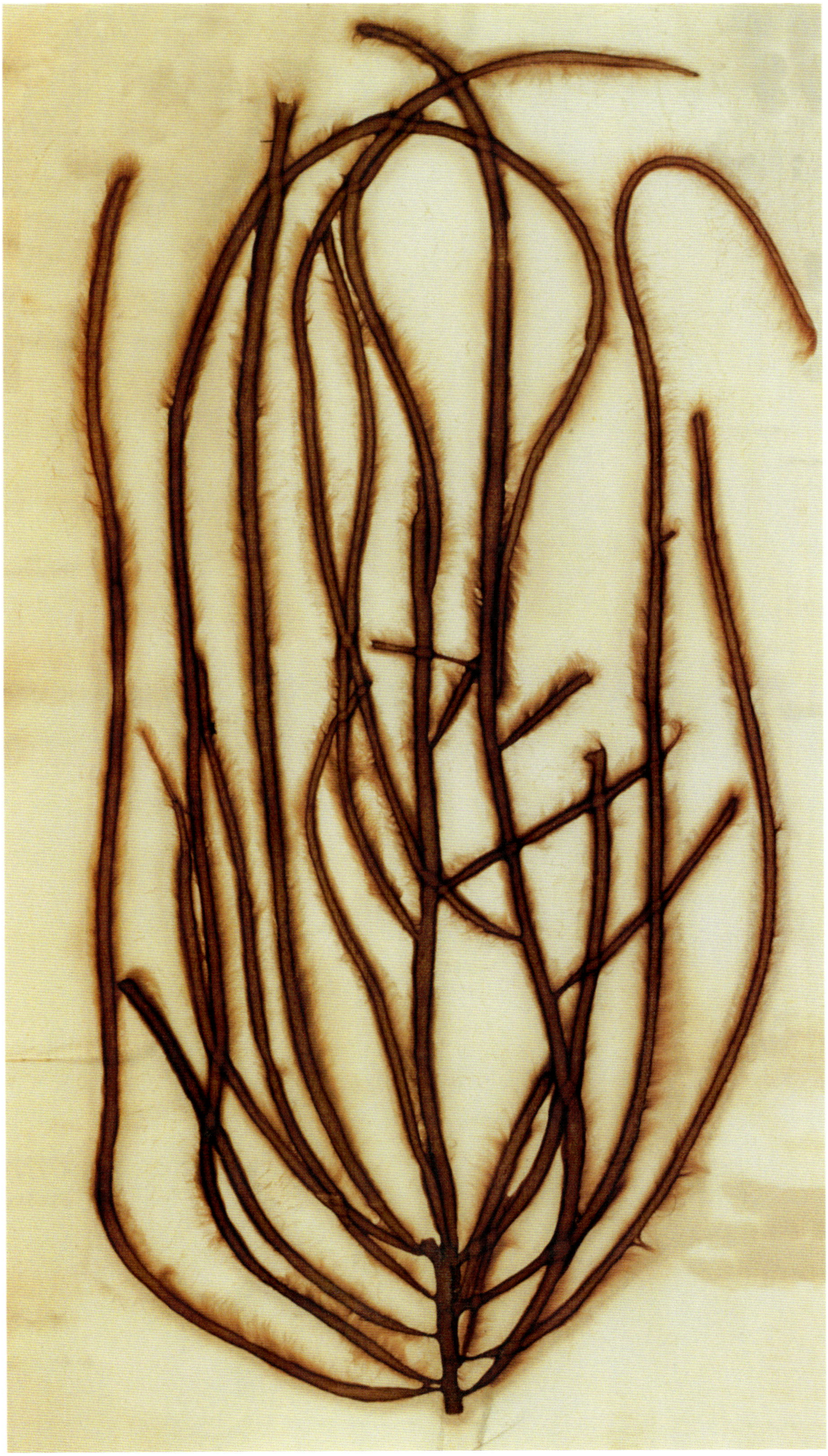

A Red Seaweed
Collected by W. H. Harvey, 1854

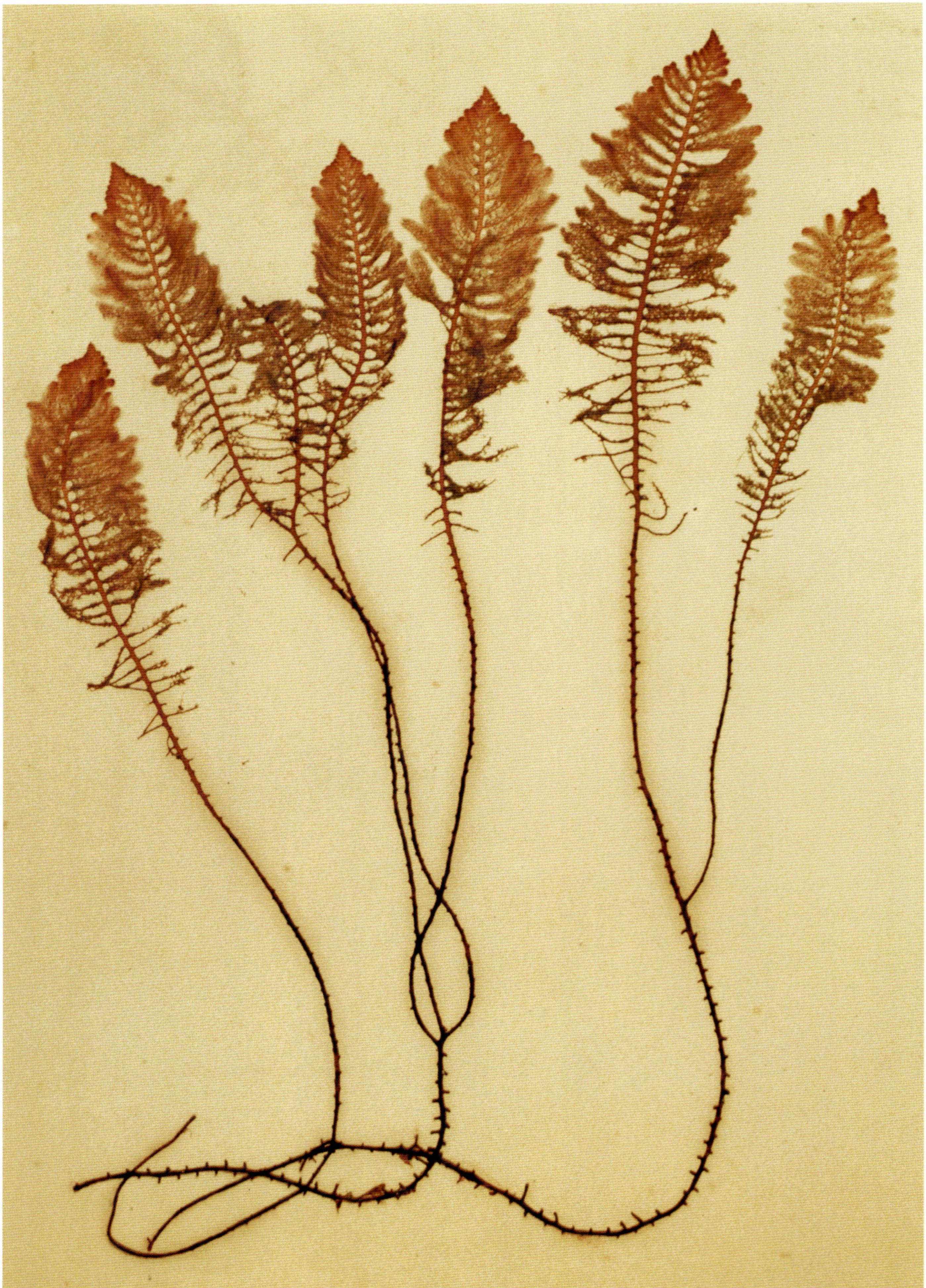

A Red Seaweed
Collected by W. H. Harvey, 1854

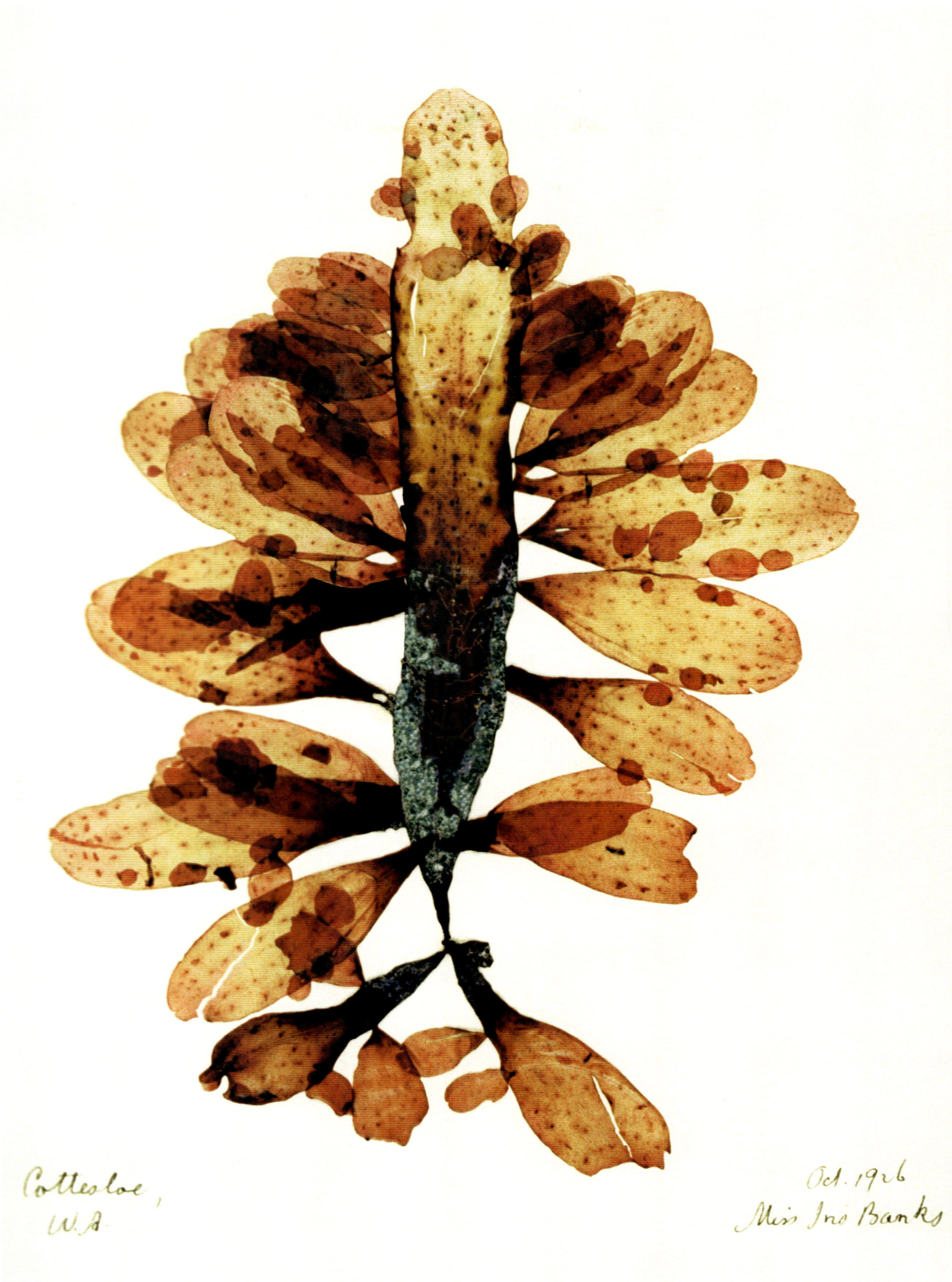

A Red Seaweed
Collected by Miss Iris Banks, 1926

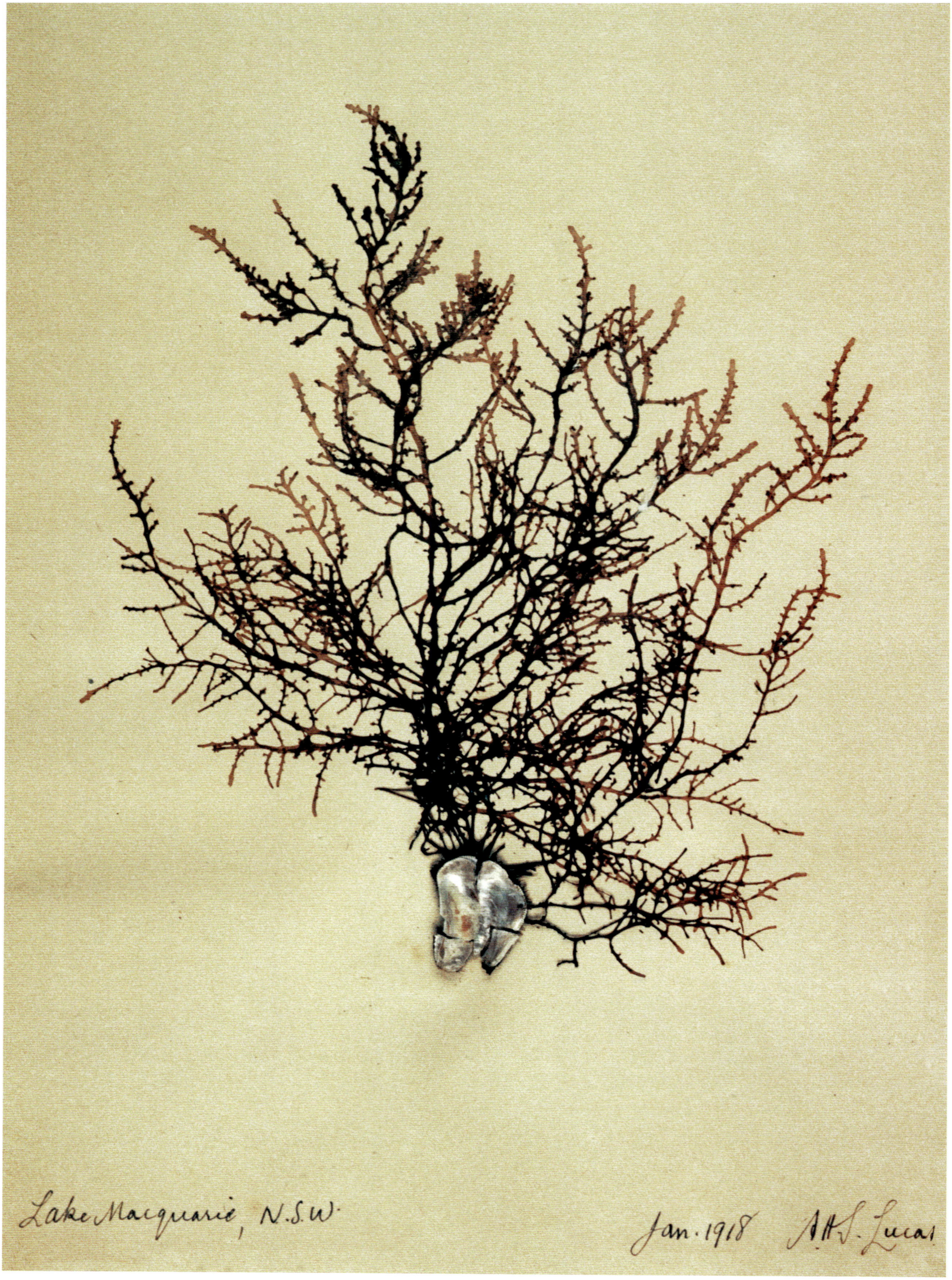

A Red Seaweed
Collected by Arthur Lucas, 1918

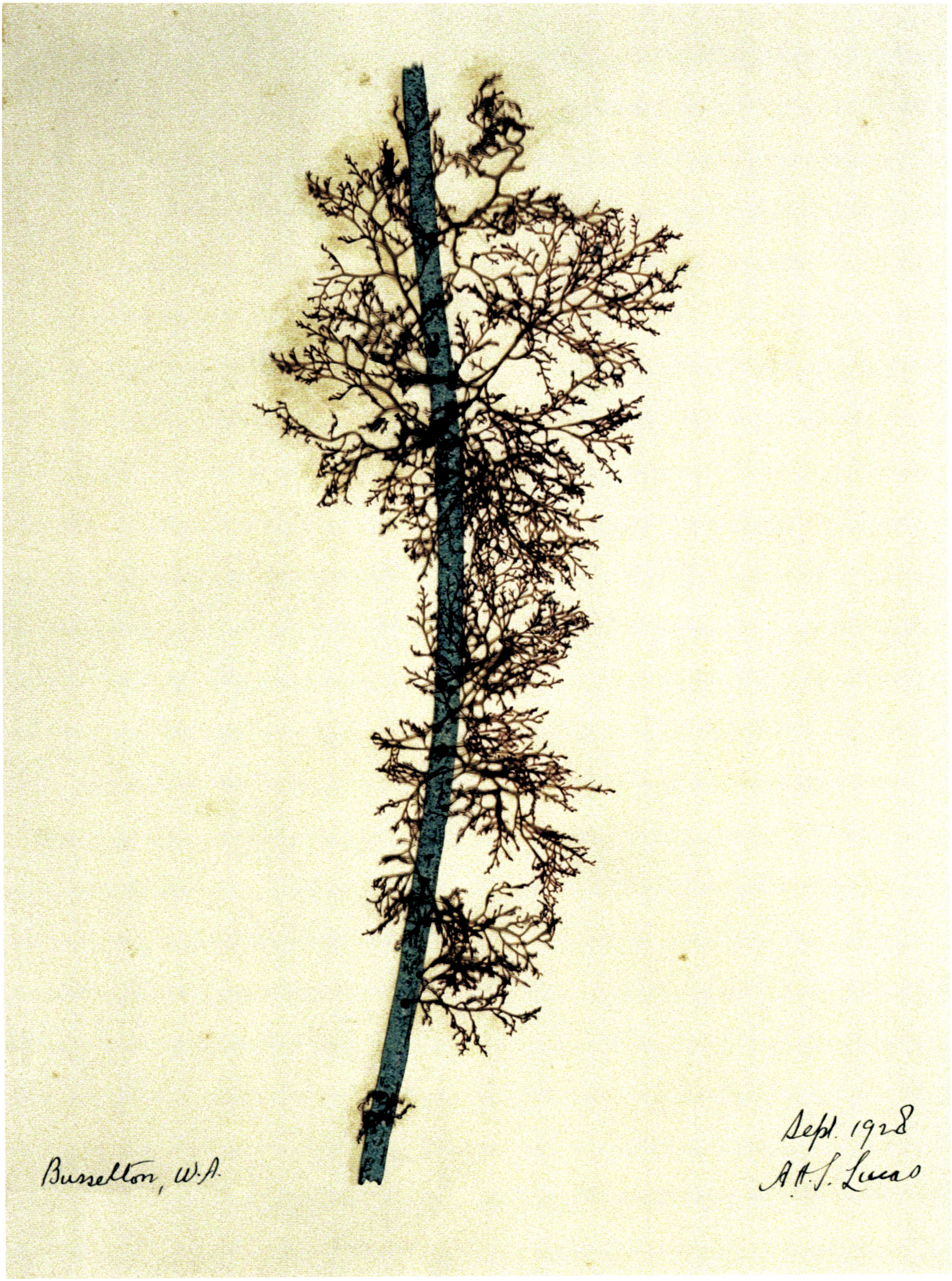

A Seagrass Leaf and a Red Alga
Collected by Arthur Lucas, 1928

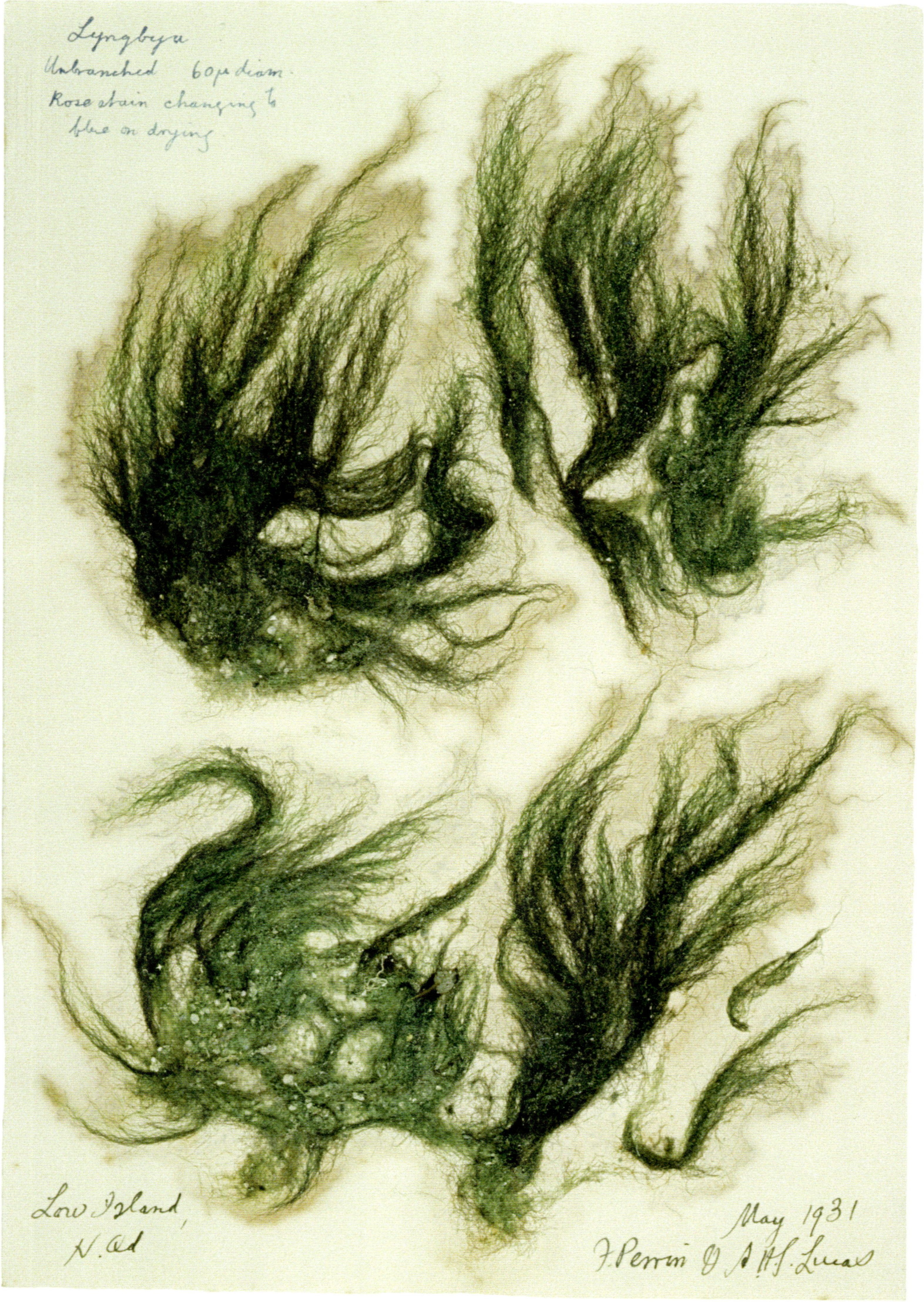

Mermaids Hair
Collected by F. Perrin and A. H. S. Lucas, 1931

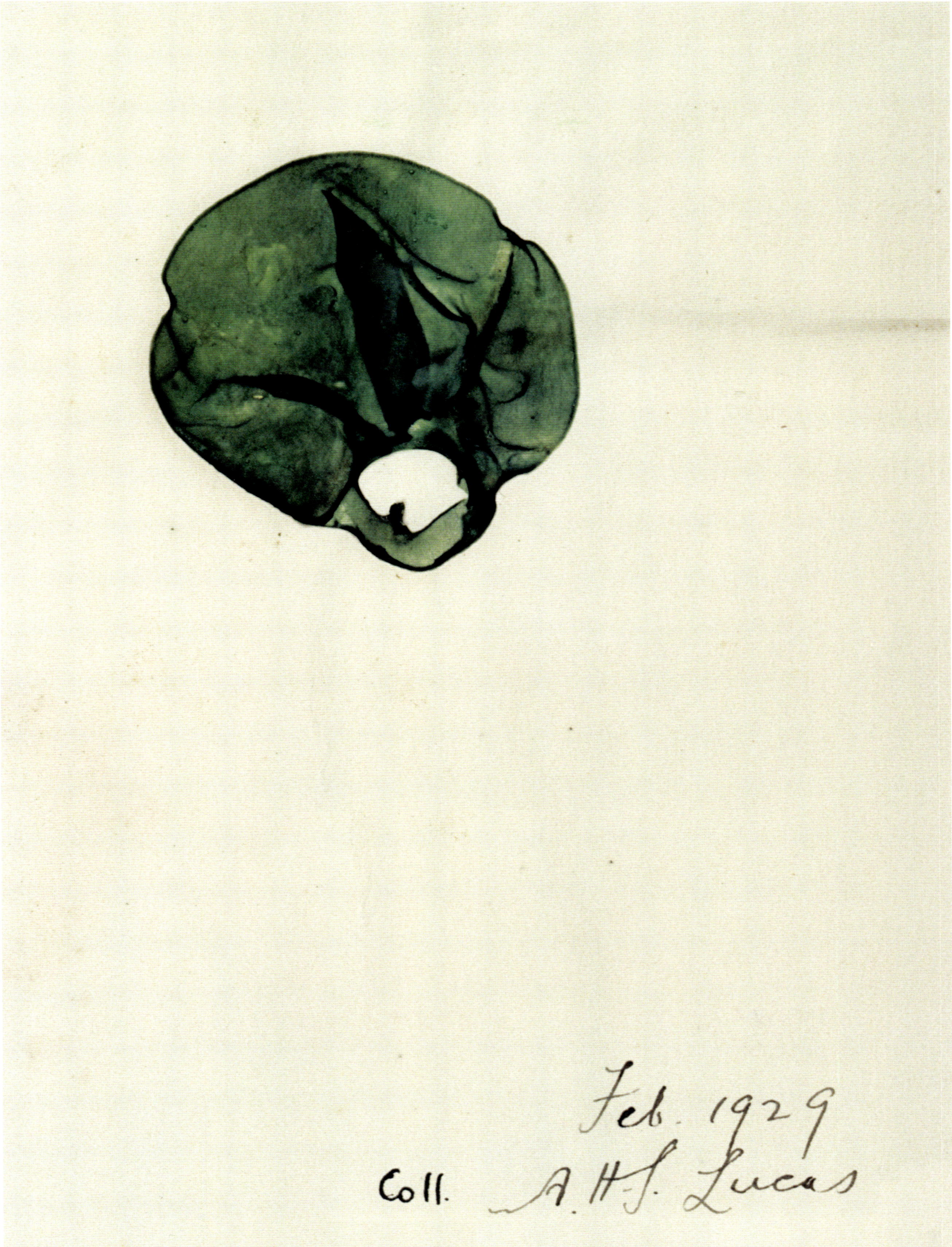

A Blue-green Seaweed
Collected by Arthur Lucas, 1929

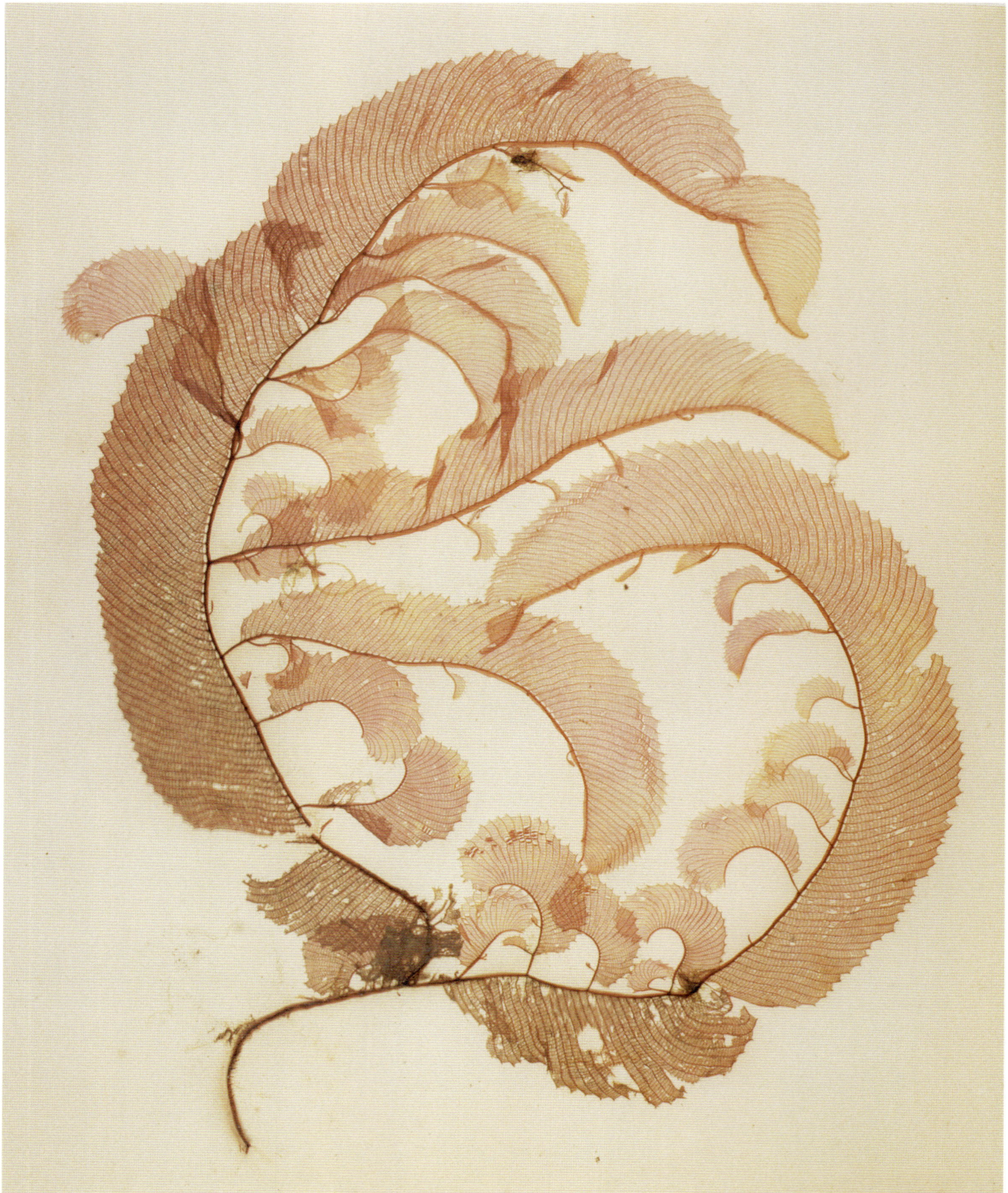

A Red Seaweed
Collected by W. H. Harvey, 1855

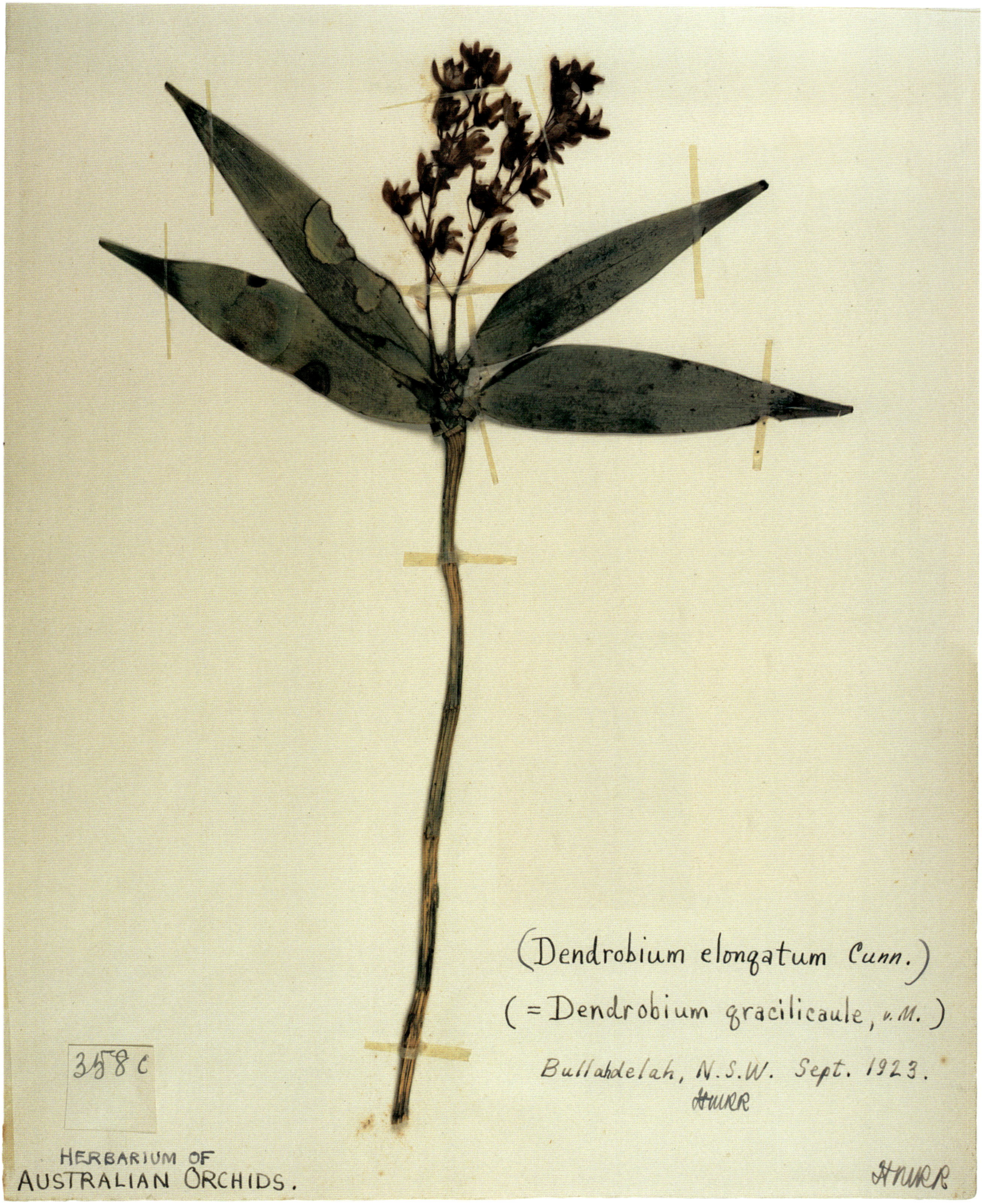

A Dendrobium Orchid
Collected by the Reverend H. M. R. Rupp, 1923

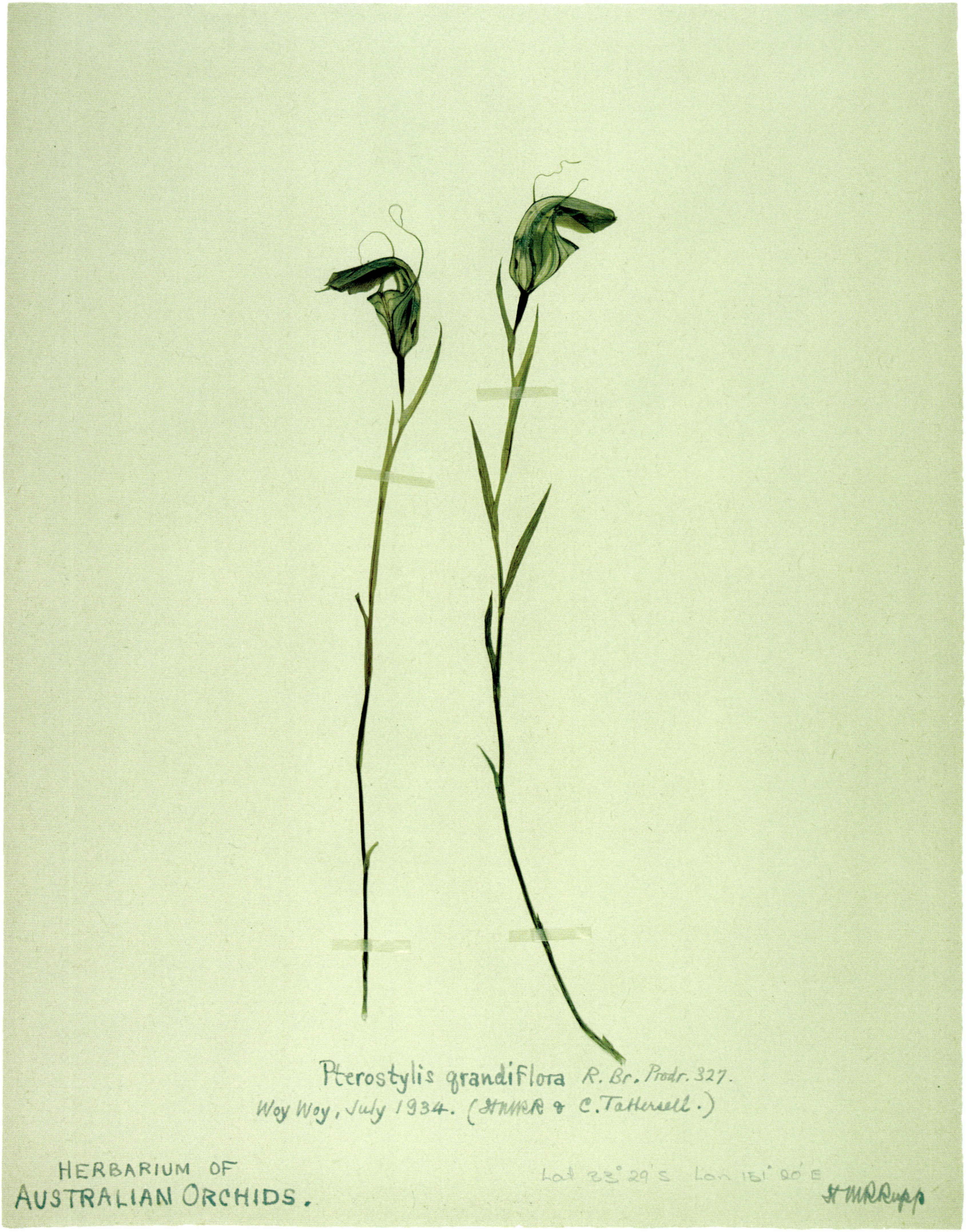

Cobra Greenhood
Collected by the Reverend H. M. R. Rupp, 1934

Hobby and decoration

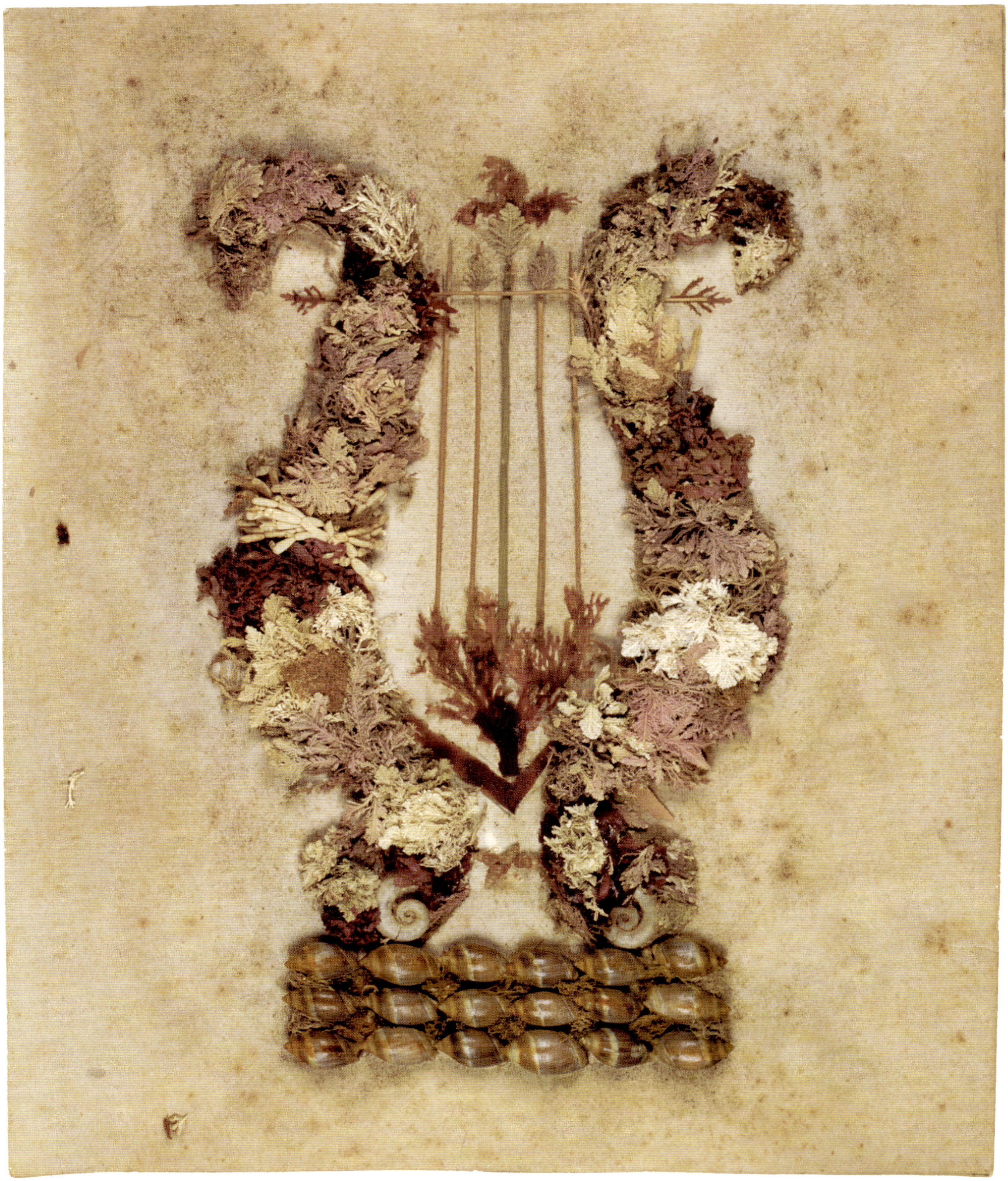

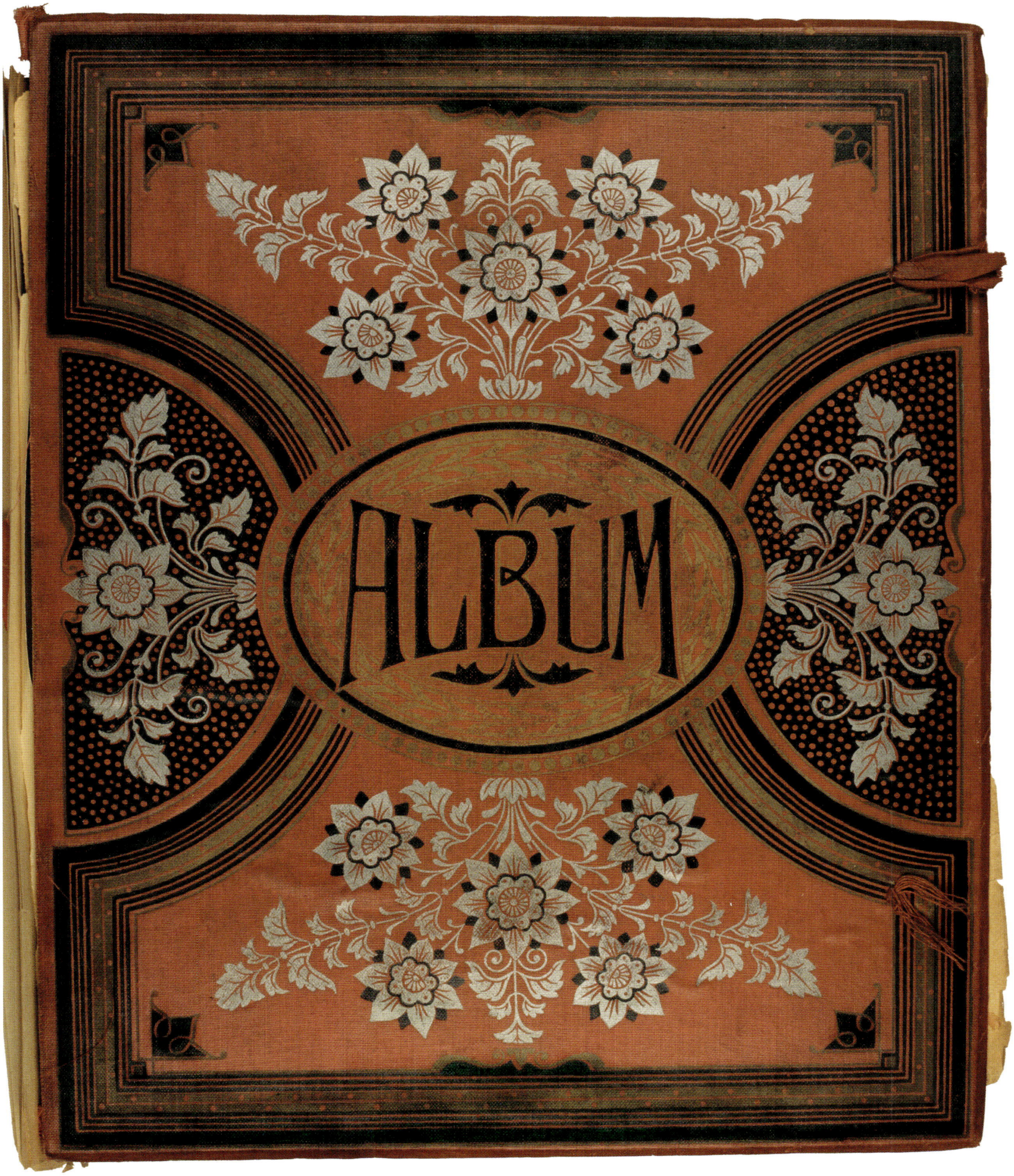

Seaweed Album
1901

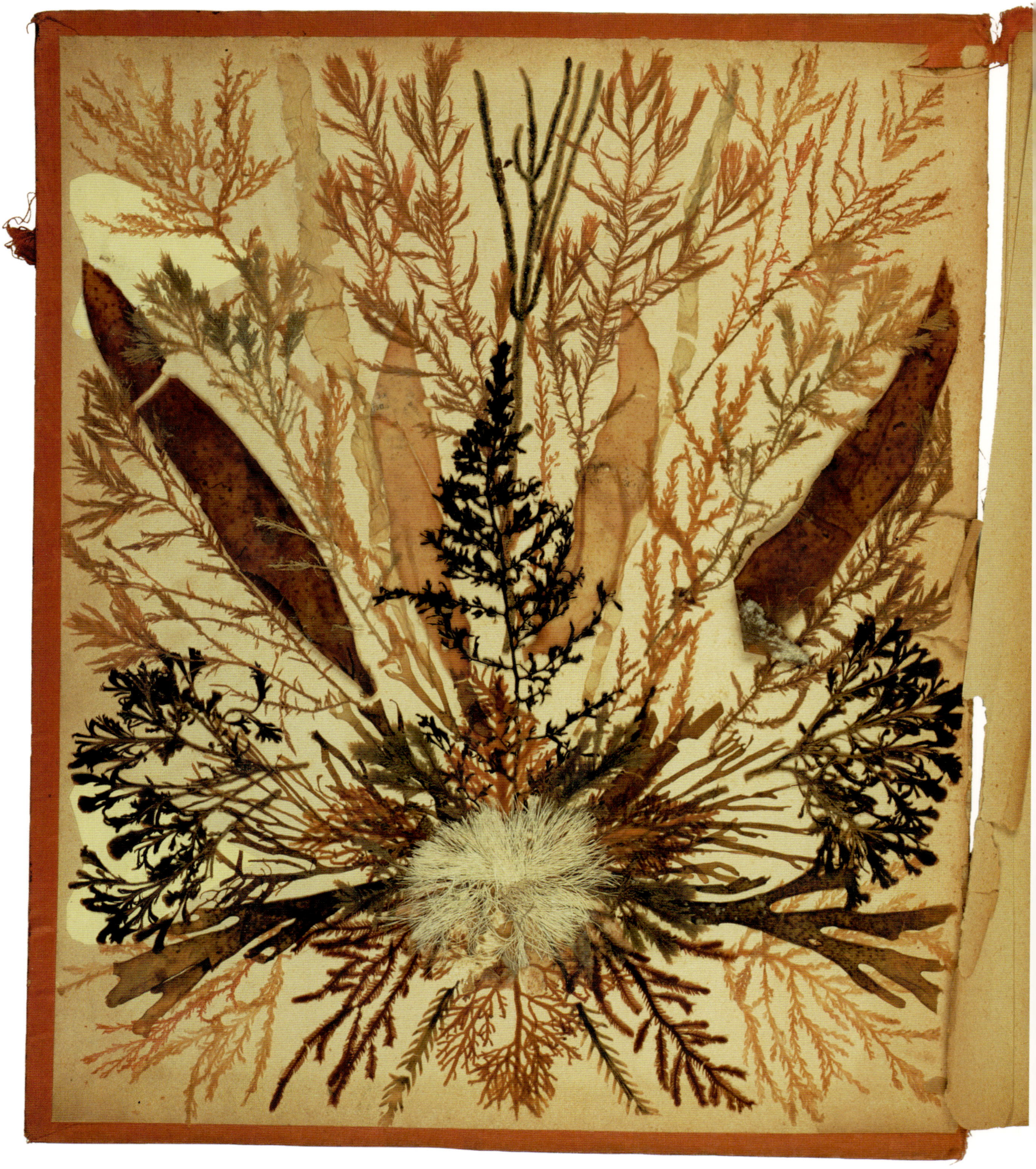

Seaweed Album Endpaper
1901

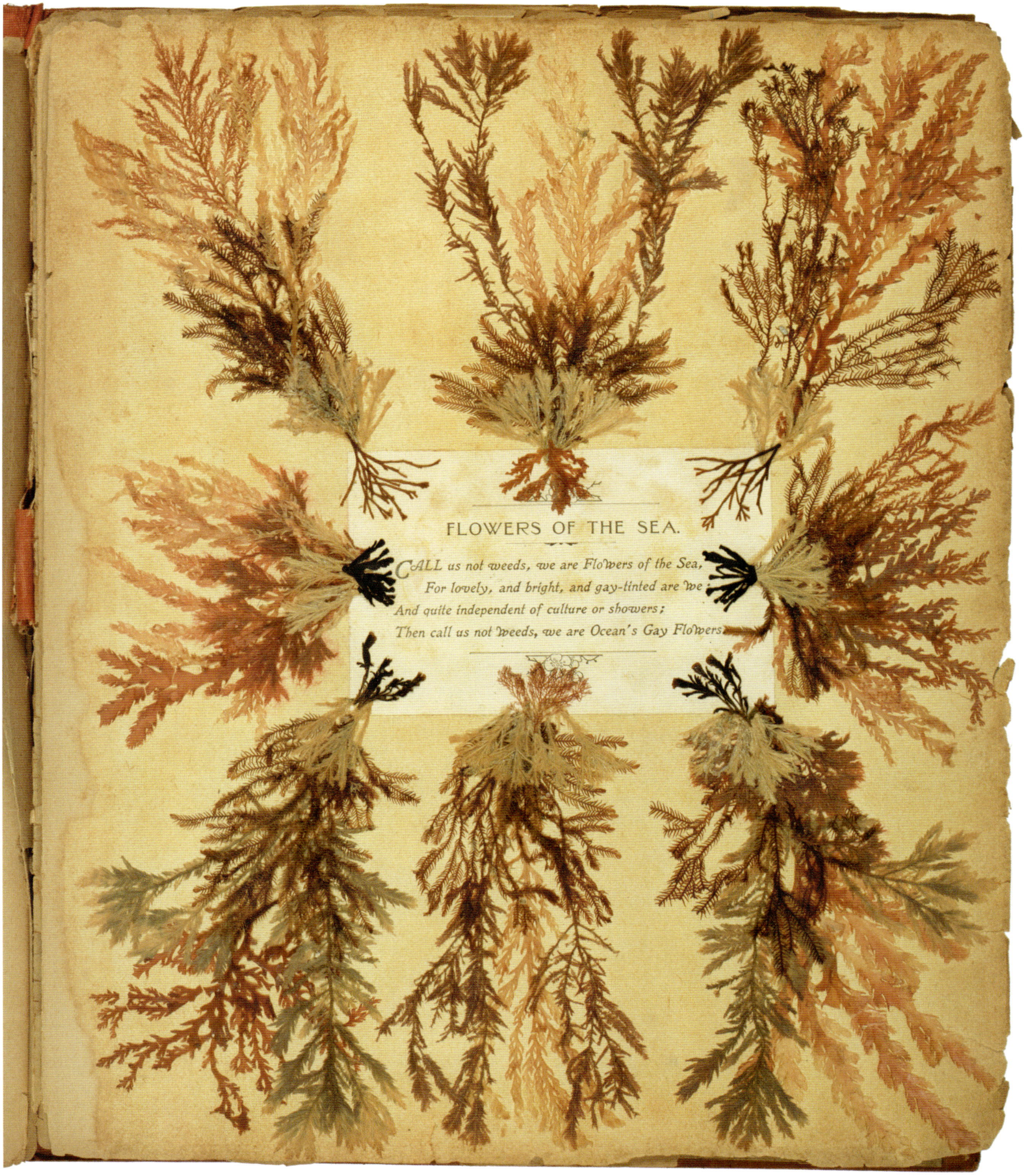

Seaweed Album Title Page
1901

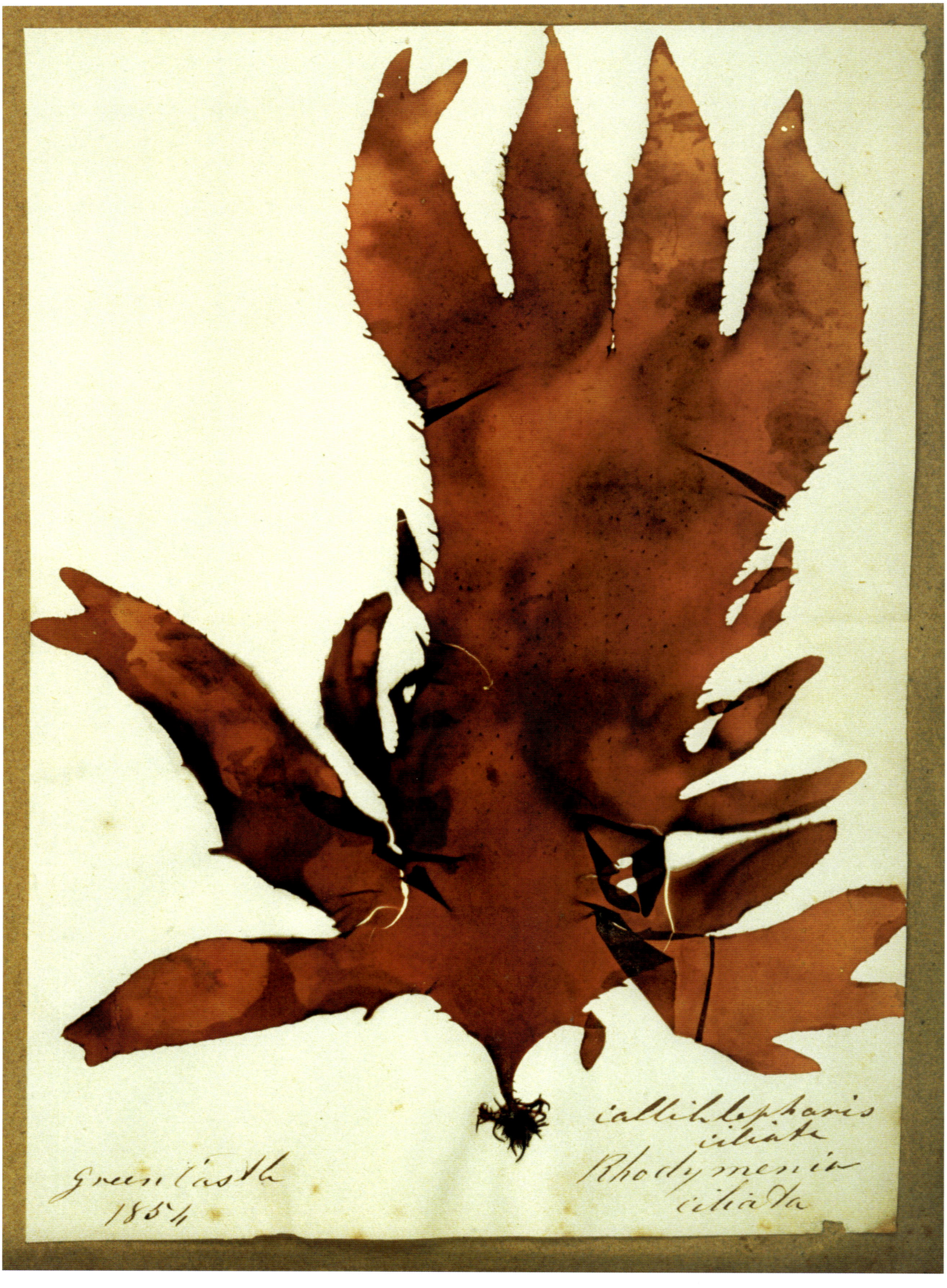

Eyelash Weed
Collected 1854

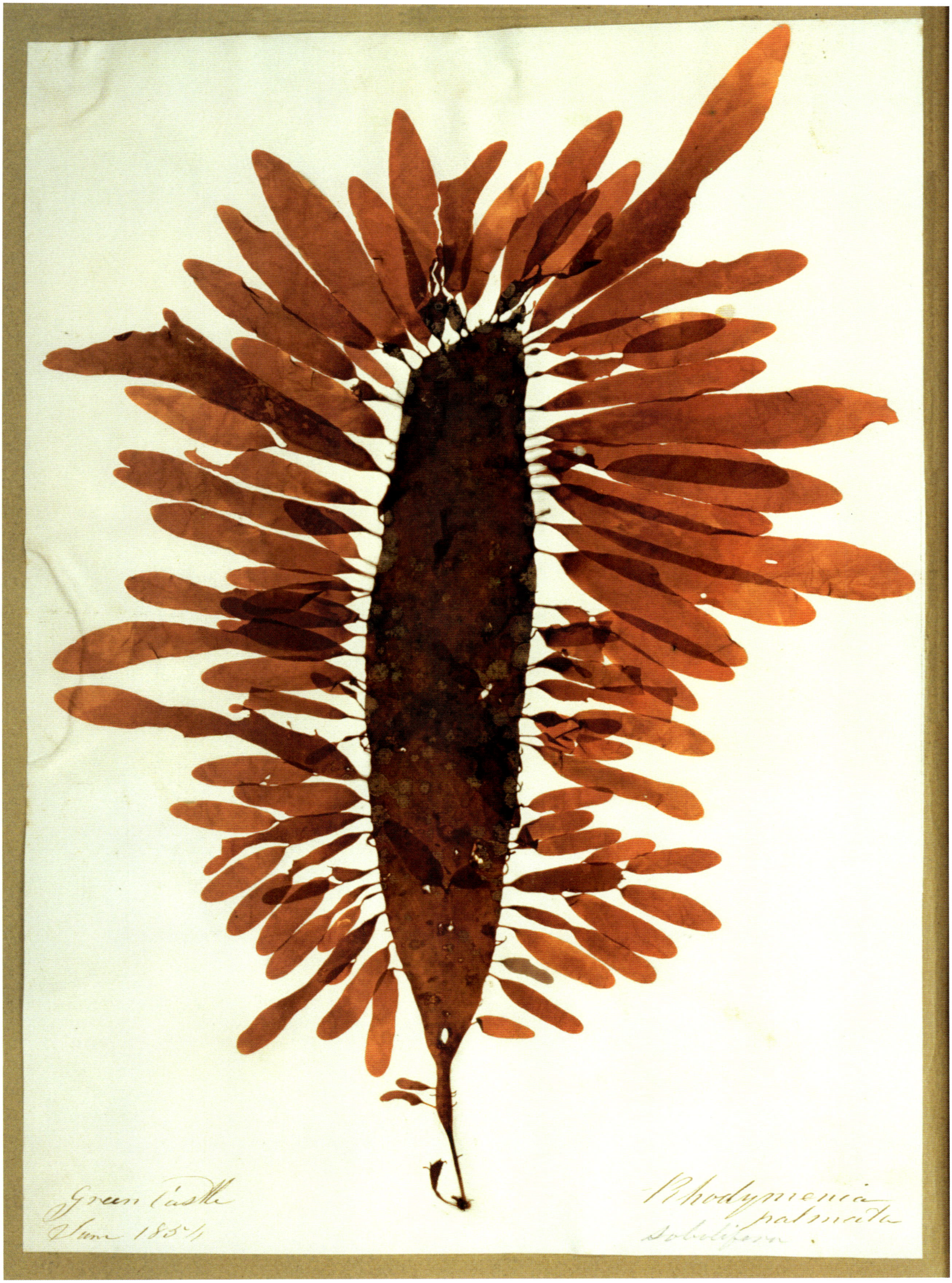

Dulse
Collected 1854

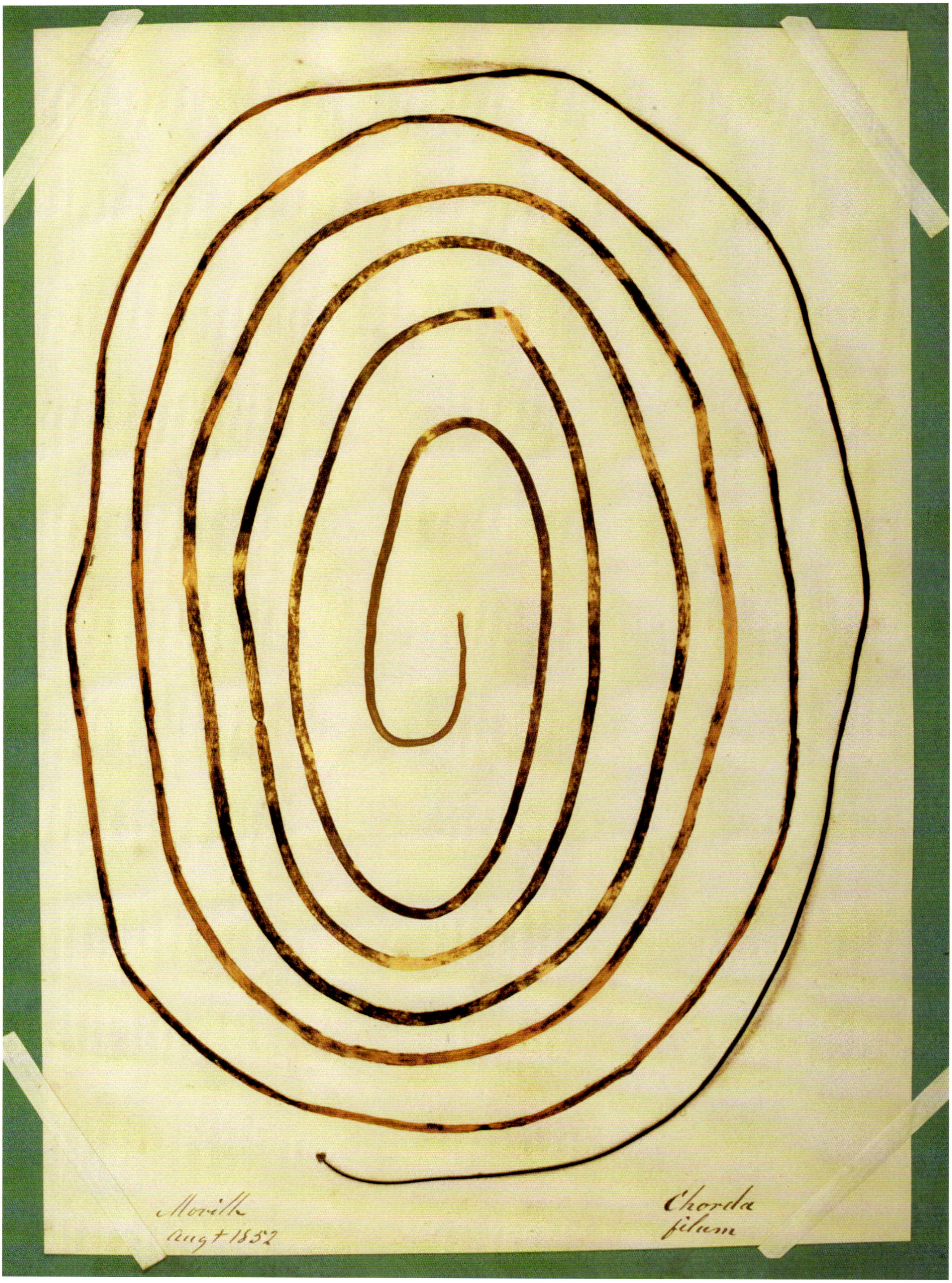

A Brown Seaweed
Collected 1852

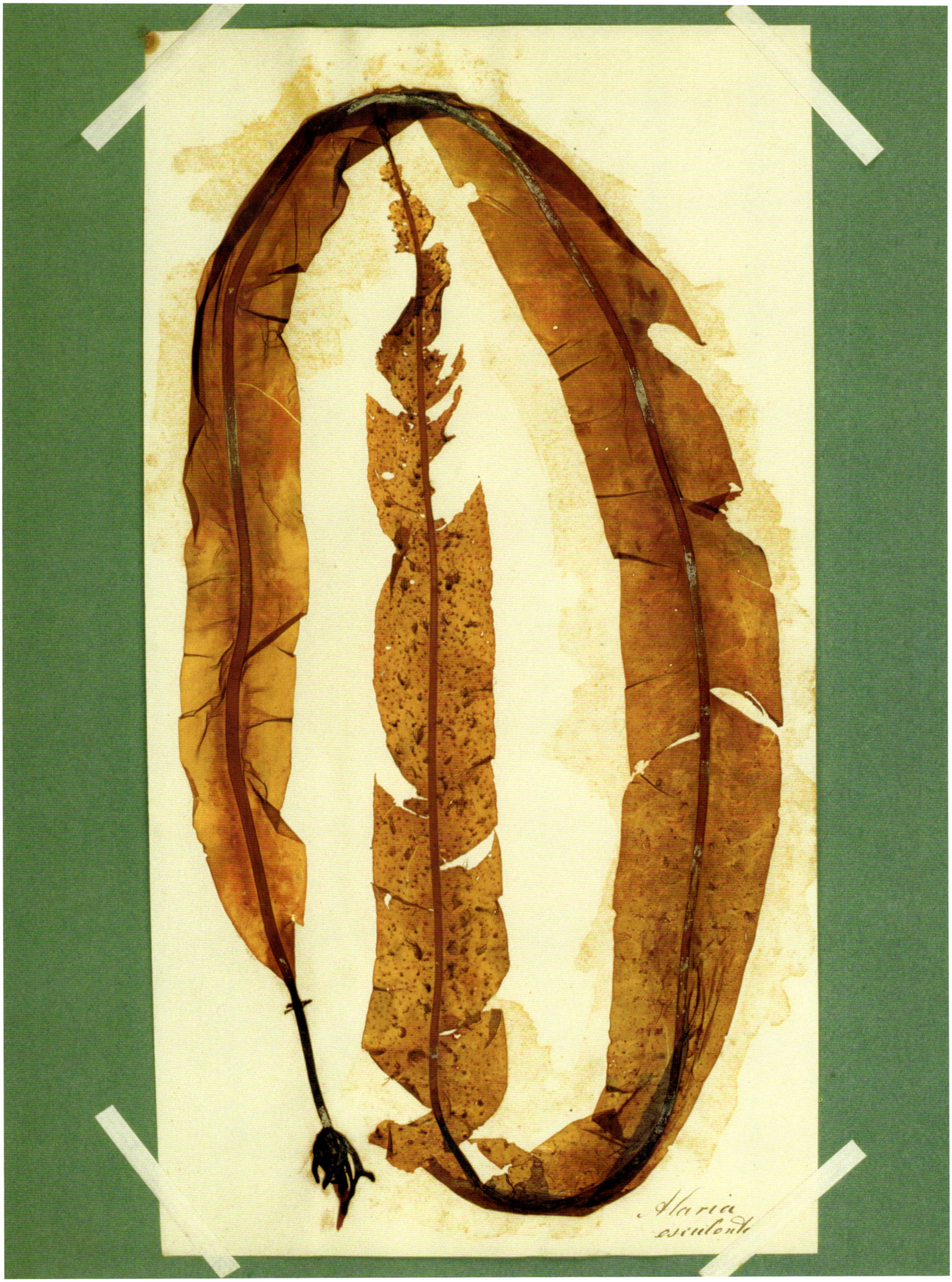

A Brown Seaweed
Collected c. 1852–1854

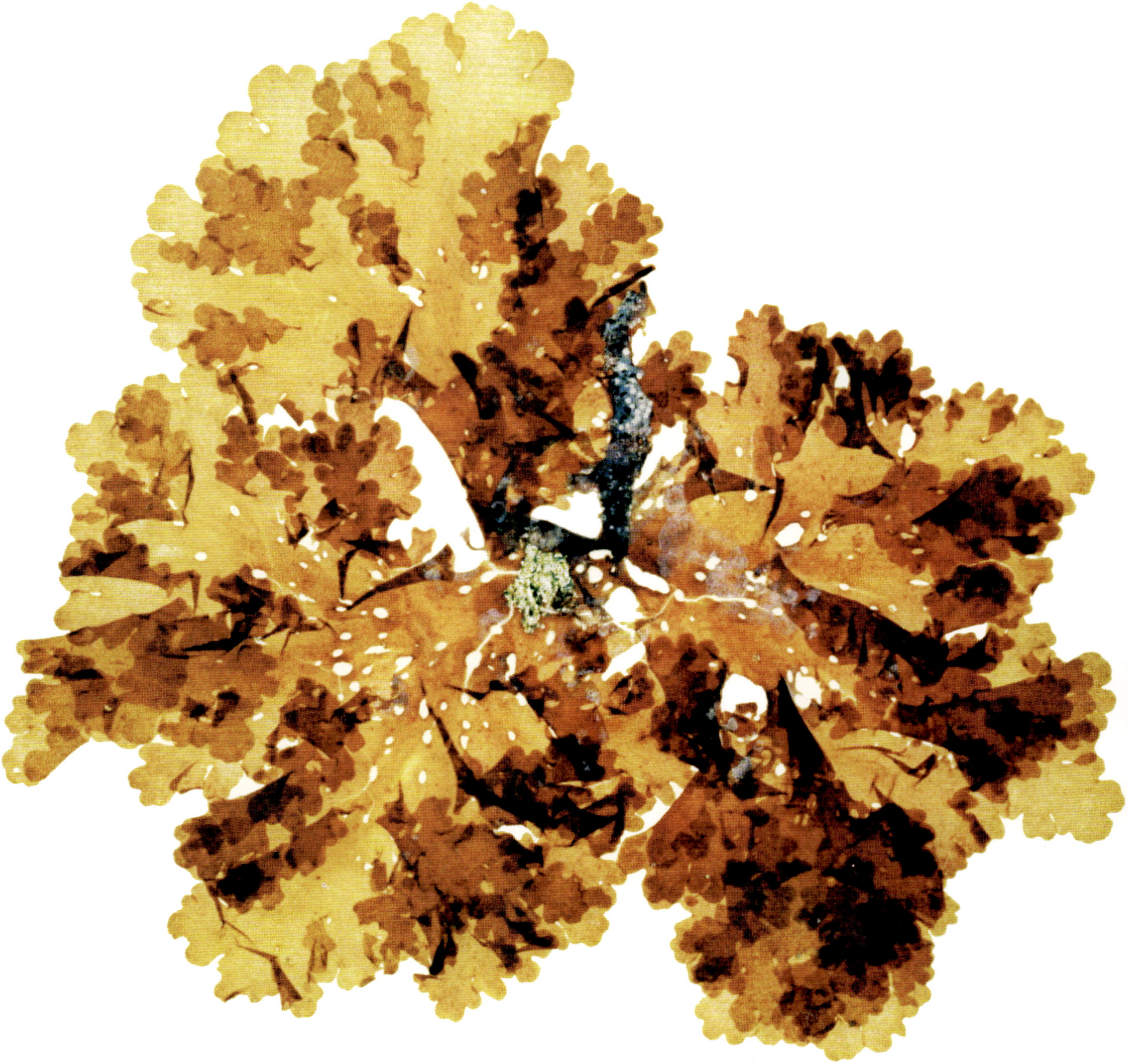

A Red Seaweed
Collected by Mrs Mary Ann Barker

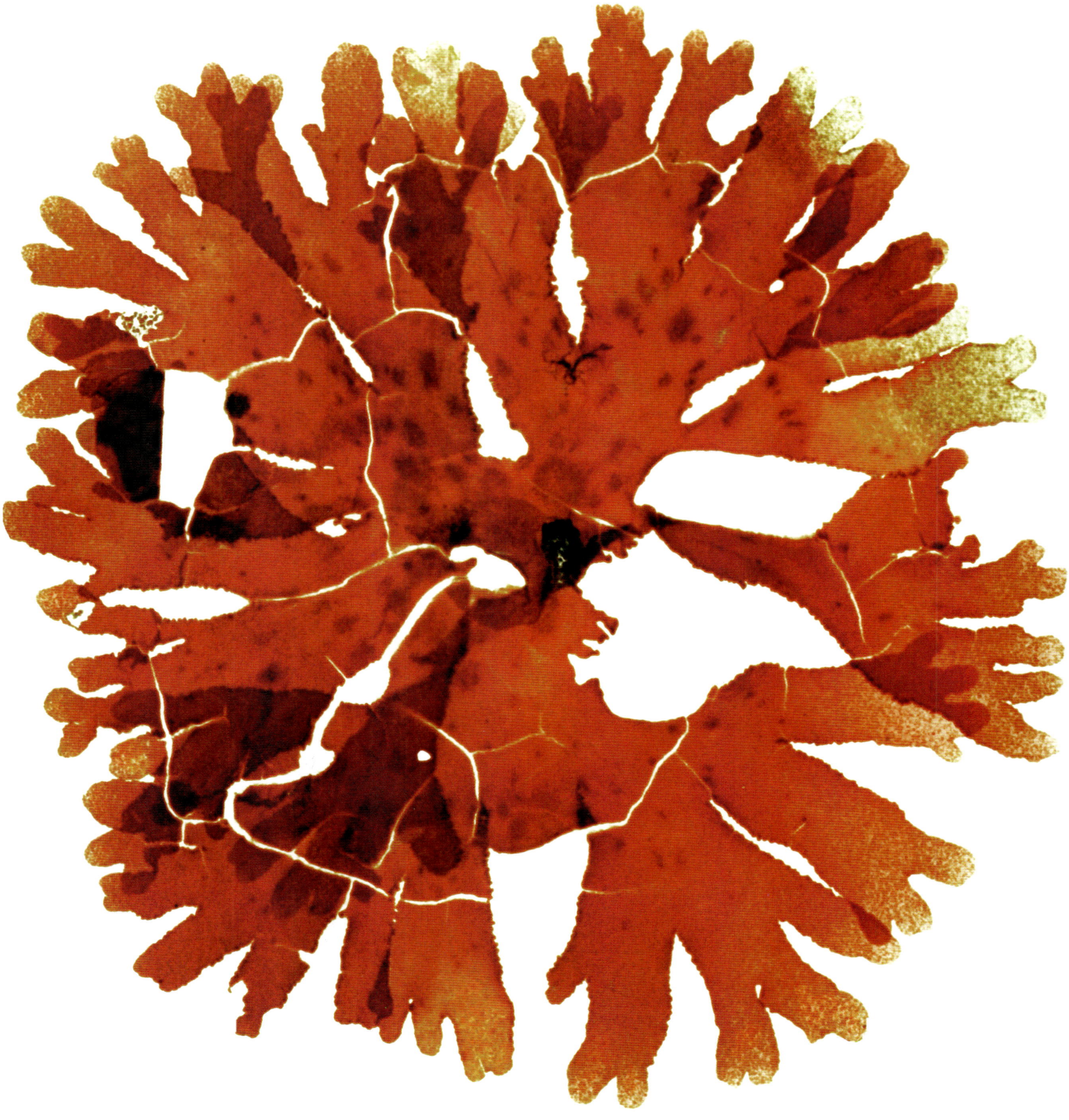

A Red Seaweed
Collected 1852

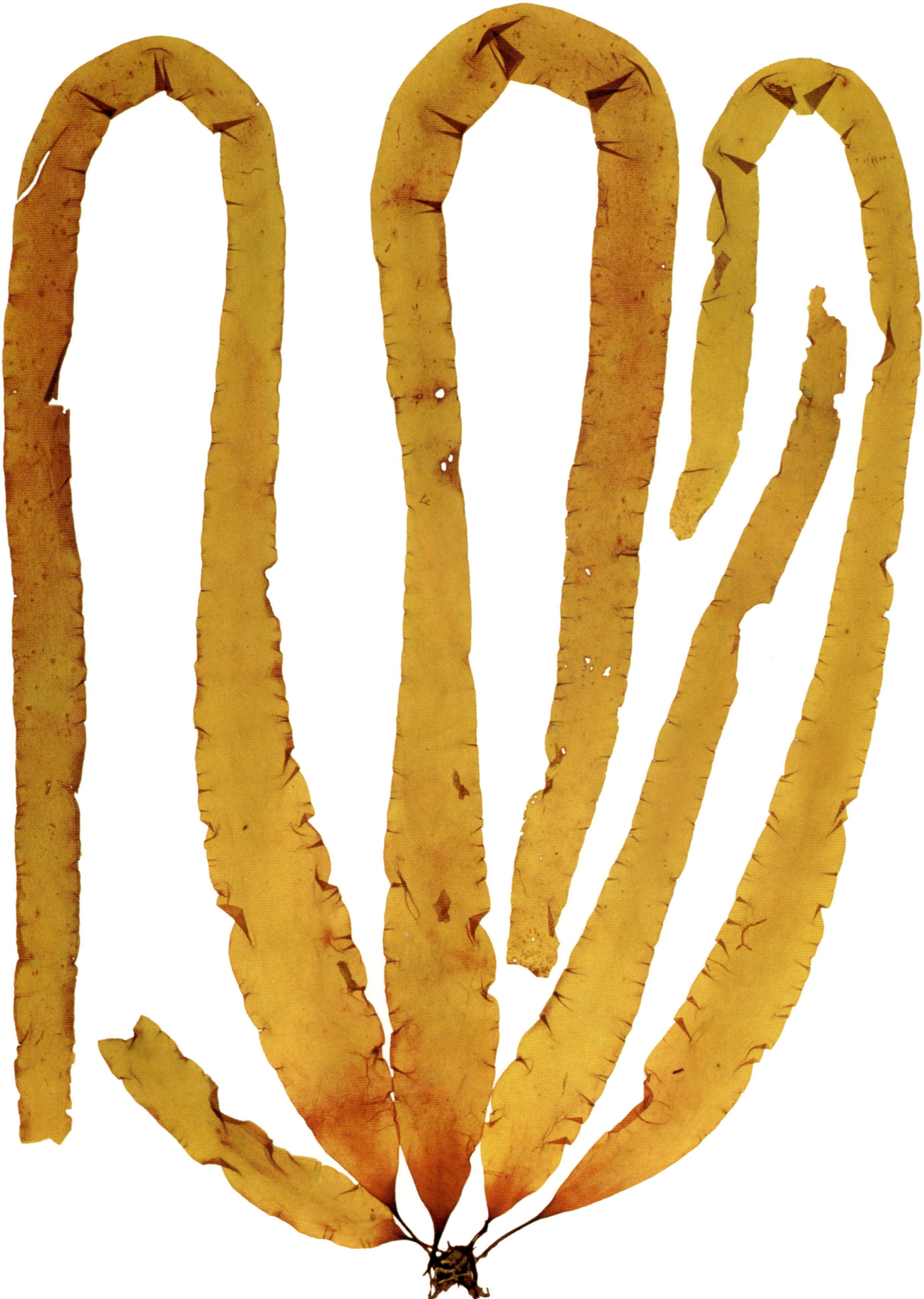

A Brown Seaweed
Collected 1853

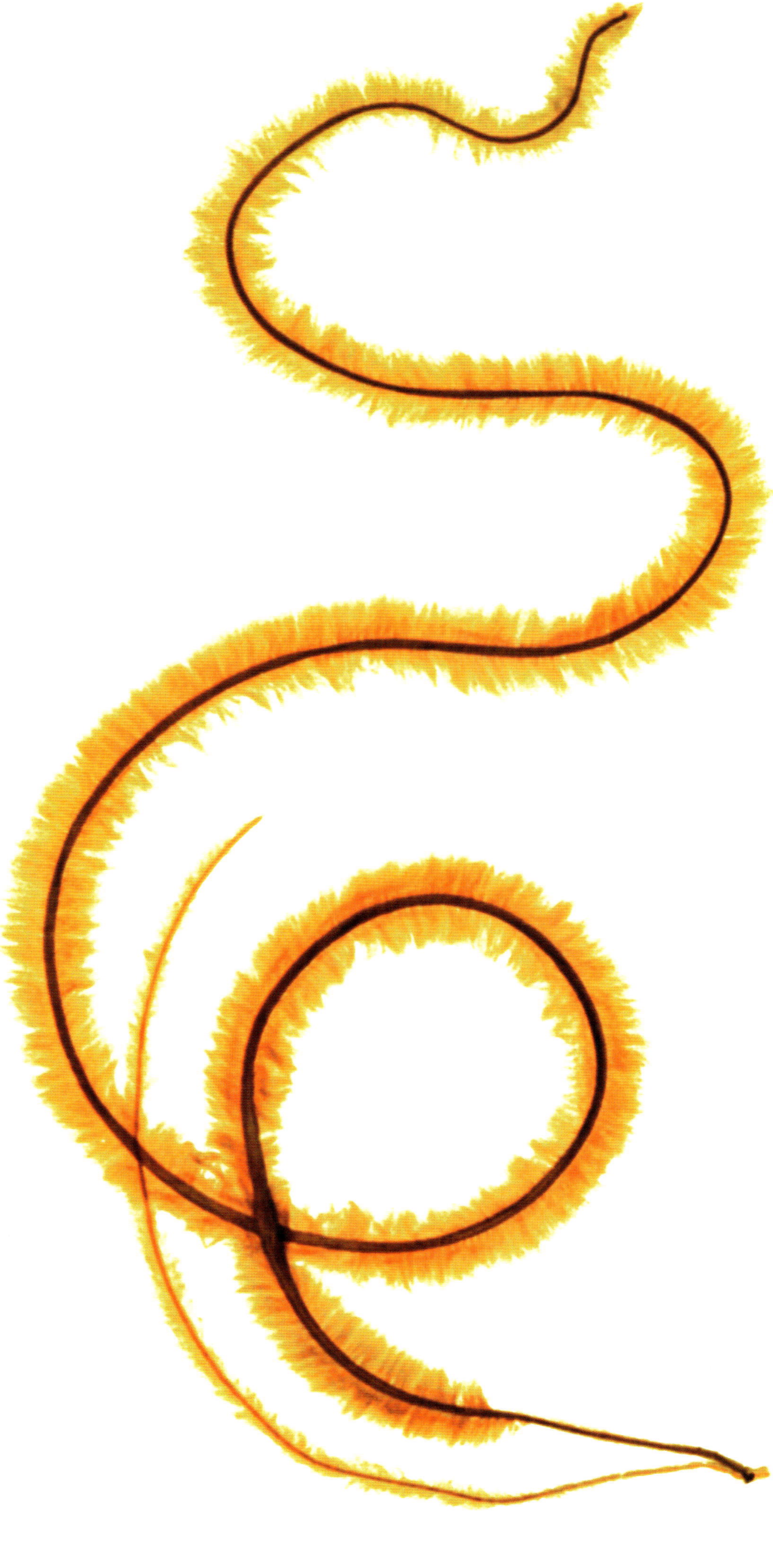

A Brown Seaweed
Collected by Dr Dickie, 1852

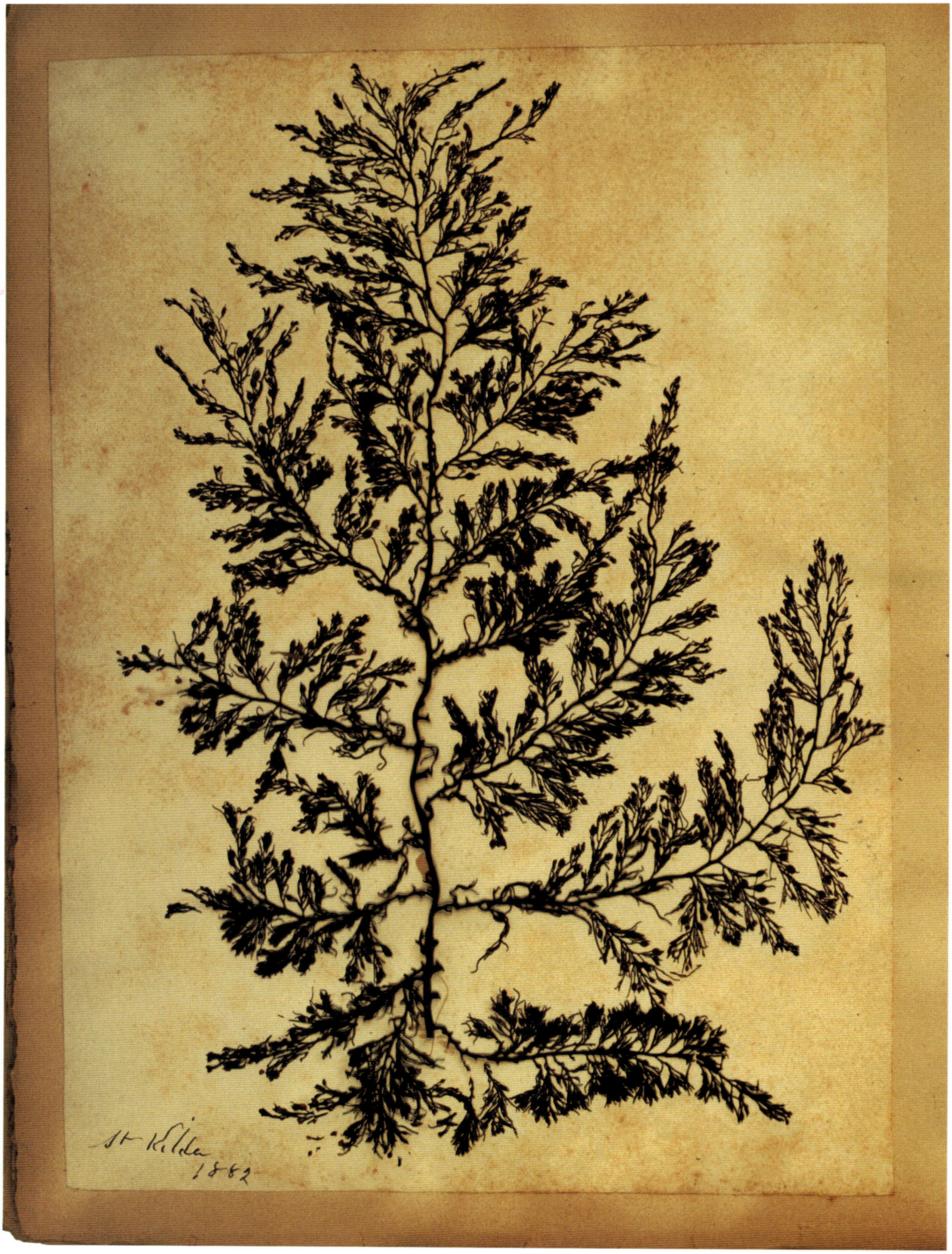

A Brown Seaweed
Collected possibly by Charles Morrison, 1882

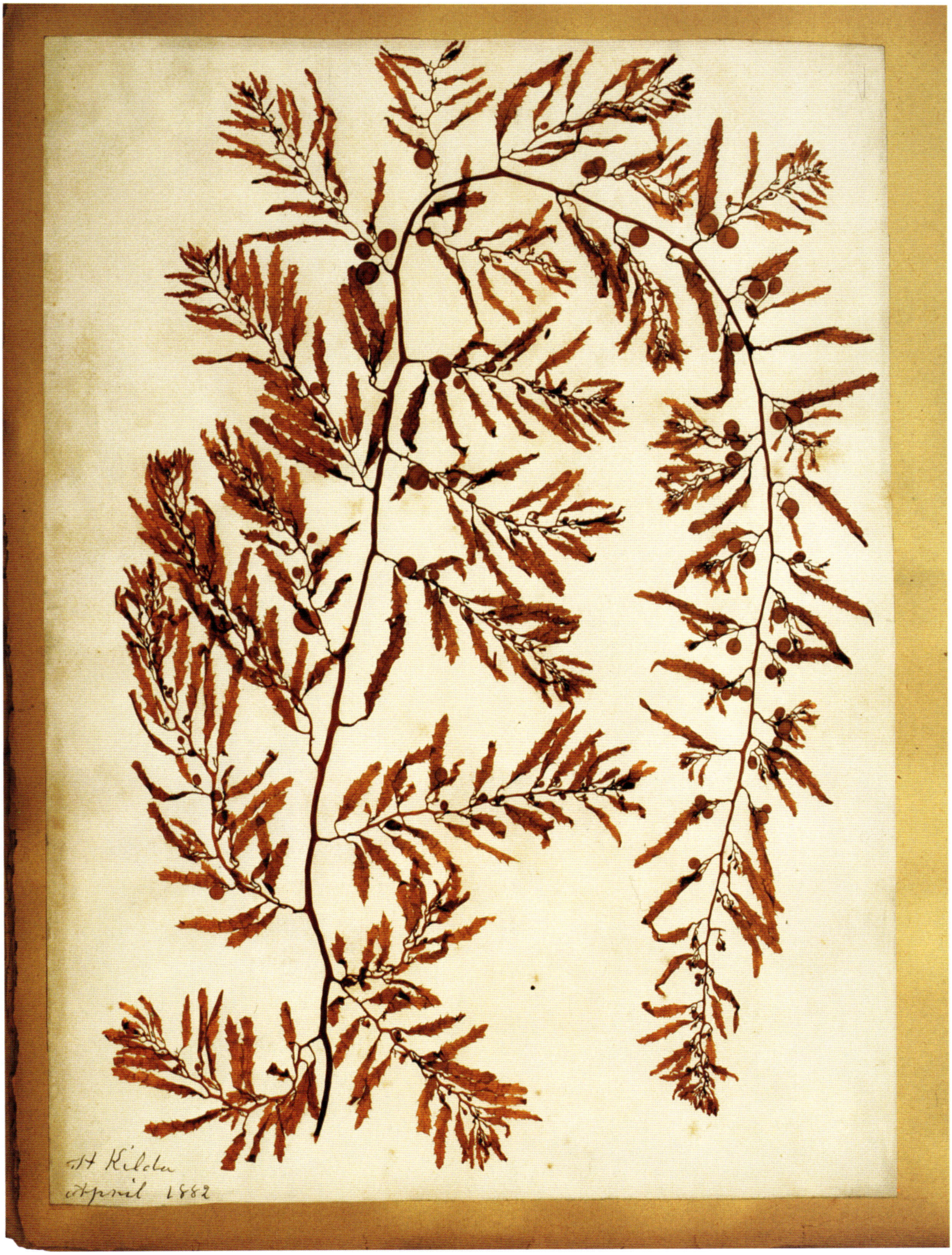

A Brown Seaweed
Collected possibly by Charles Morrison, 1882

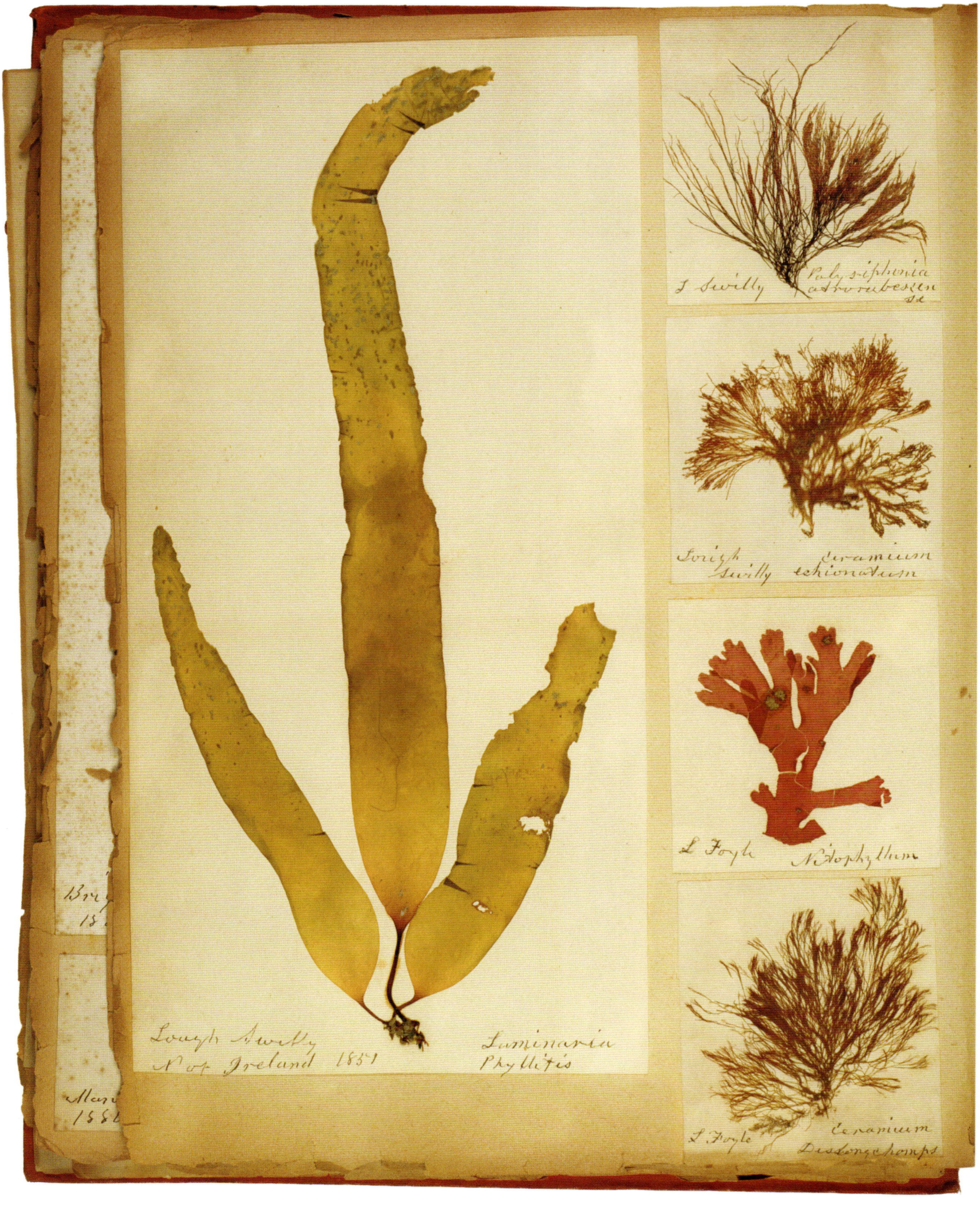

Typical Page of Seaweed Album
1901

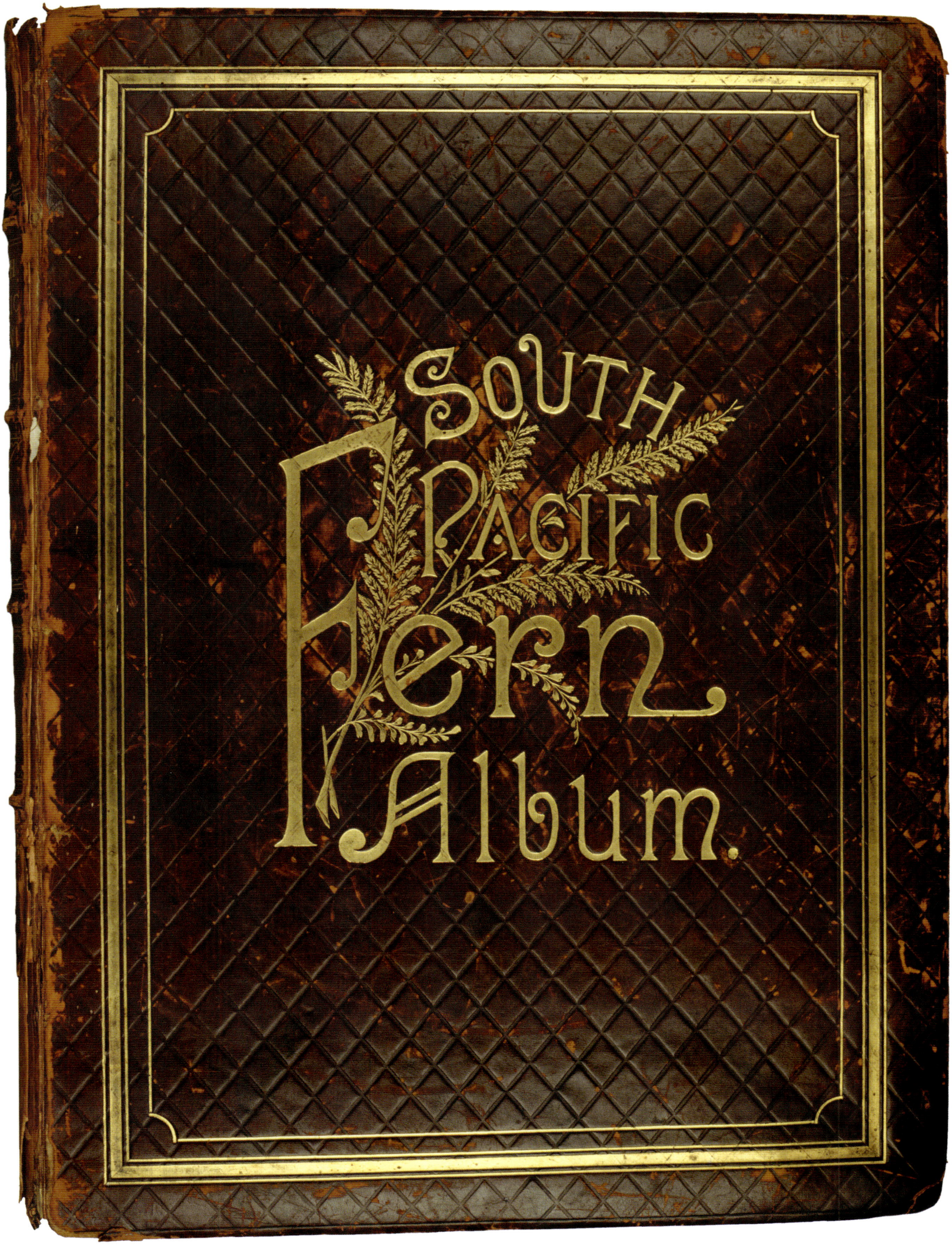

South Pacific Fern Album 'New Zealand Section'
1895

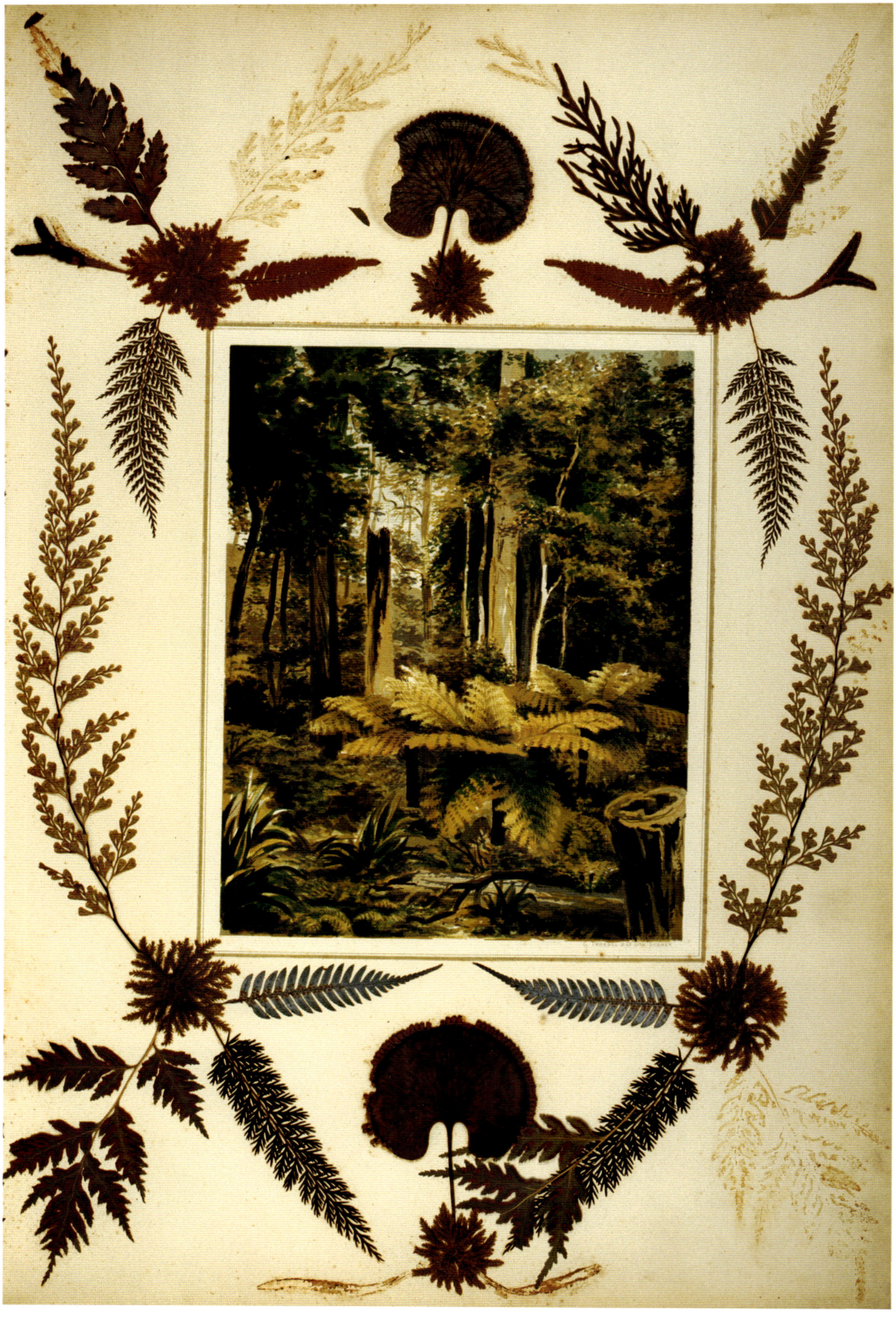

South Pacific Fern Album 'New Zealand Section'
1895

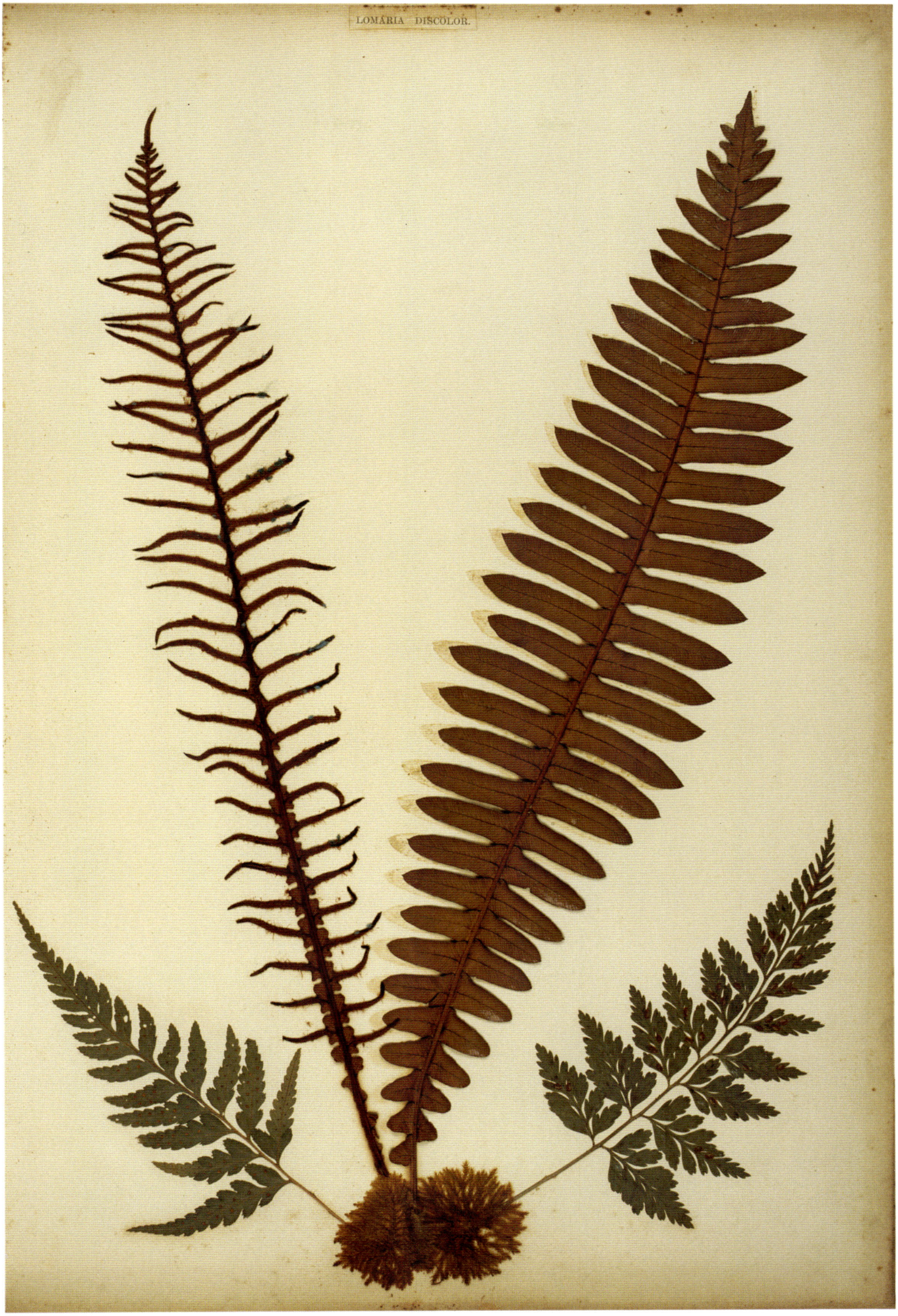

South Pacific Fern Album 'New Zealand Section'
1895

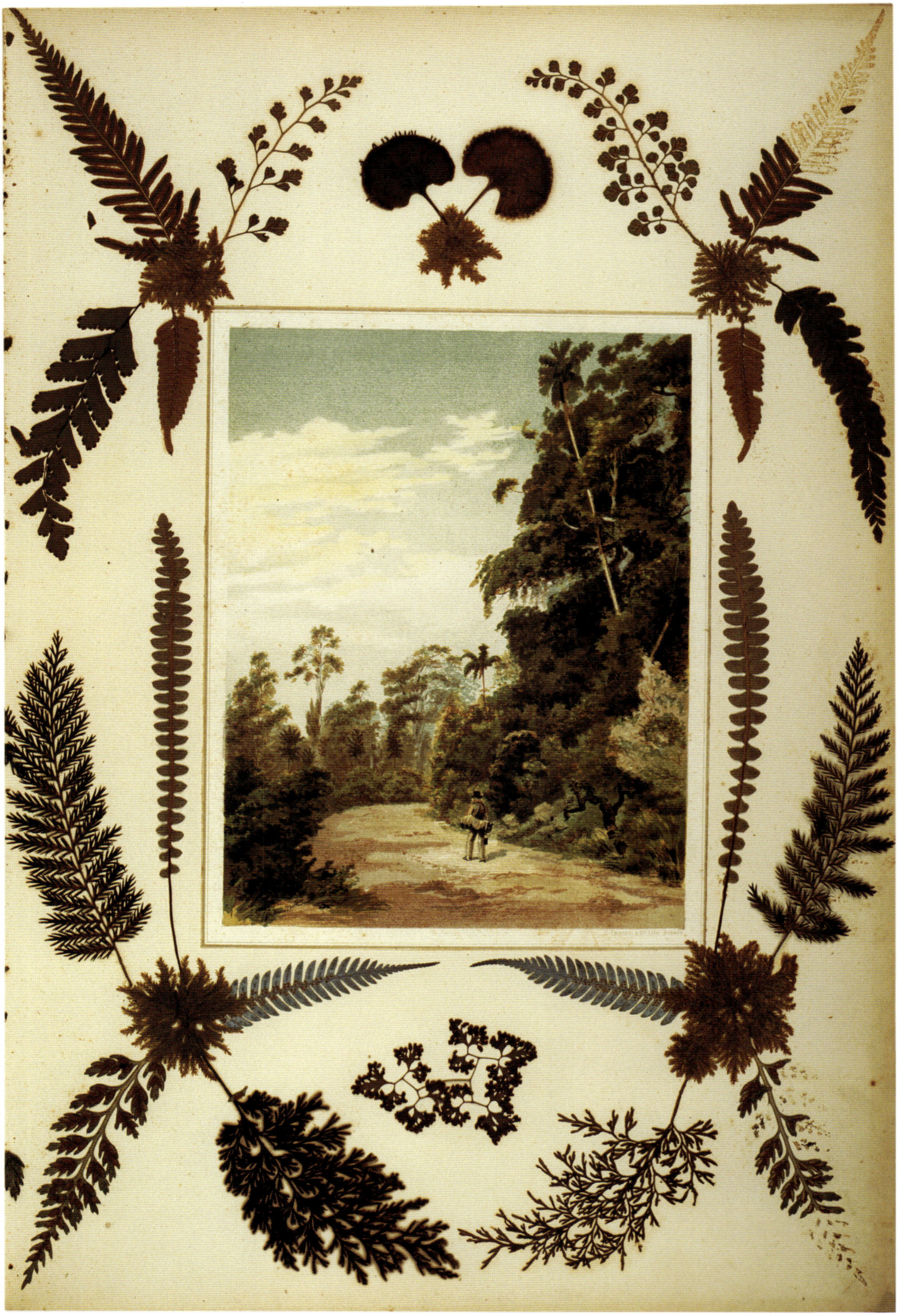

South Pacific Fern Album 'New Zealand Section'
1895

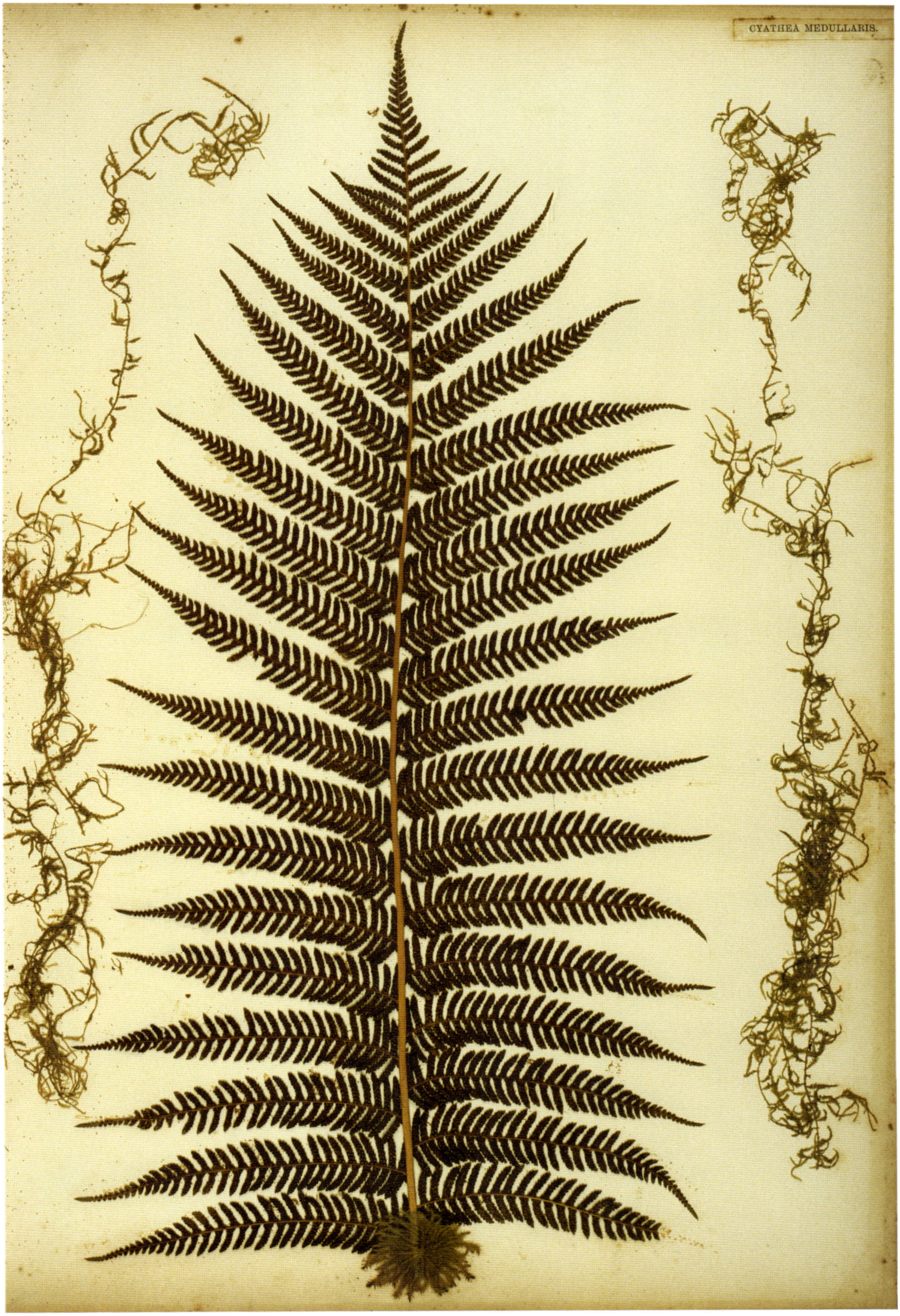

South Pacific Fern Album 'New Zealand Section'
1895

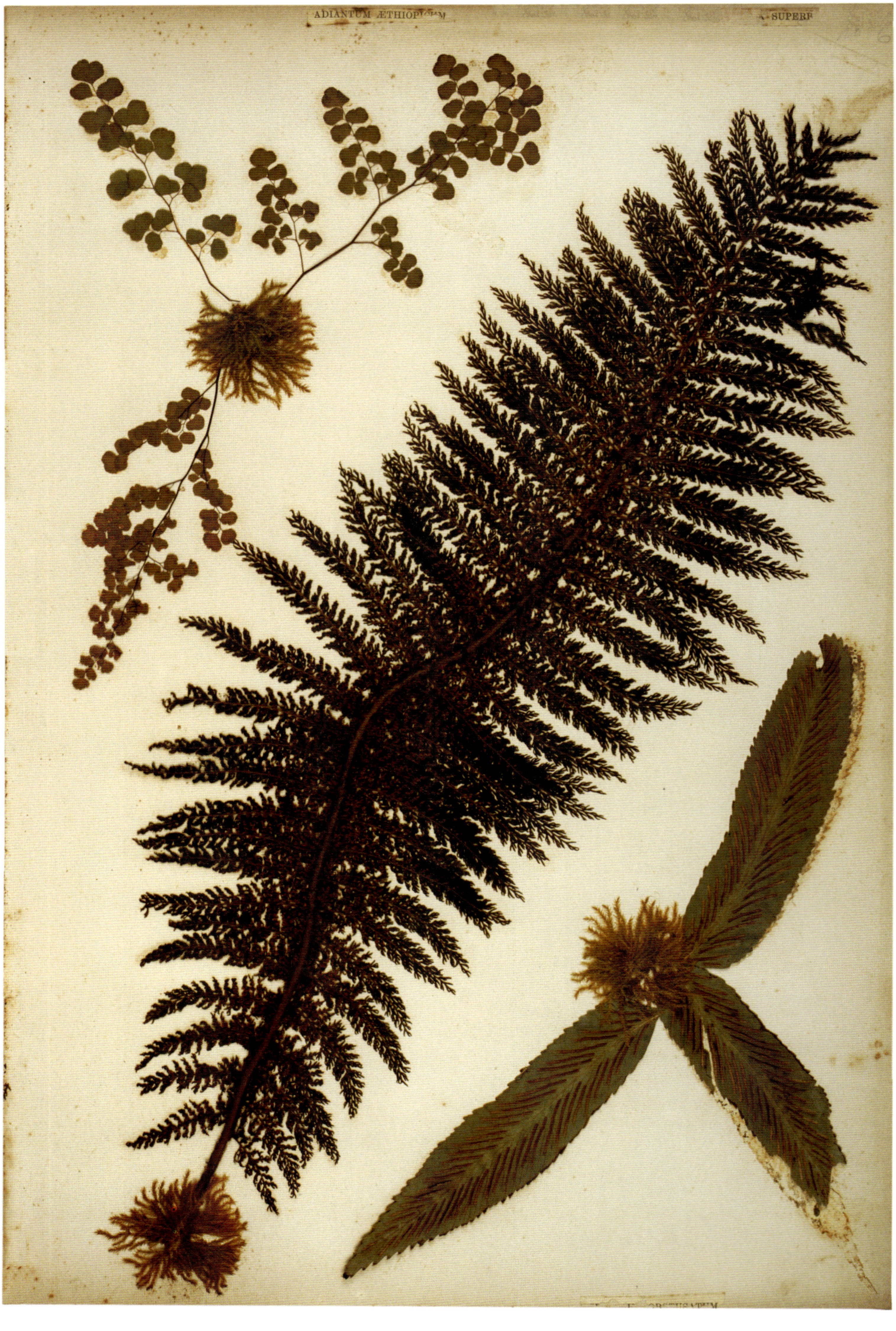

South Pacific Fern Album 'New Zealand Section'
1895

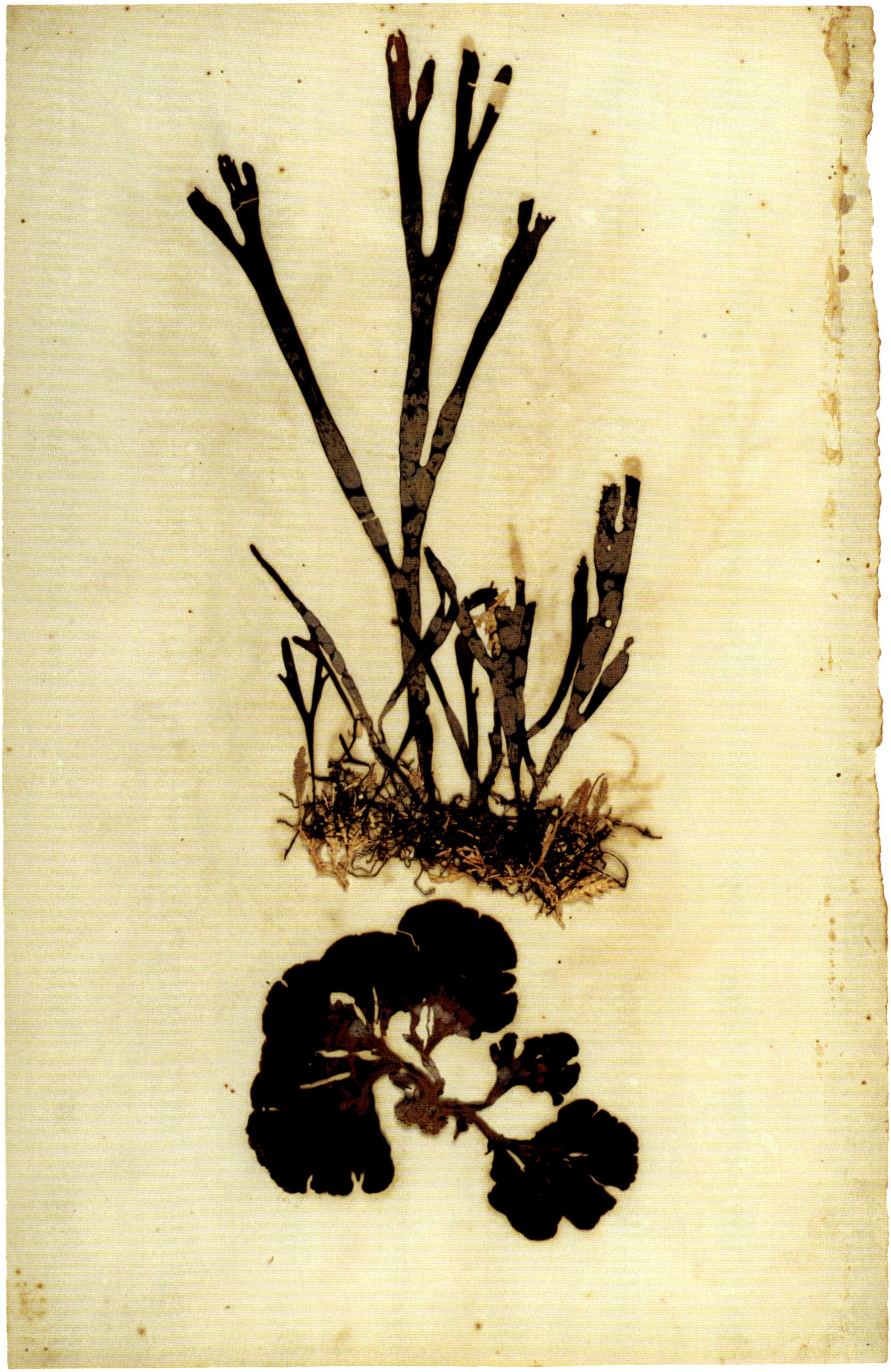

Brown Seaweeds
Collected by John Thompson, 1860s

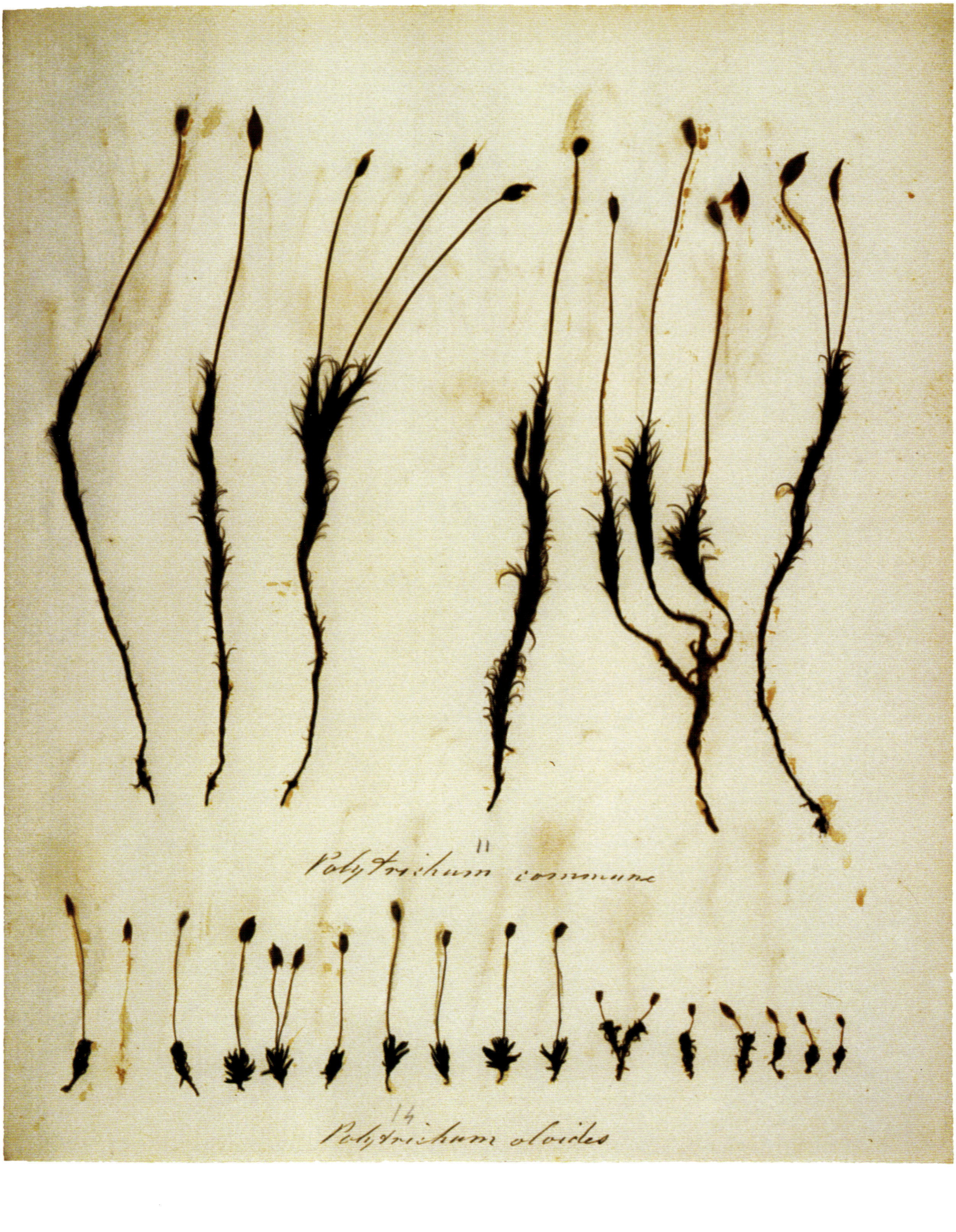

Haircap Mosses

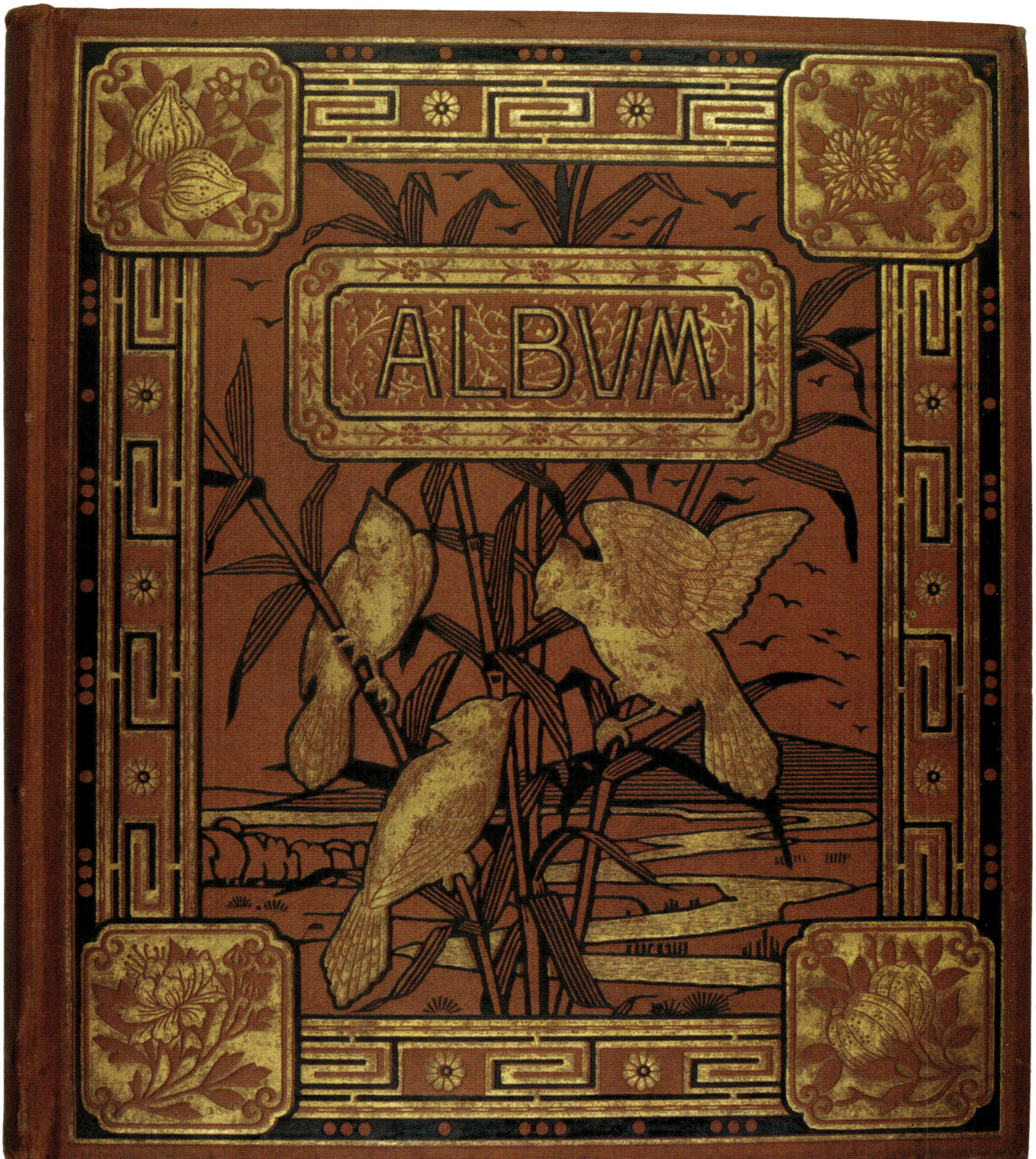

Victorian Fern Album

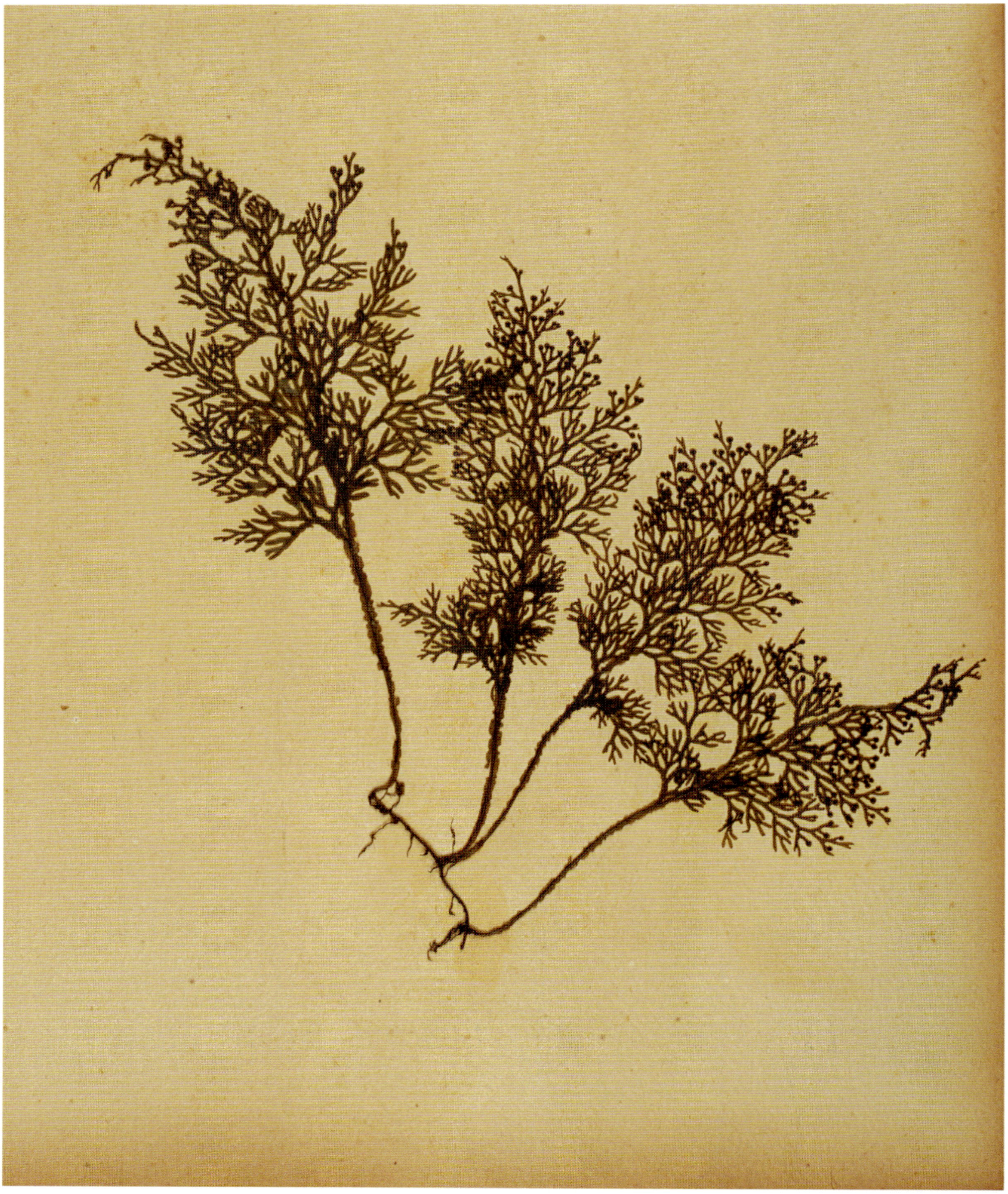

A Filmy Fern

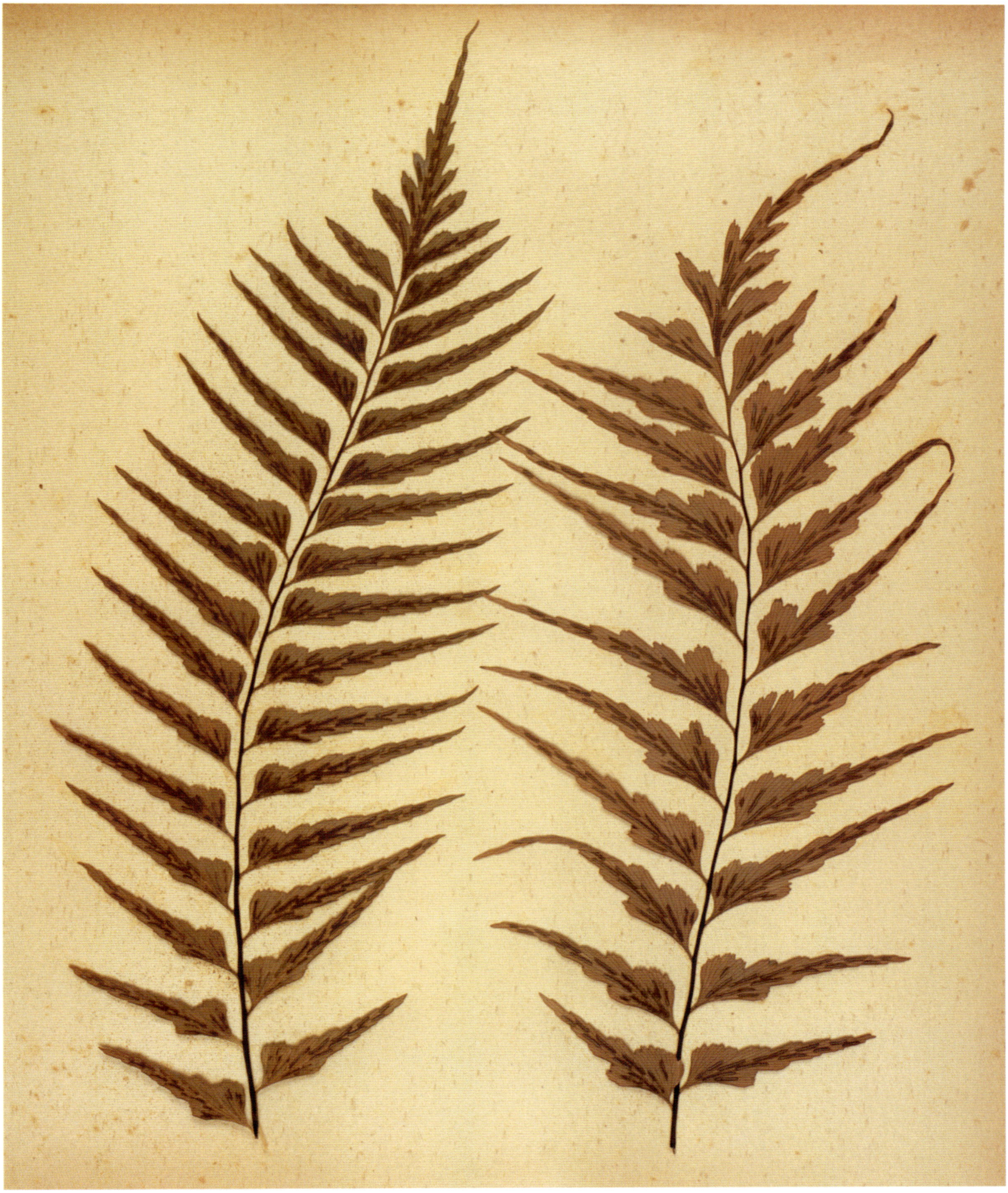

Asplenium Ferns

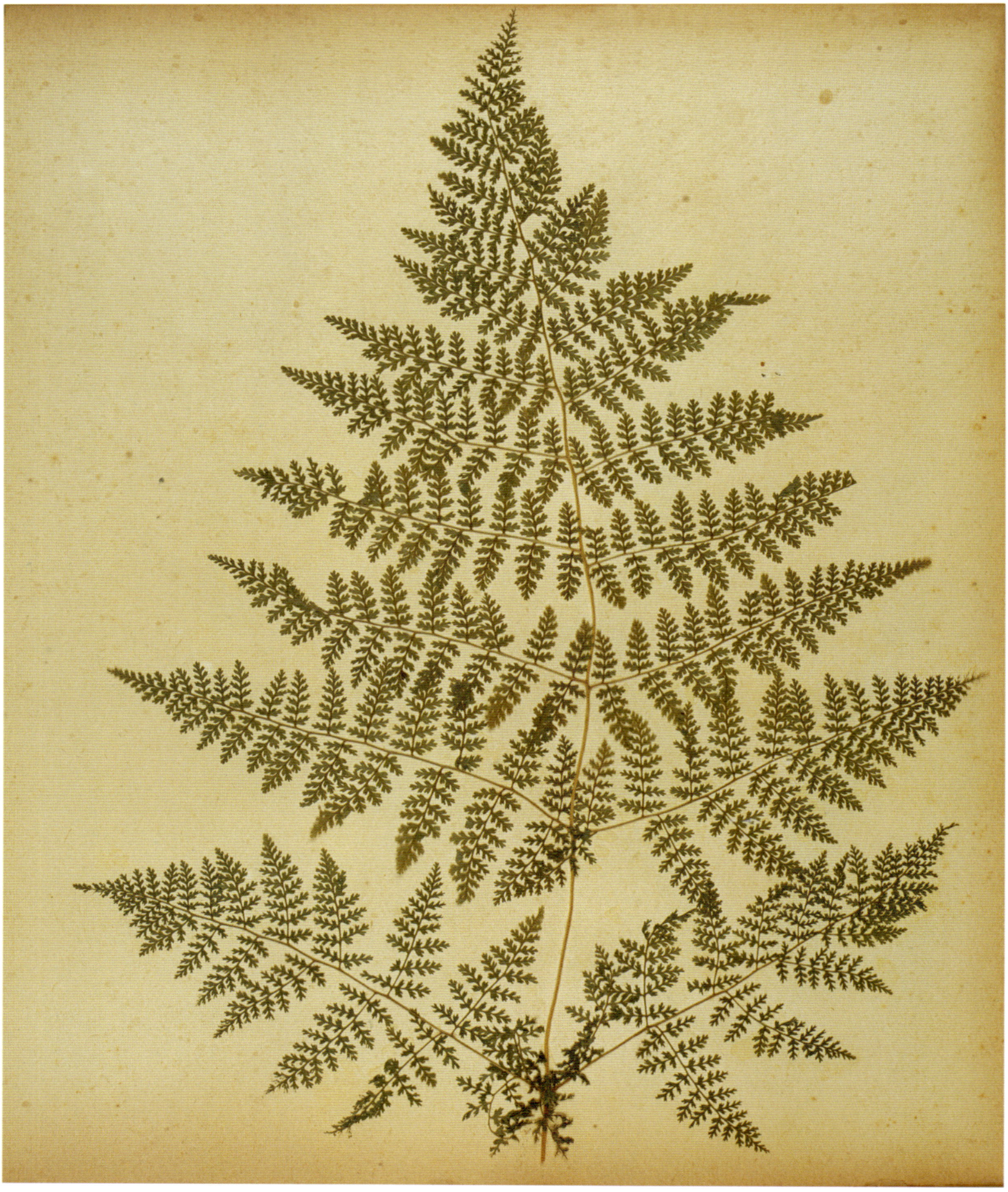

A Hypolepis Fern

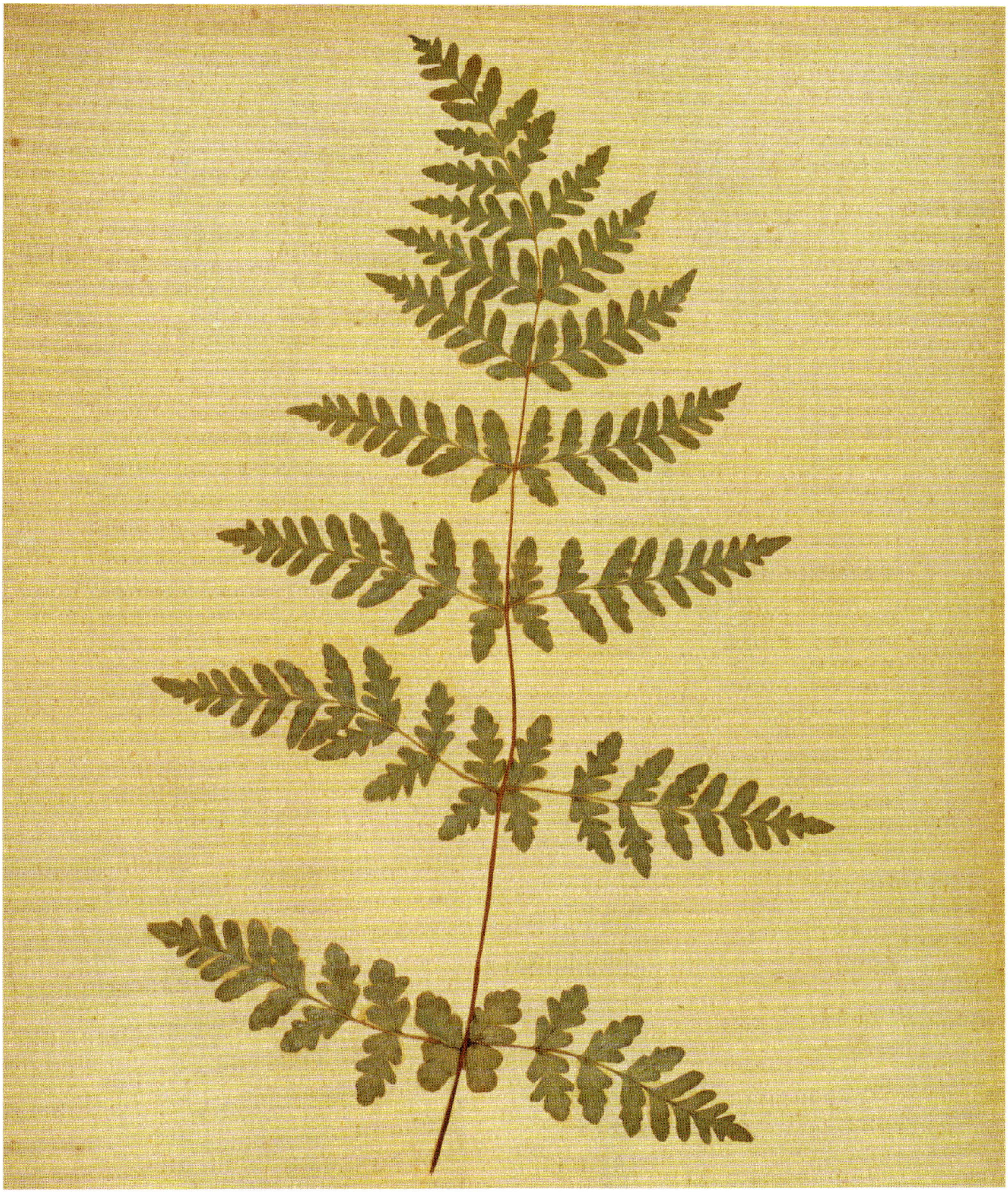

A Histiopteris Fern

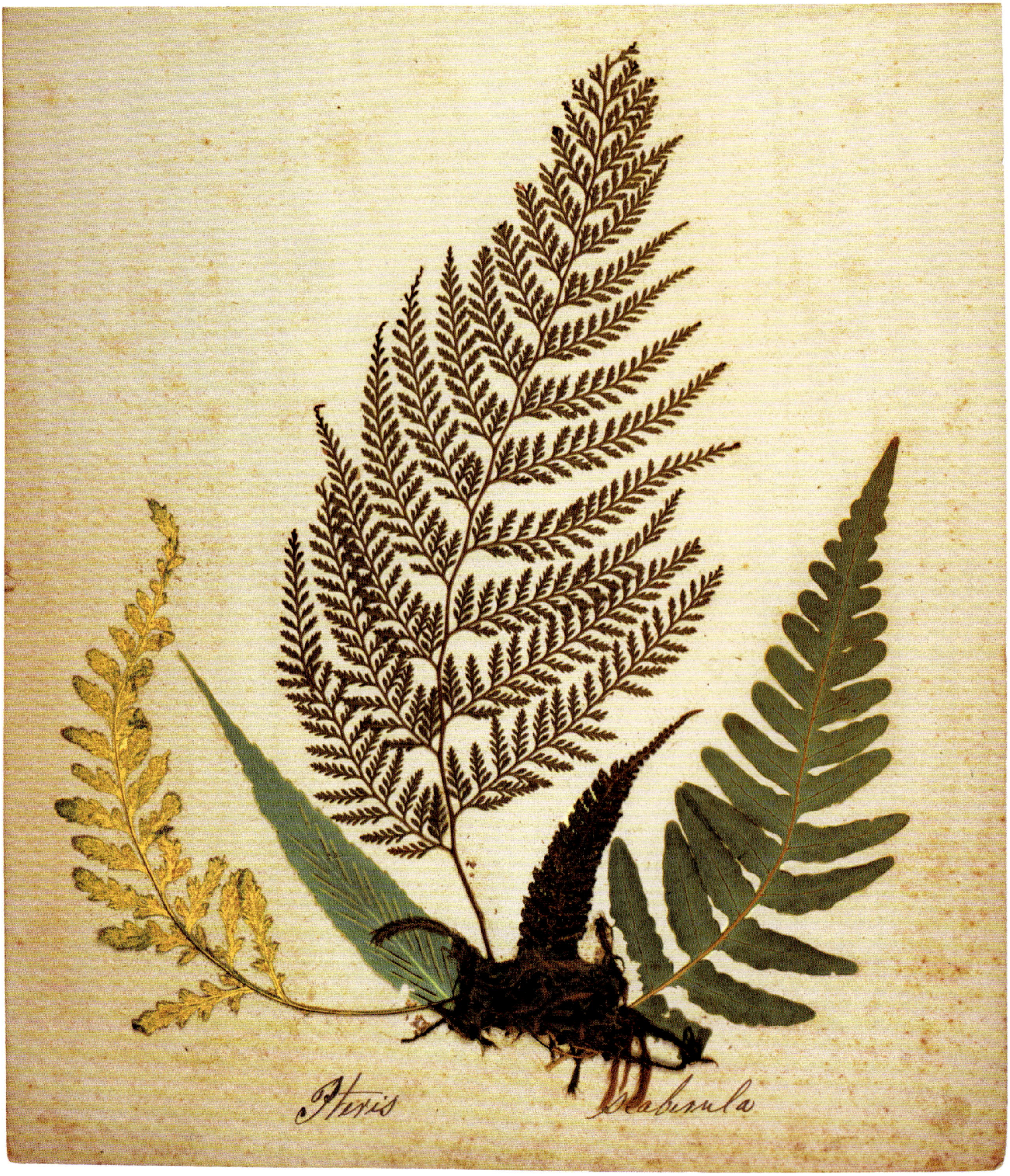

Louisa Atkinson's Fern Pressings

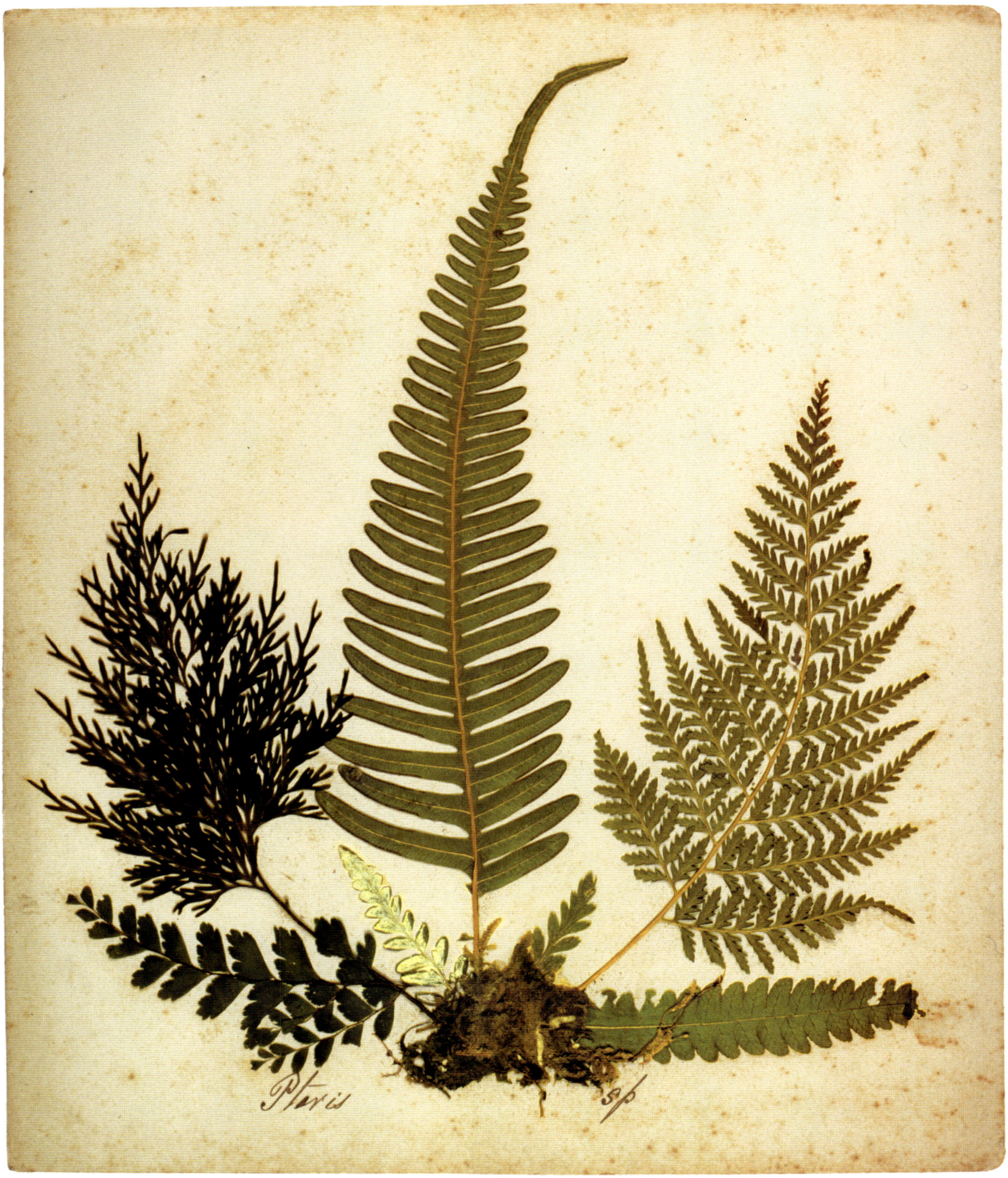

Louisa Atkinson's Fern Pressings

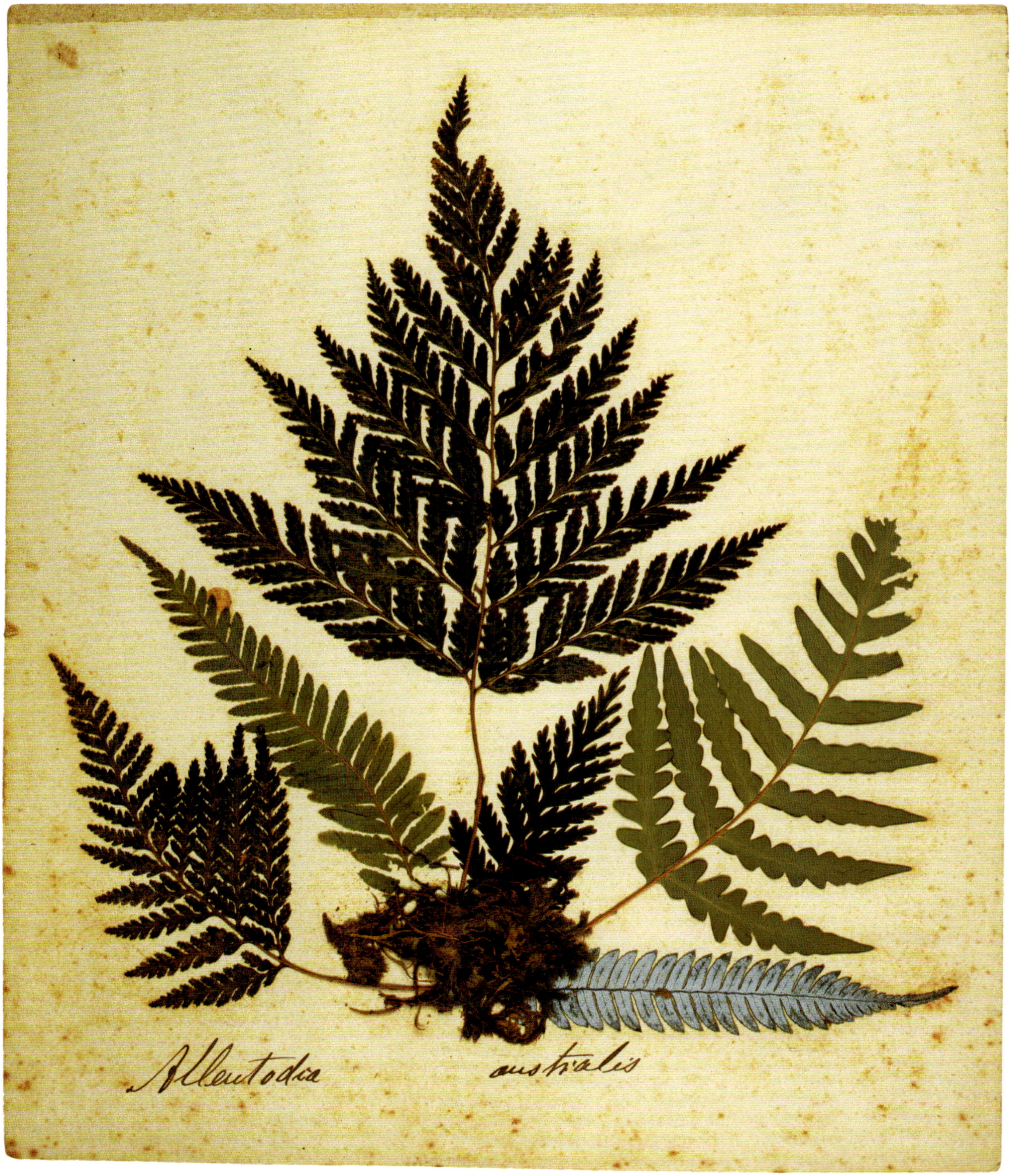

Louisa Atkinson's Fern Pressings

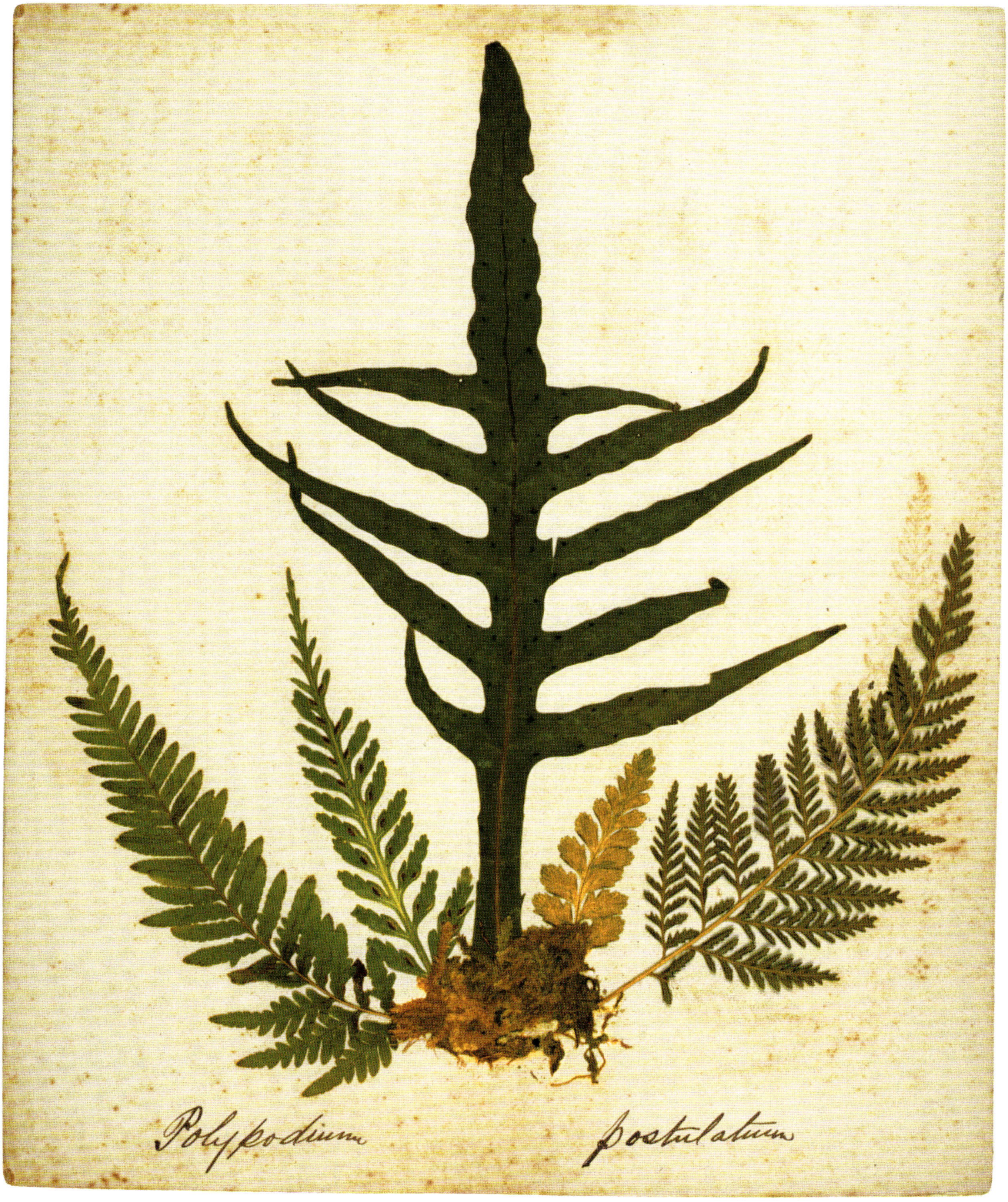

Louisa Atkinson's Fern Pressings

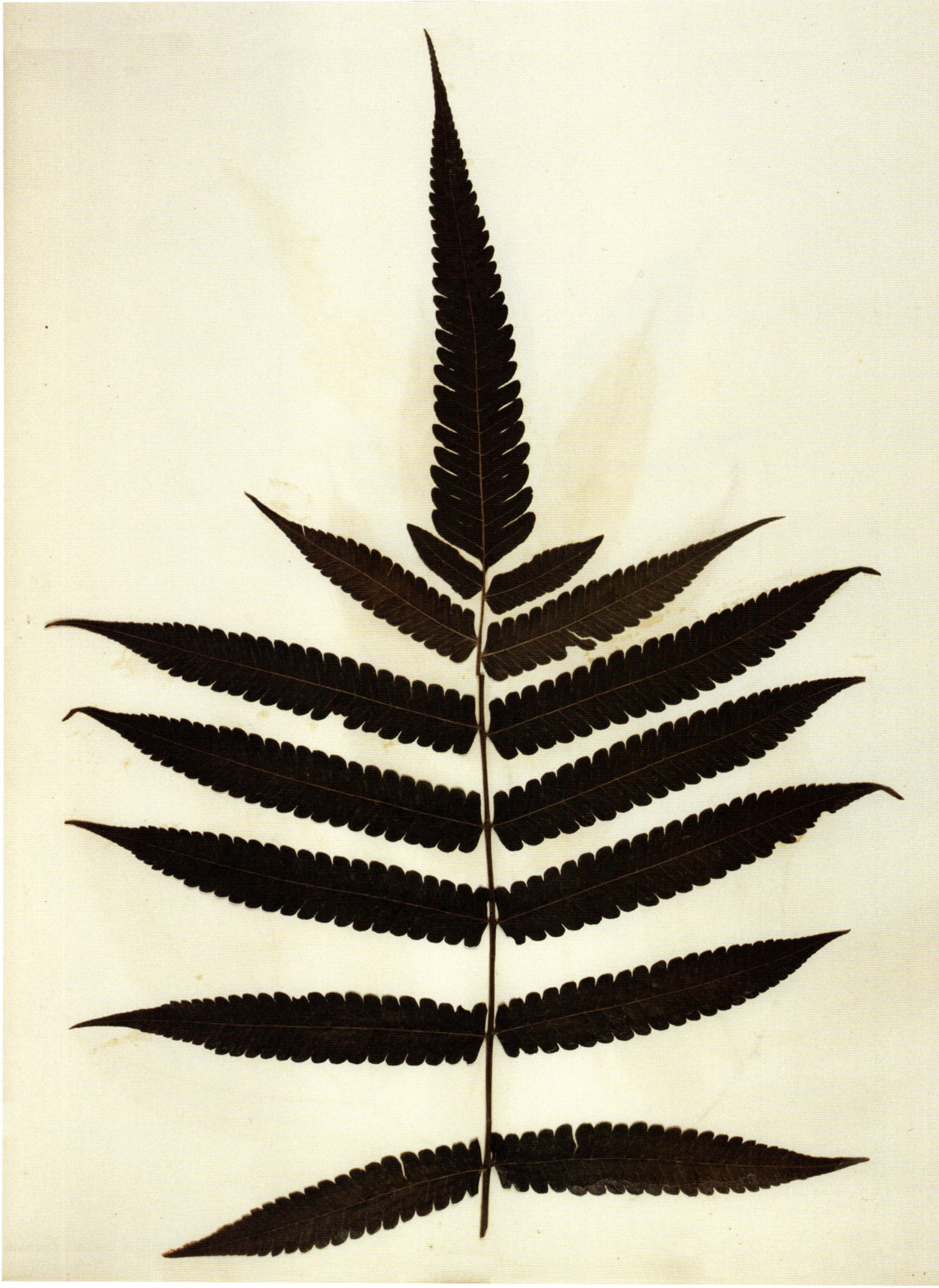

A Tropical Fern

Exotics

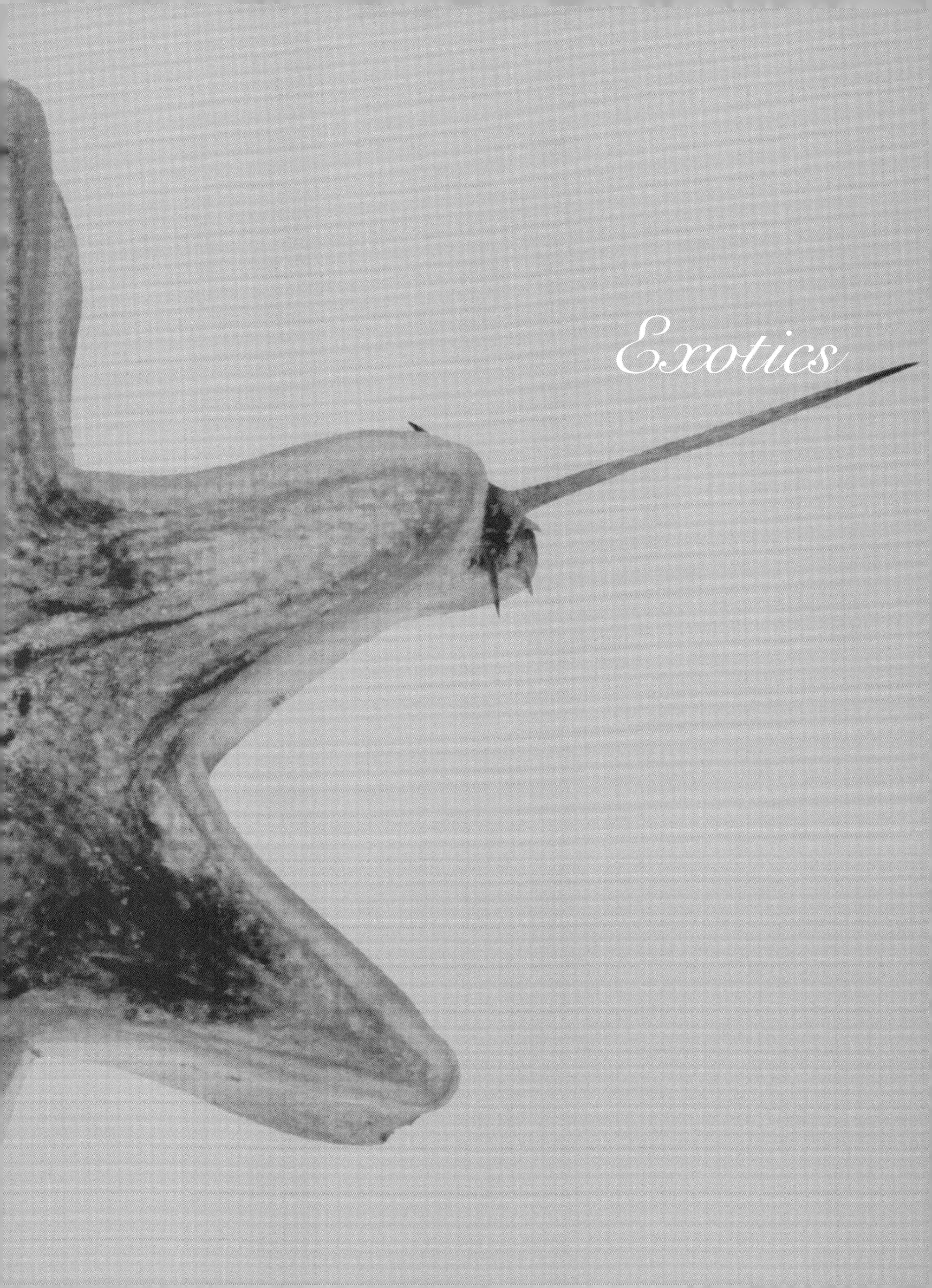

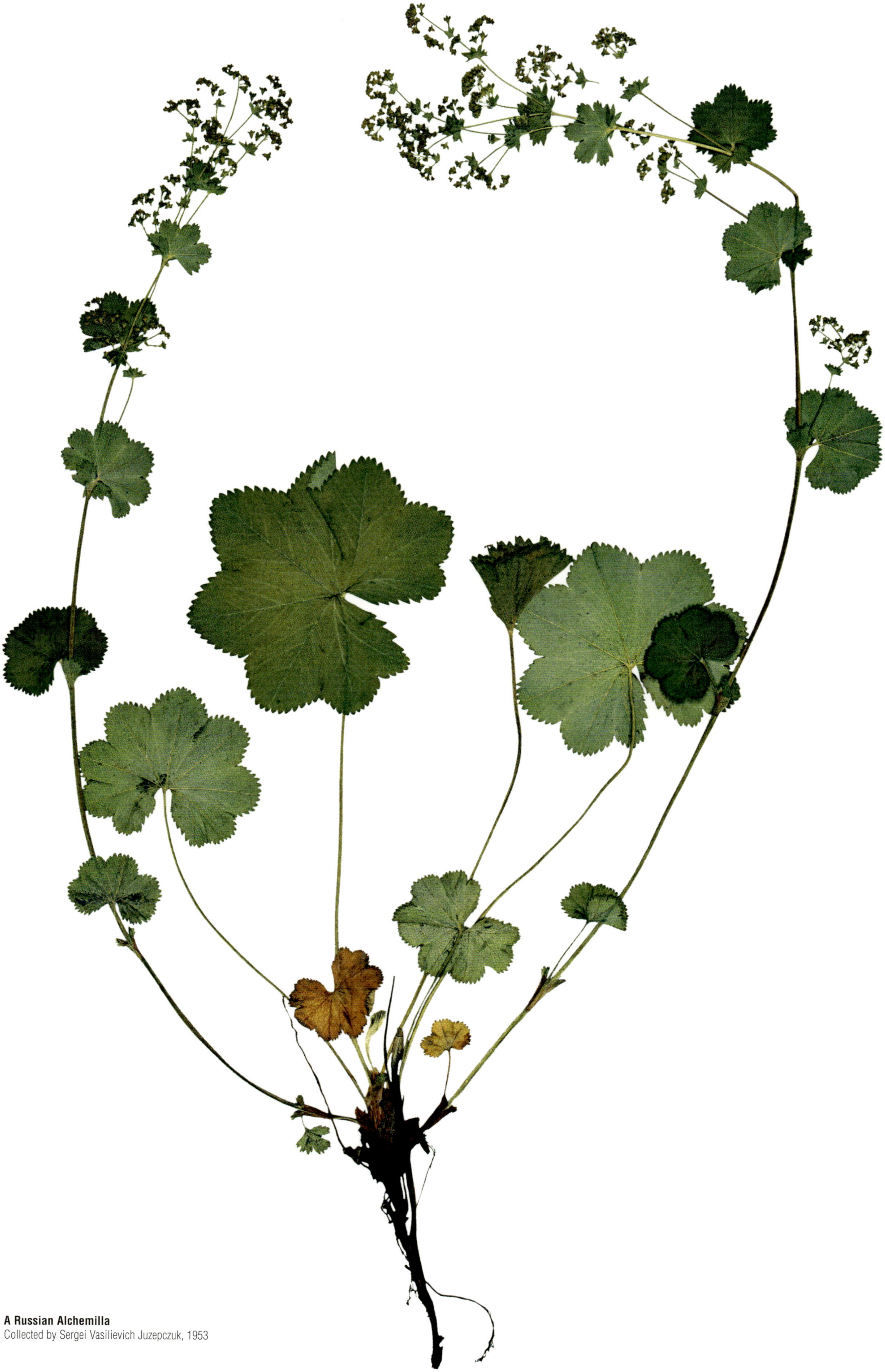

A Russian Alchemilla
Collected by Sergei Vasilievich Juzepczuk, 1953

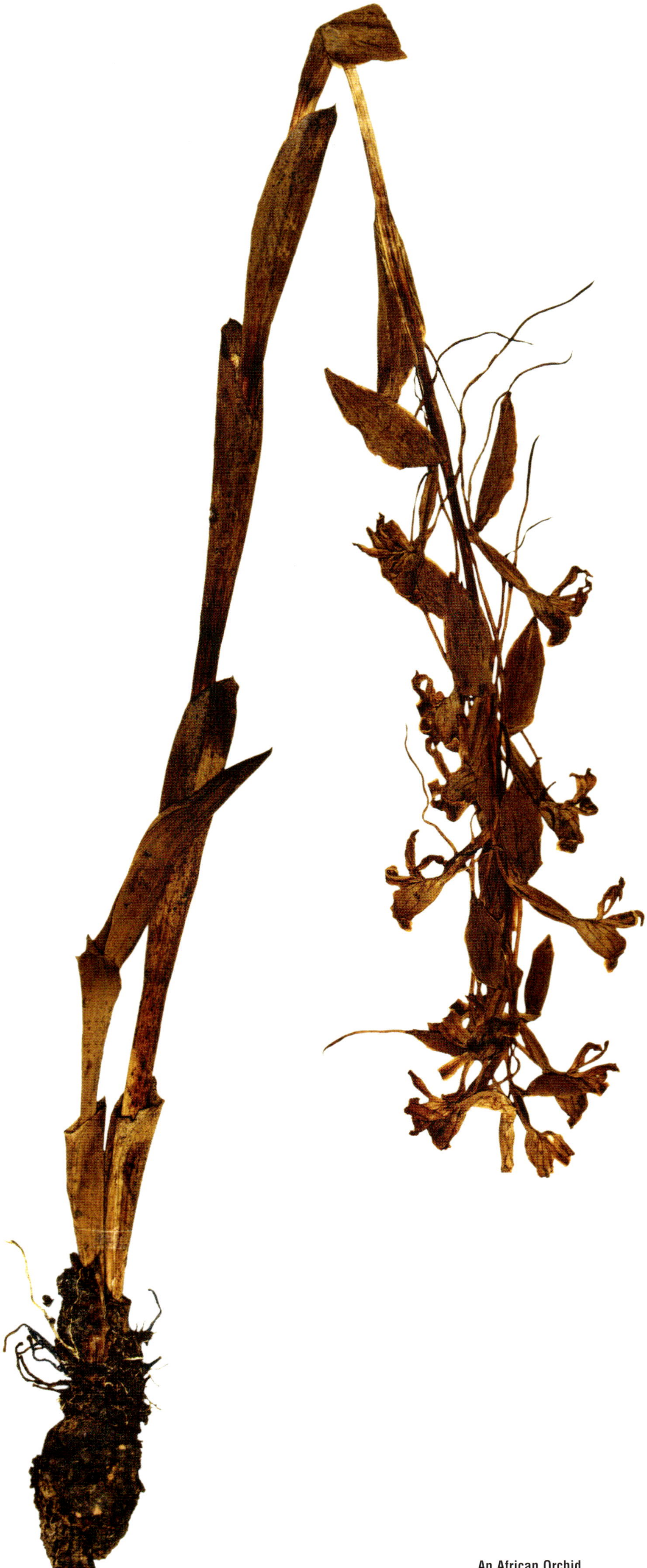

An African Orchid
Collected by R. K. Brummitt and R. M. Polhill, 1975

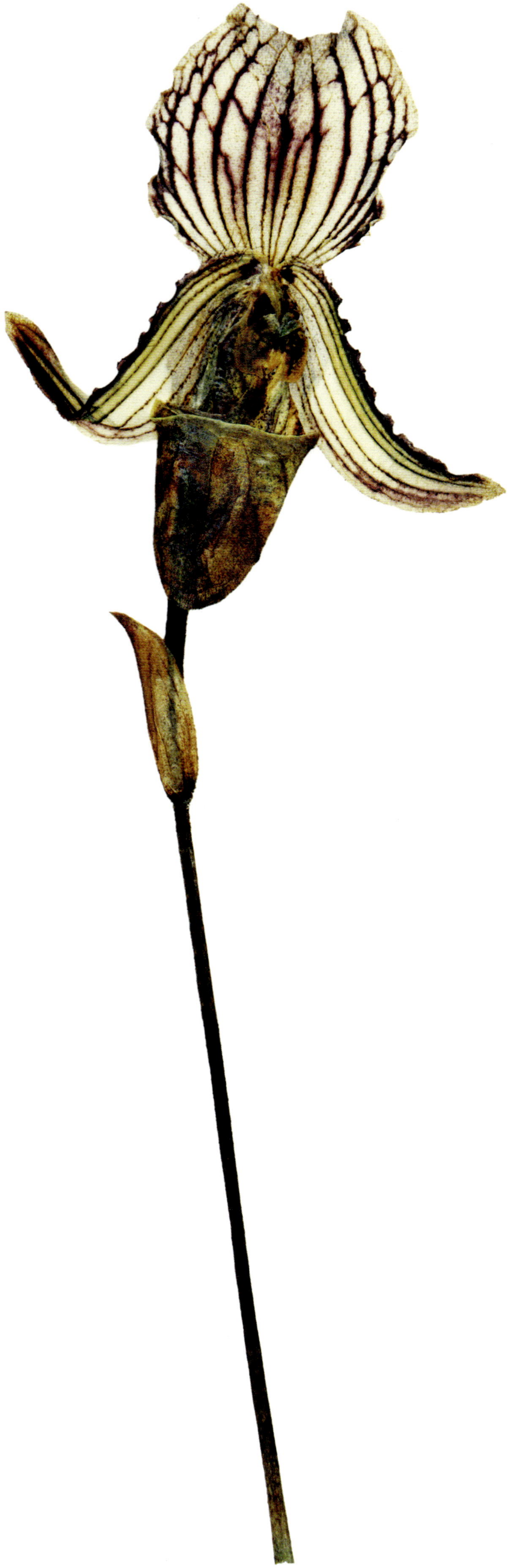

A Slipper Orchid
Collected by Dianne C. Evans, 1988

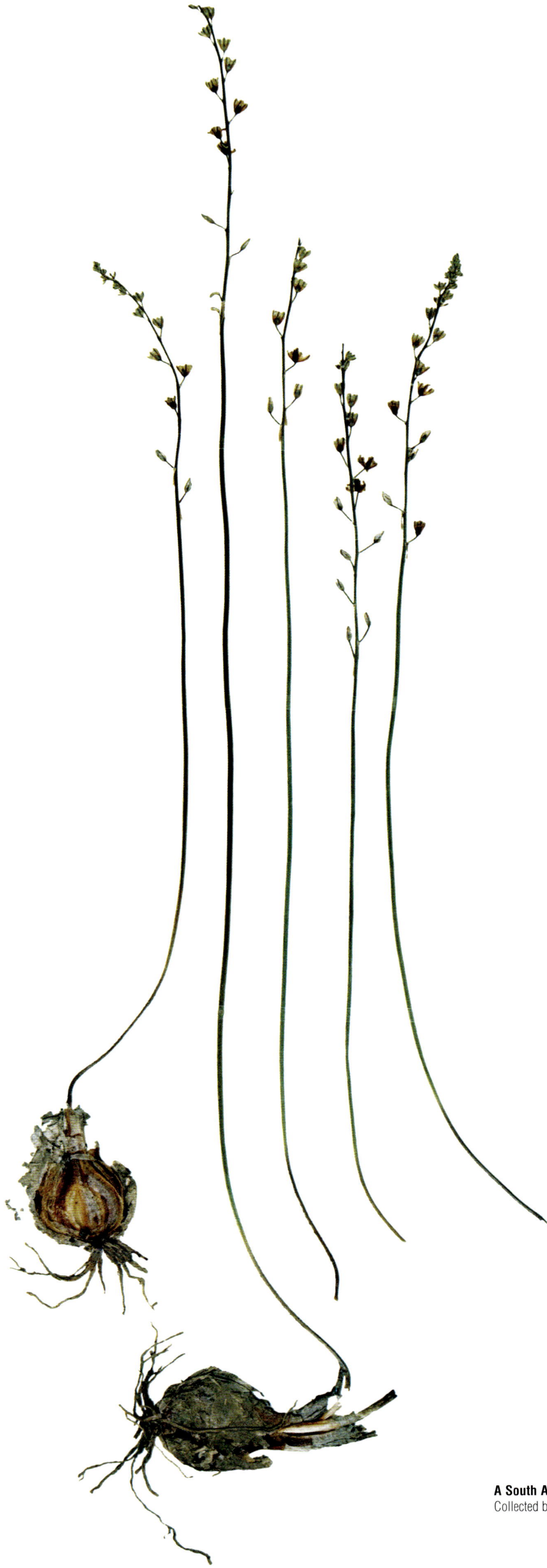

A South African Drimia
Collected by W. J. Hanekom, 1972

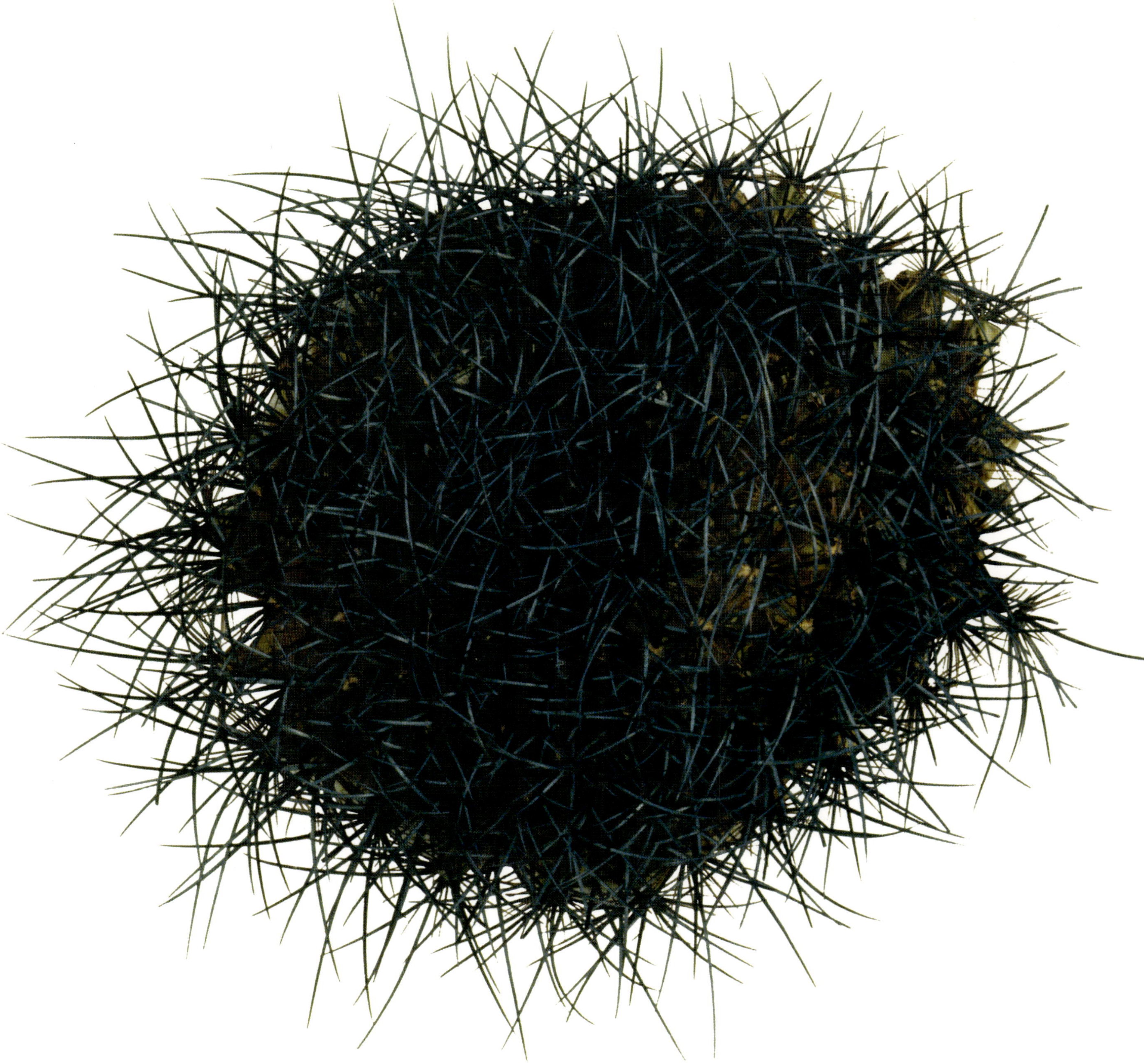

A Mexican Cactus
Collected by C. G. Pringle, 1901

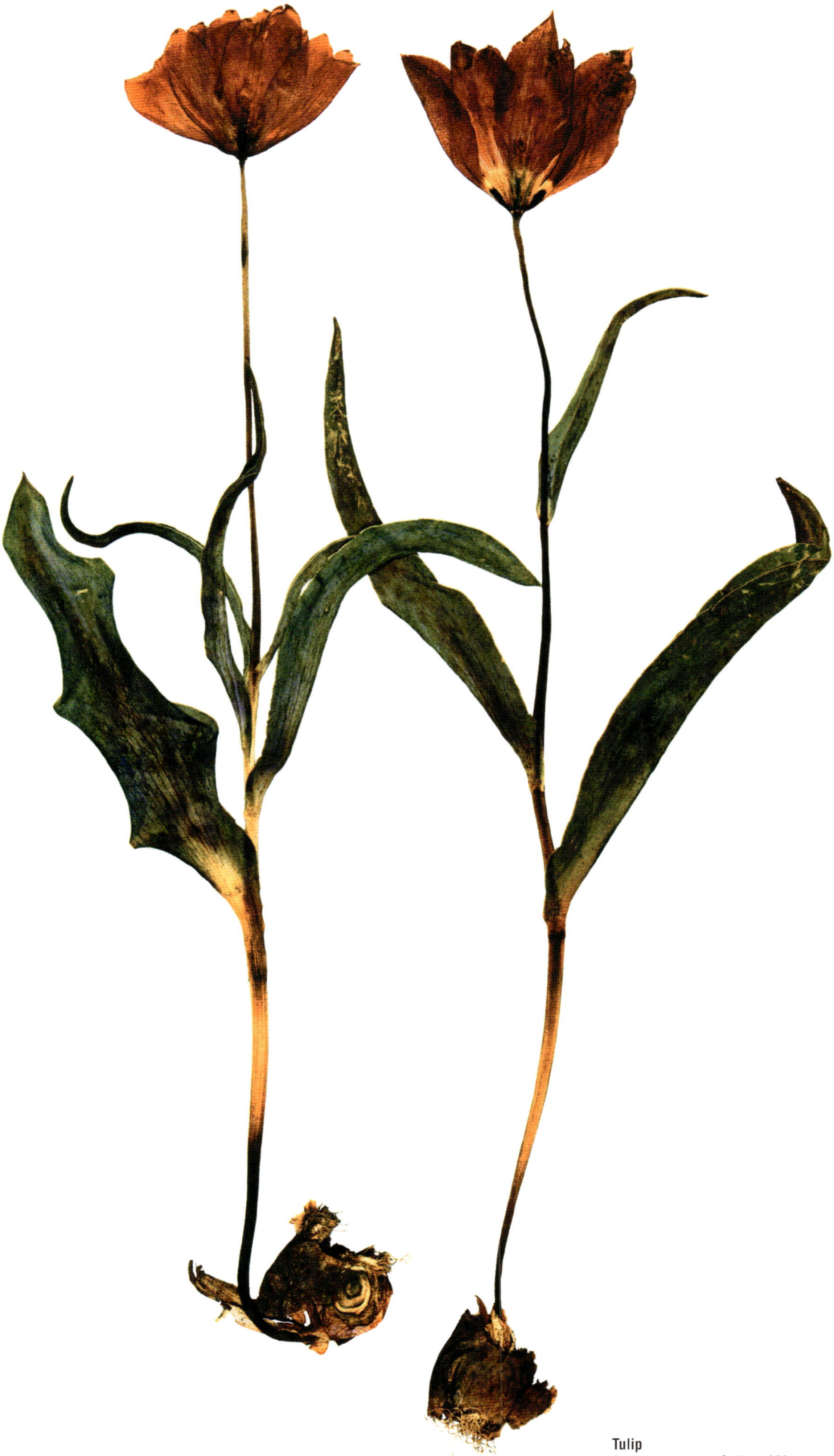

Tulip
Collected by Alexis Callier, 1900

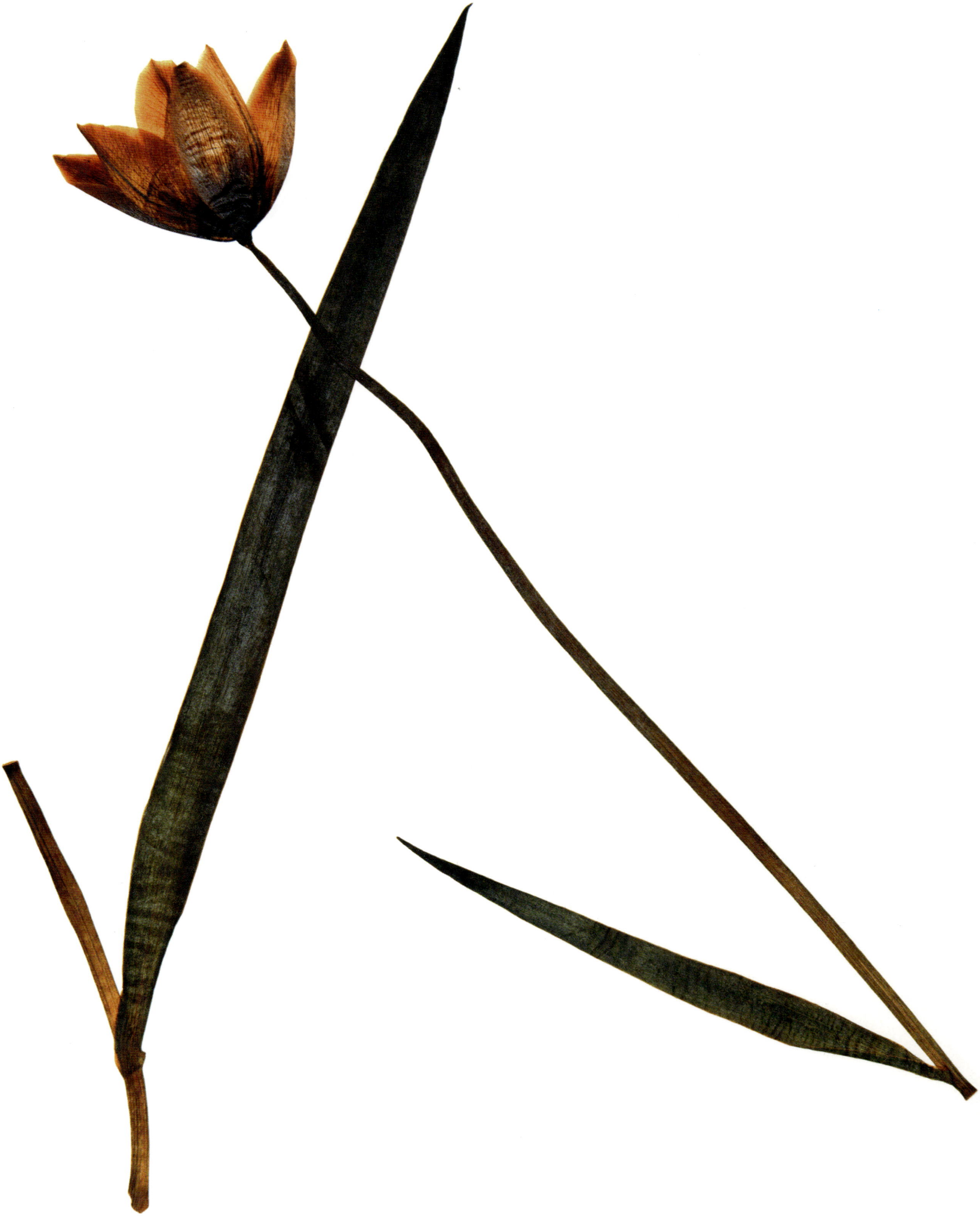

A Romanian Tulip
Collected by A. Richter, 1908

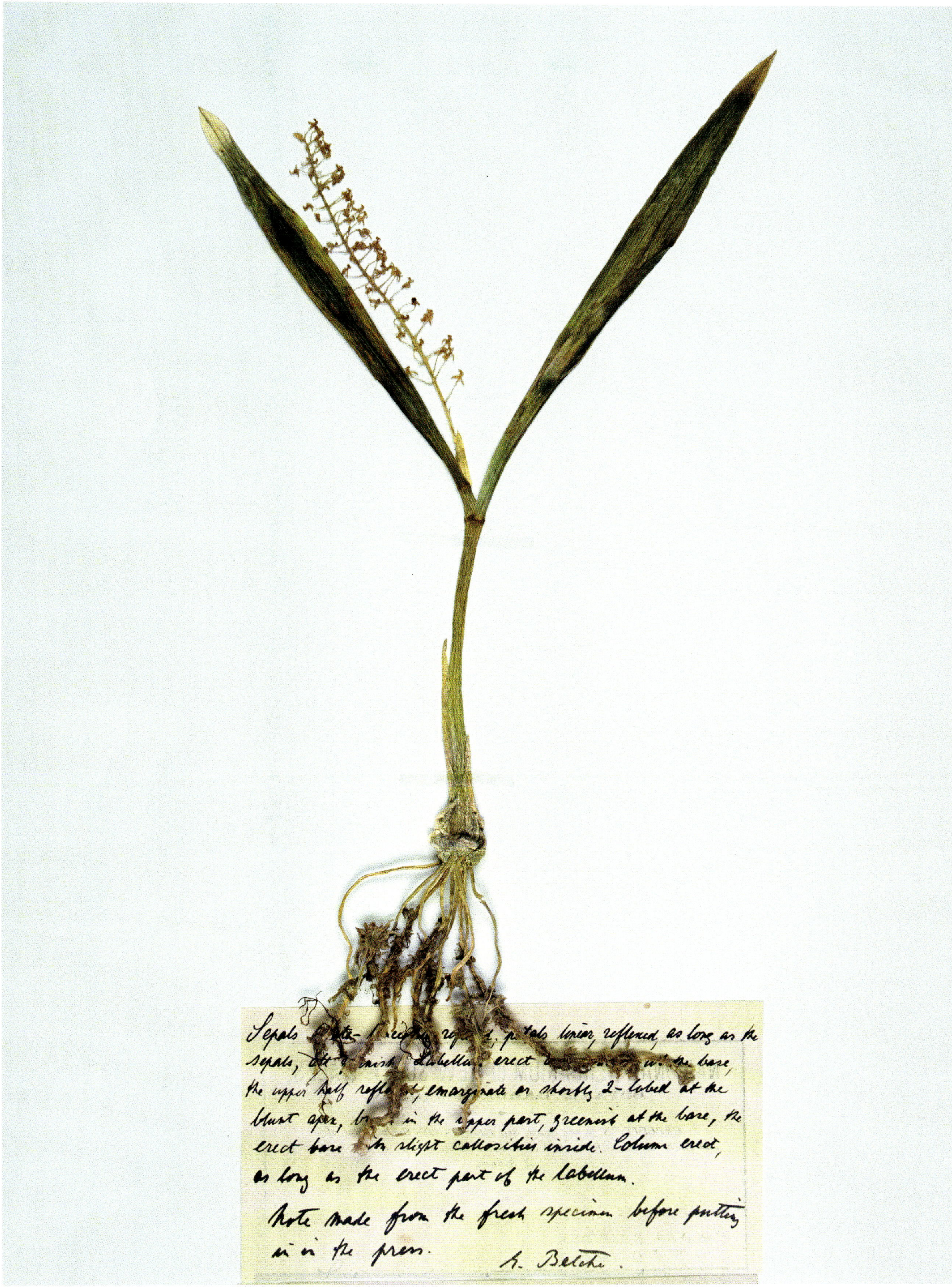

A Vanuatu Liparis
Collected by Dr Marriso, 1897

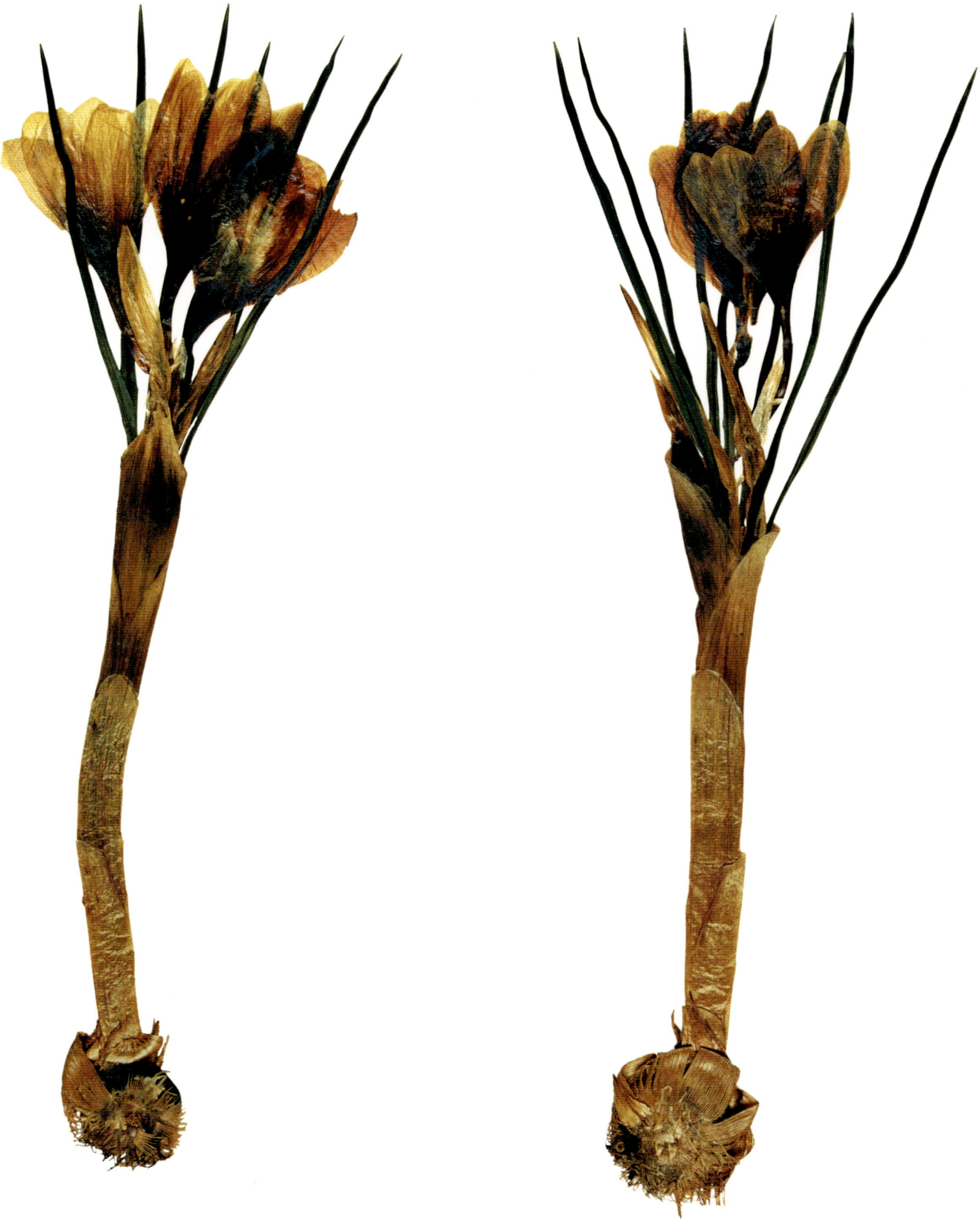

Yellow Crocus
Collected 1844

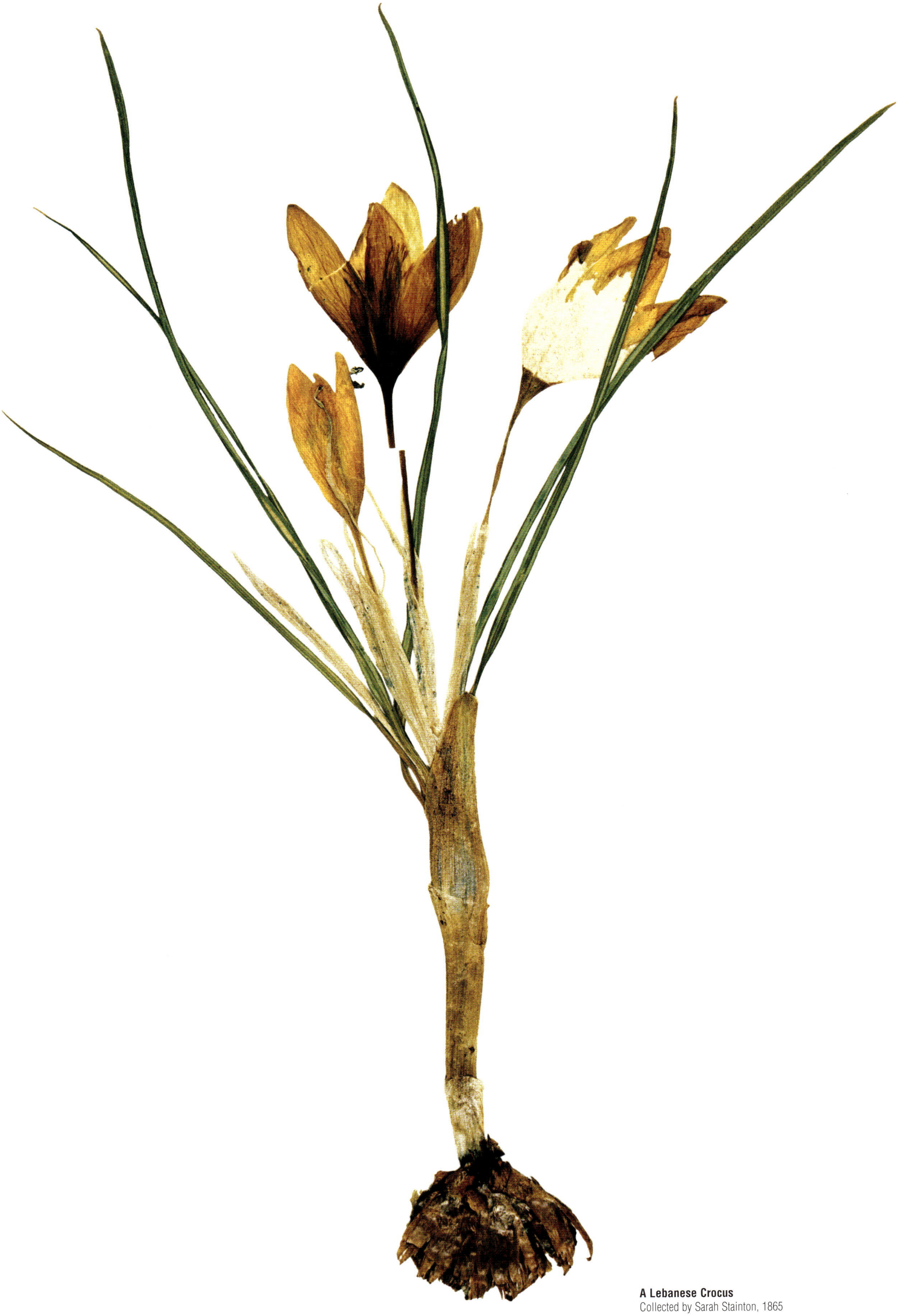

A Lebanese Crocus
Collected by Sarah Stainton, 1865

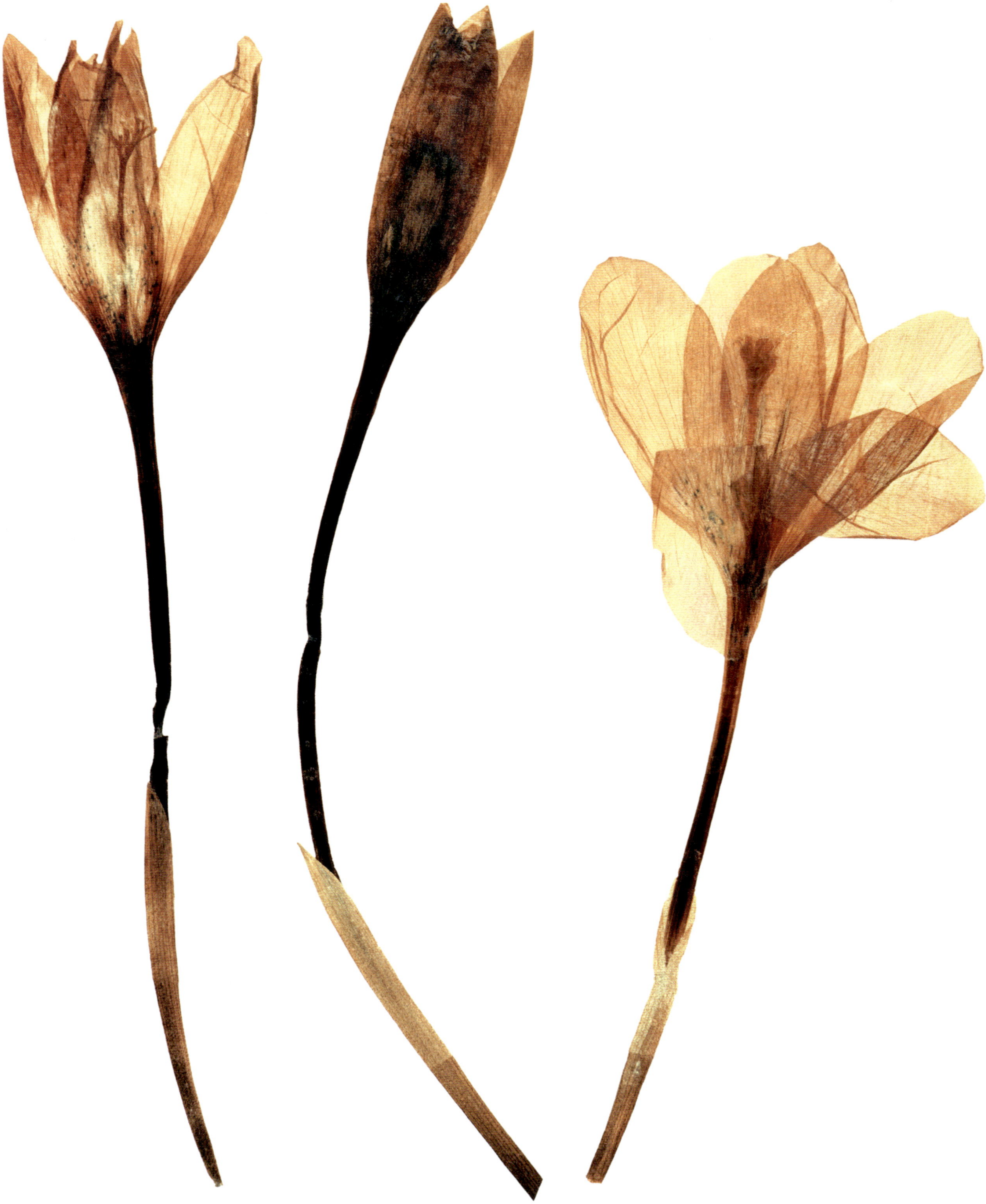

Naked-flowering Crocus
Collected 1858

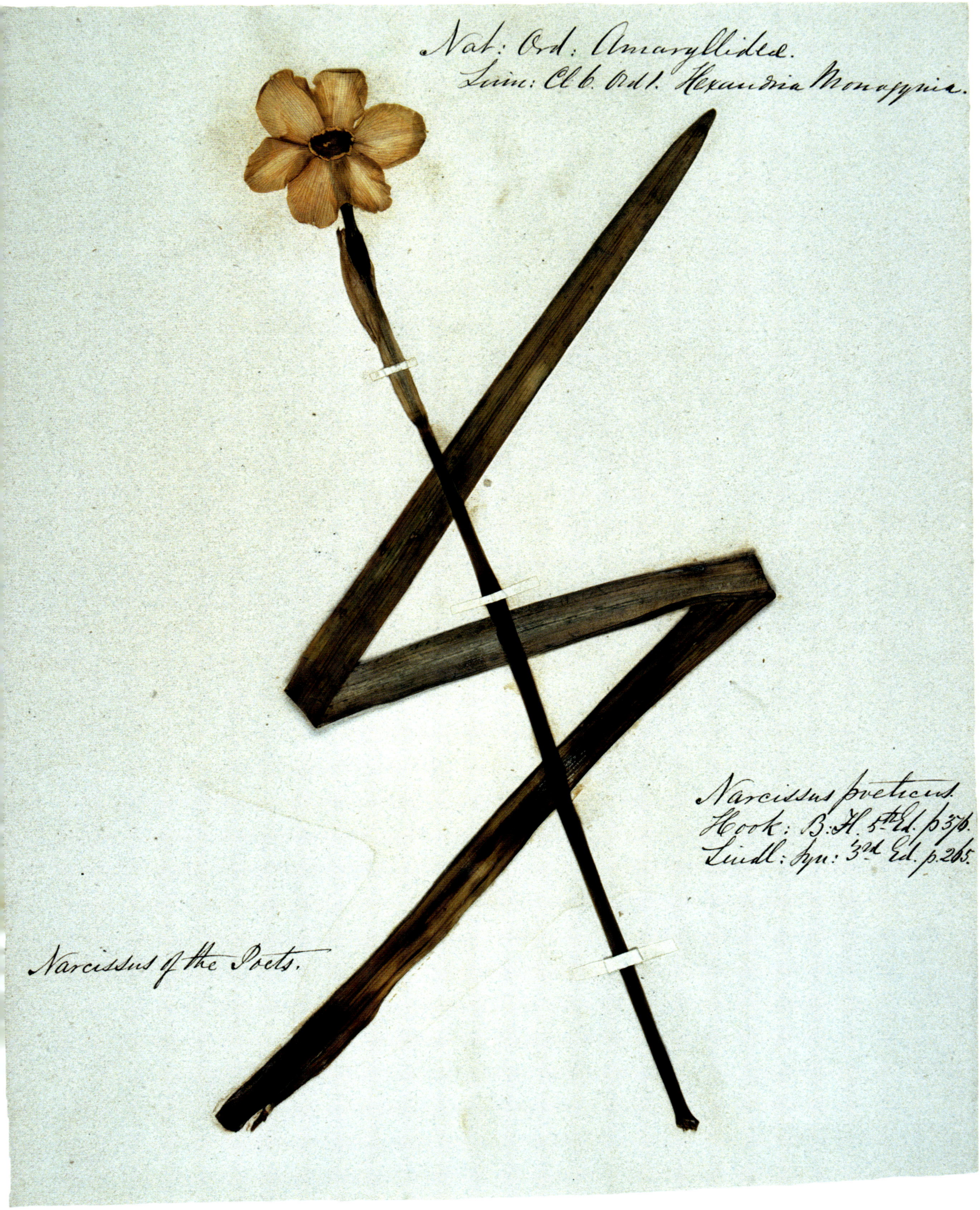

Pheasants Eye Narcissus

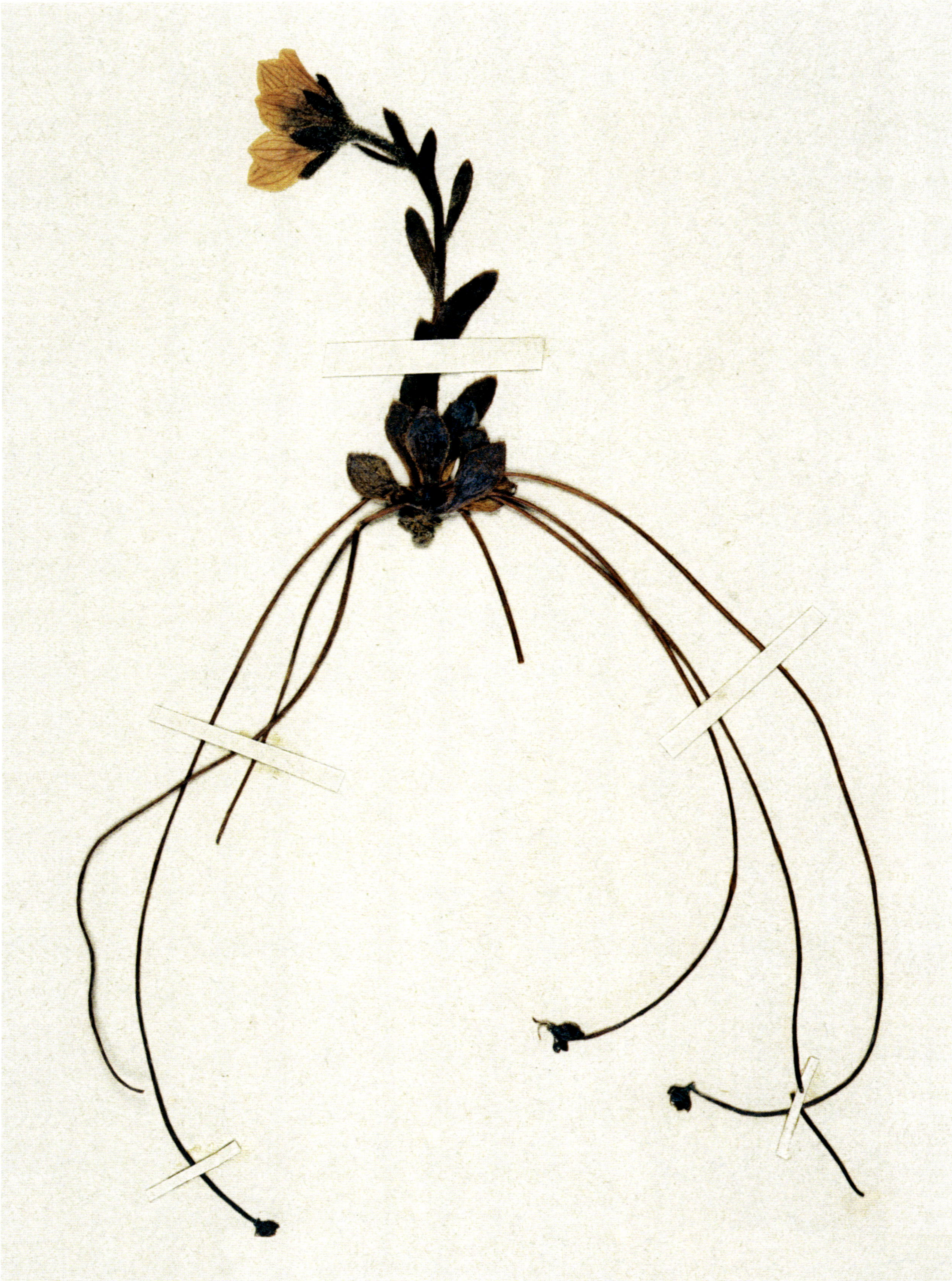

Spiderplant
Collected 1819–1825

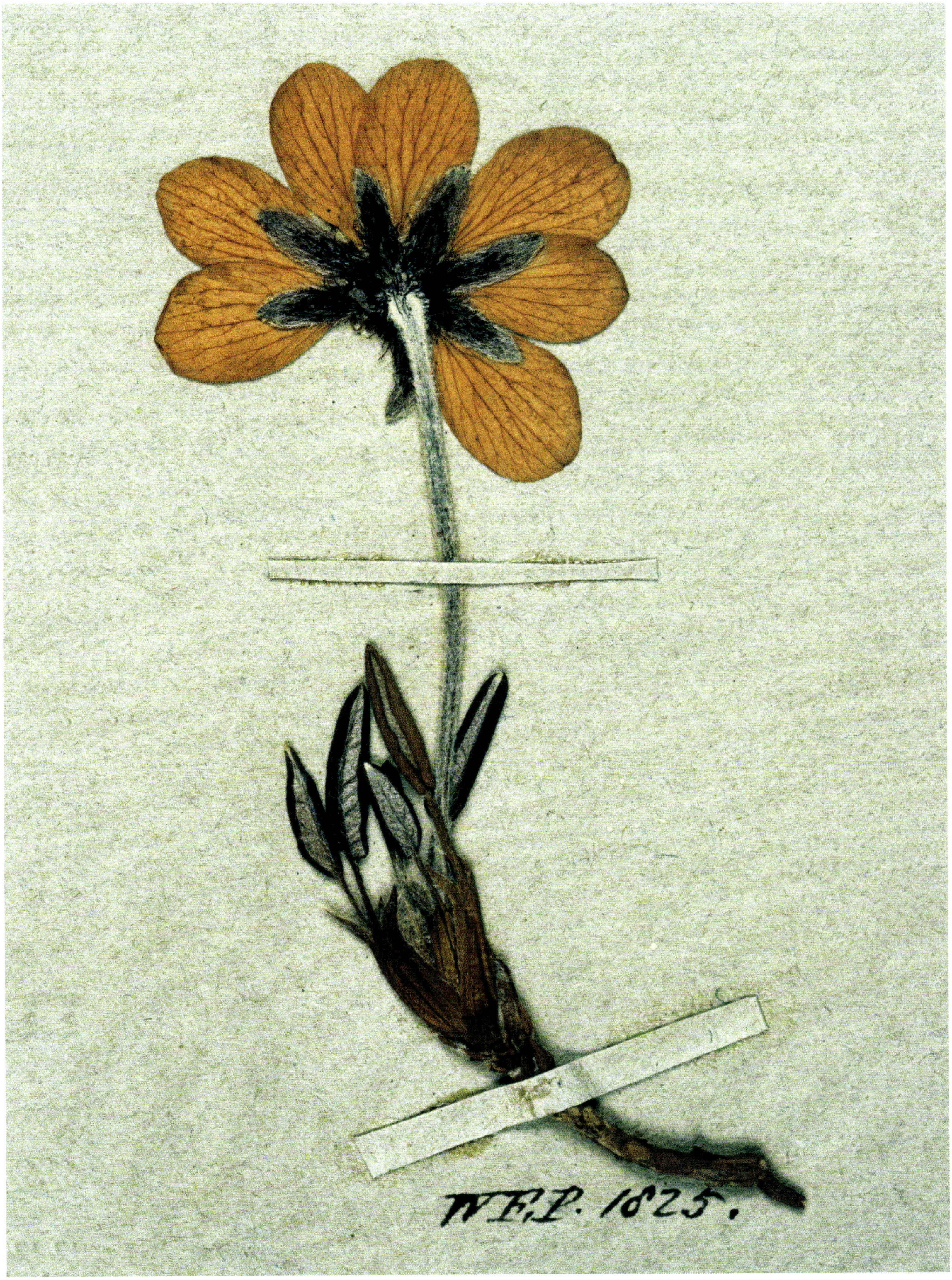

An Arctic Dryas
Collected 1825

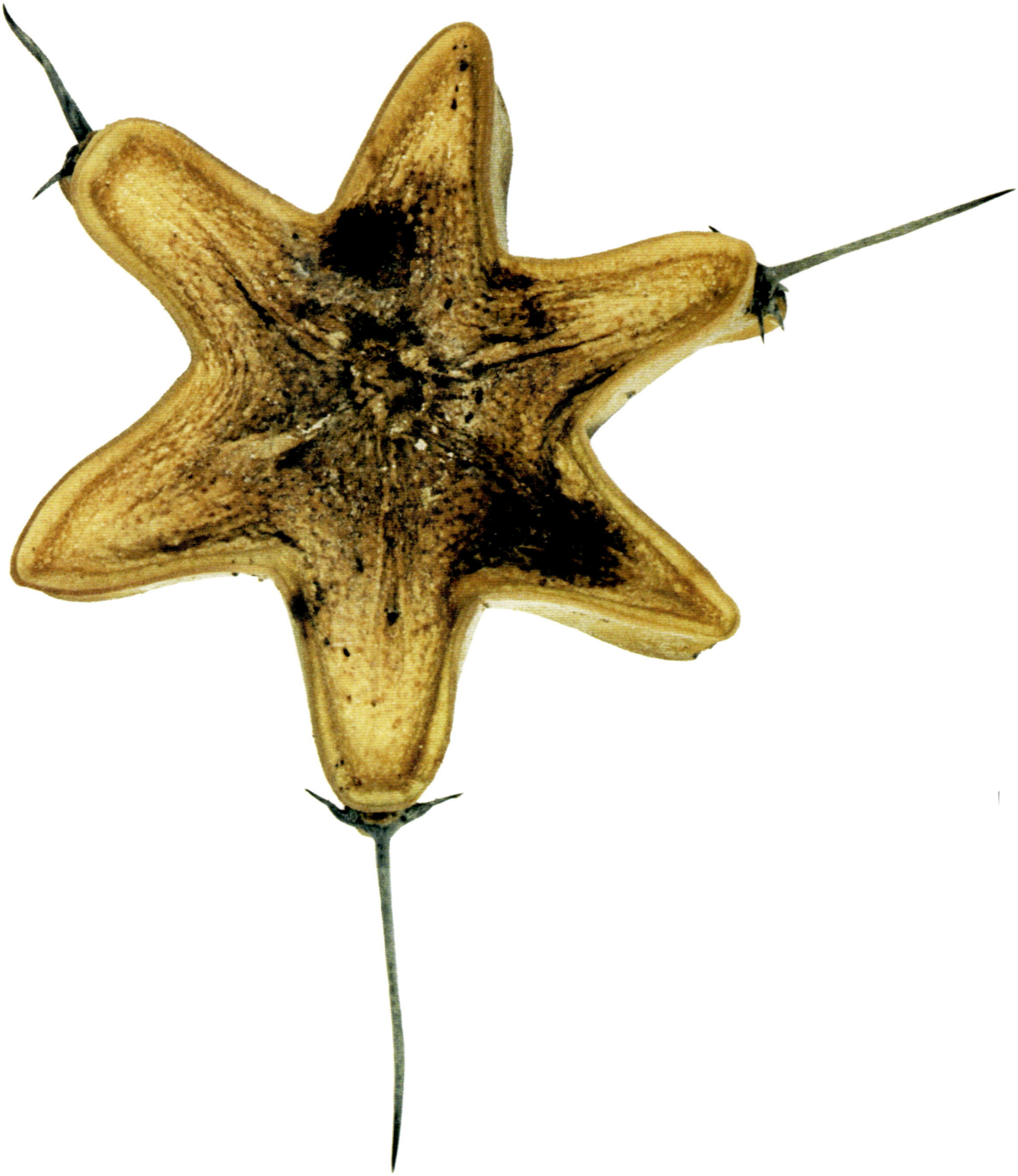

Blue Candle
Collected by C. G. Pringle, 1891

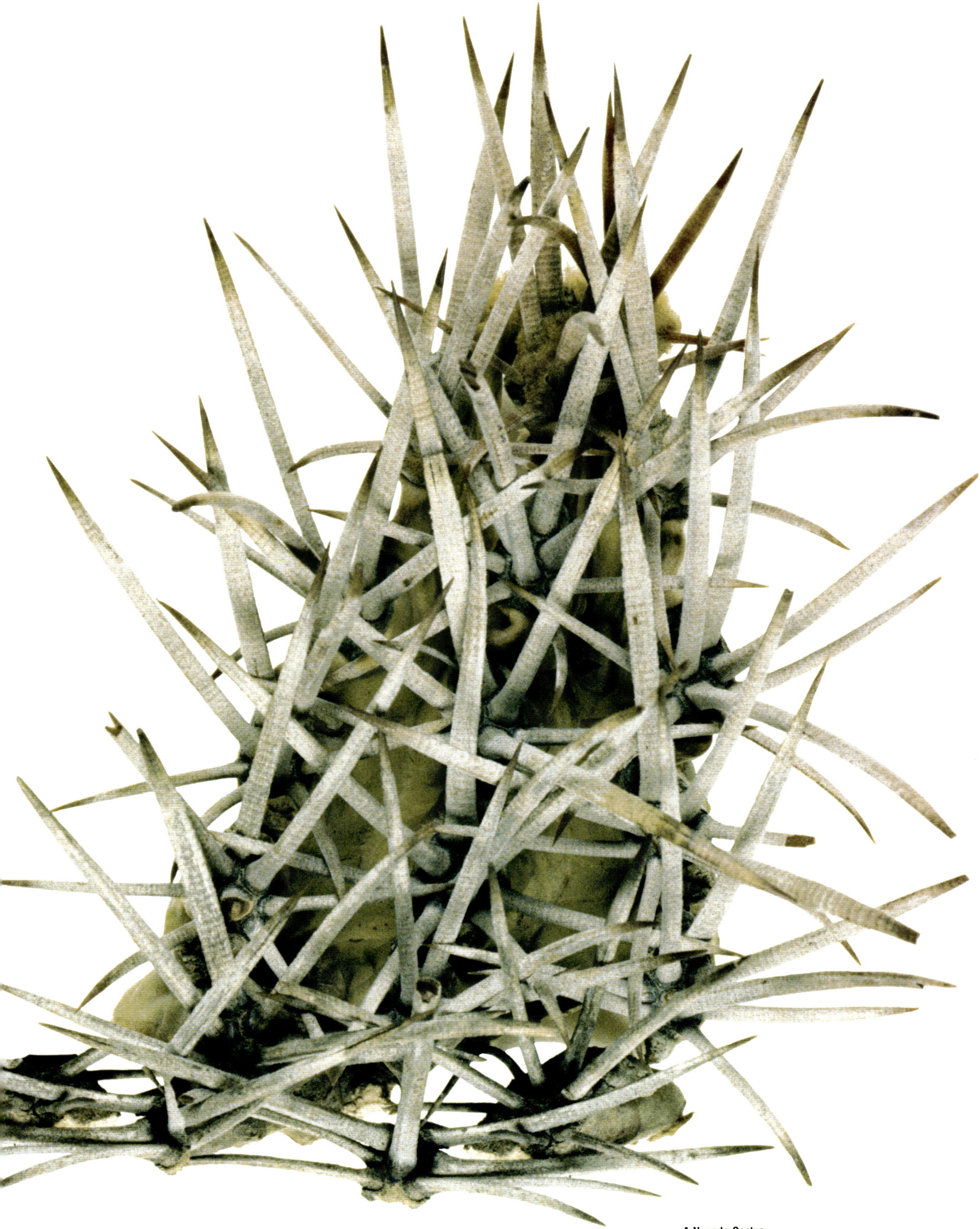

A Nevada Cactus
Collected by I. W. Clokey, 1939

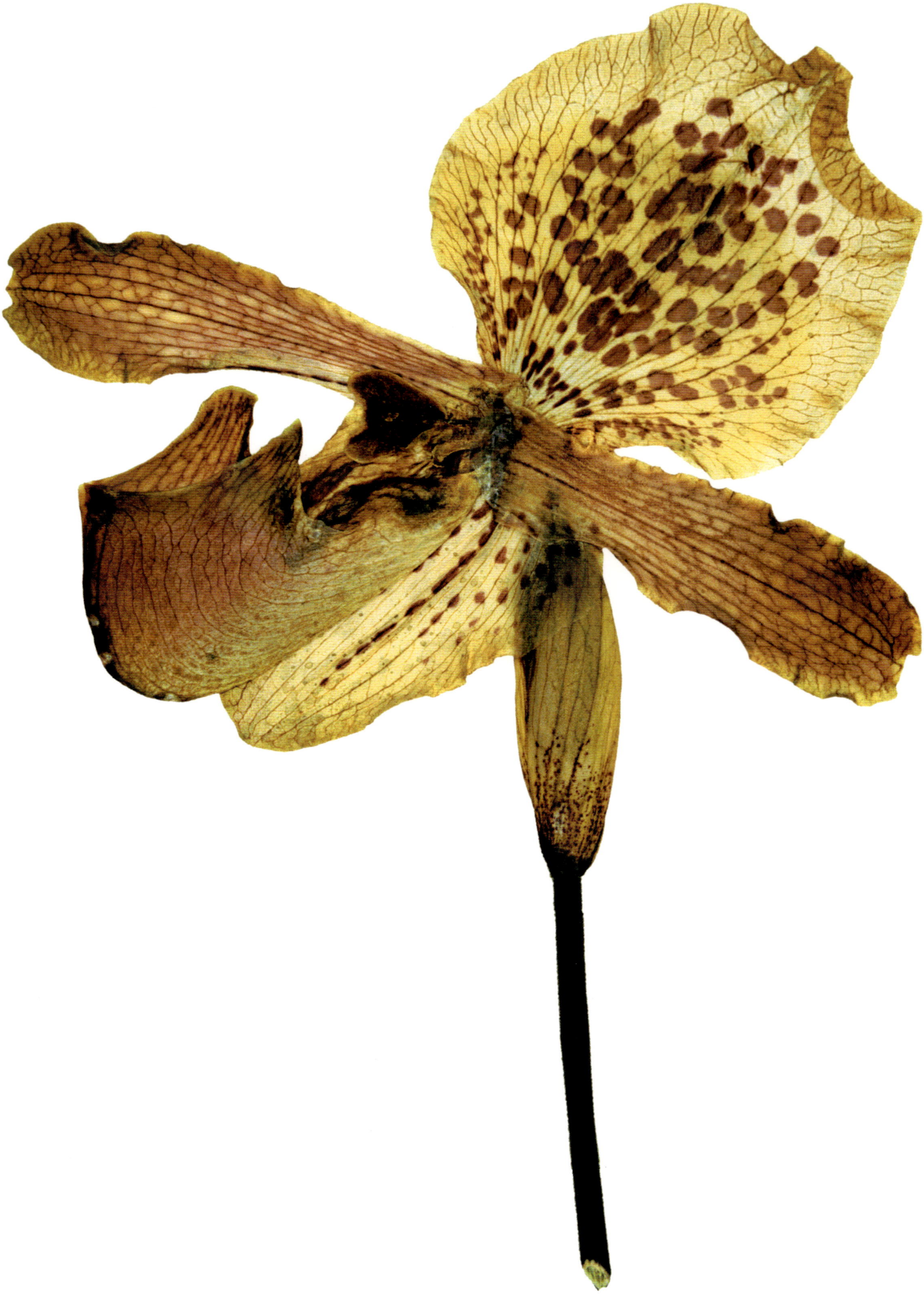

A Slipper Orchid
Collected by Dianne Evans, 1988

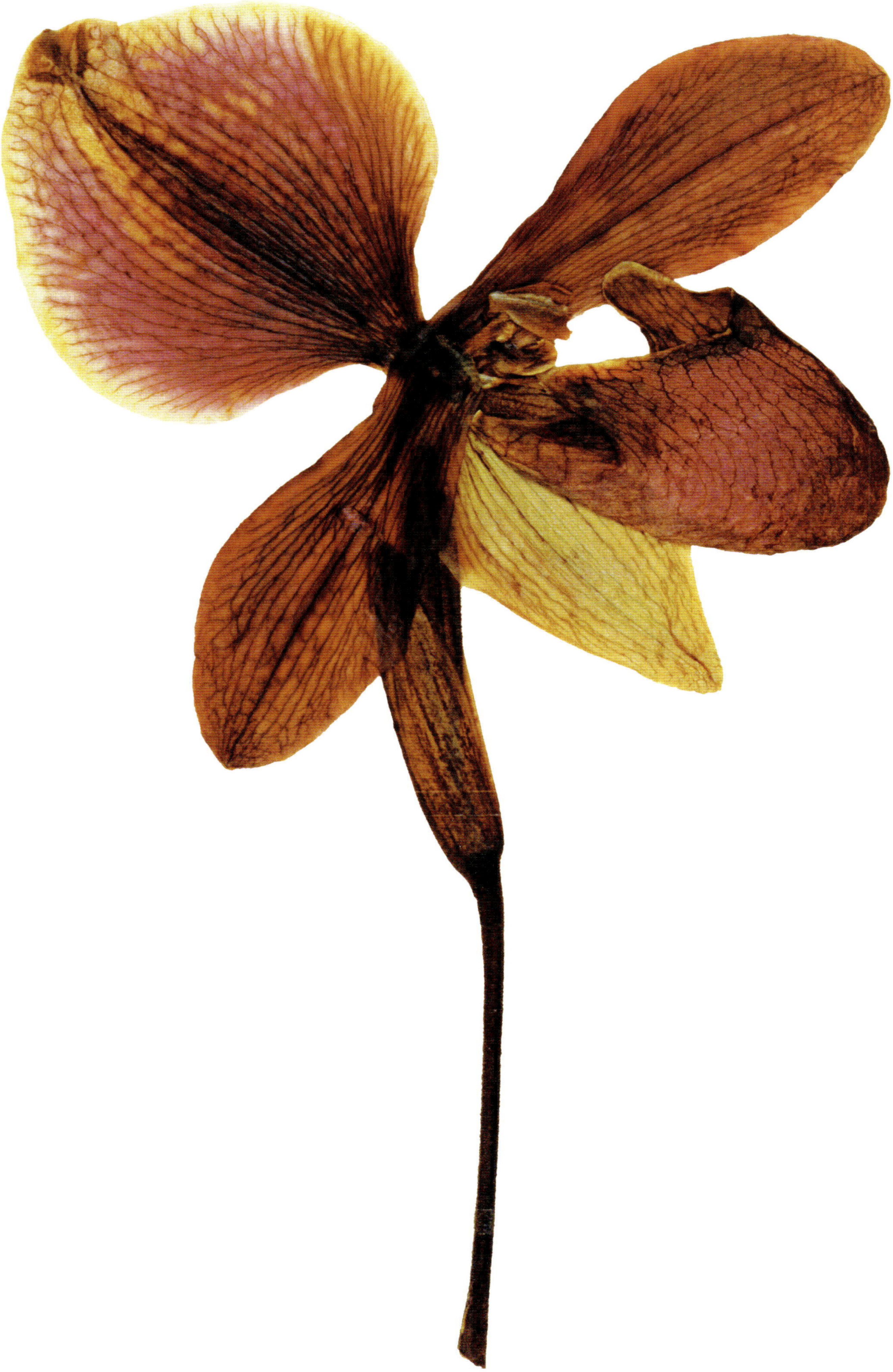

A Slipper Orchid Hybrid
Collected by Dianne Evans, 1988

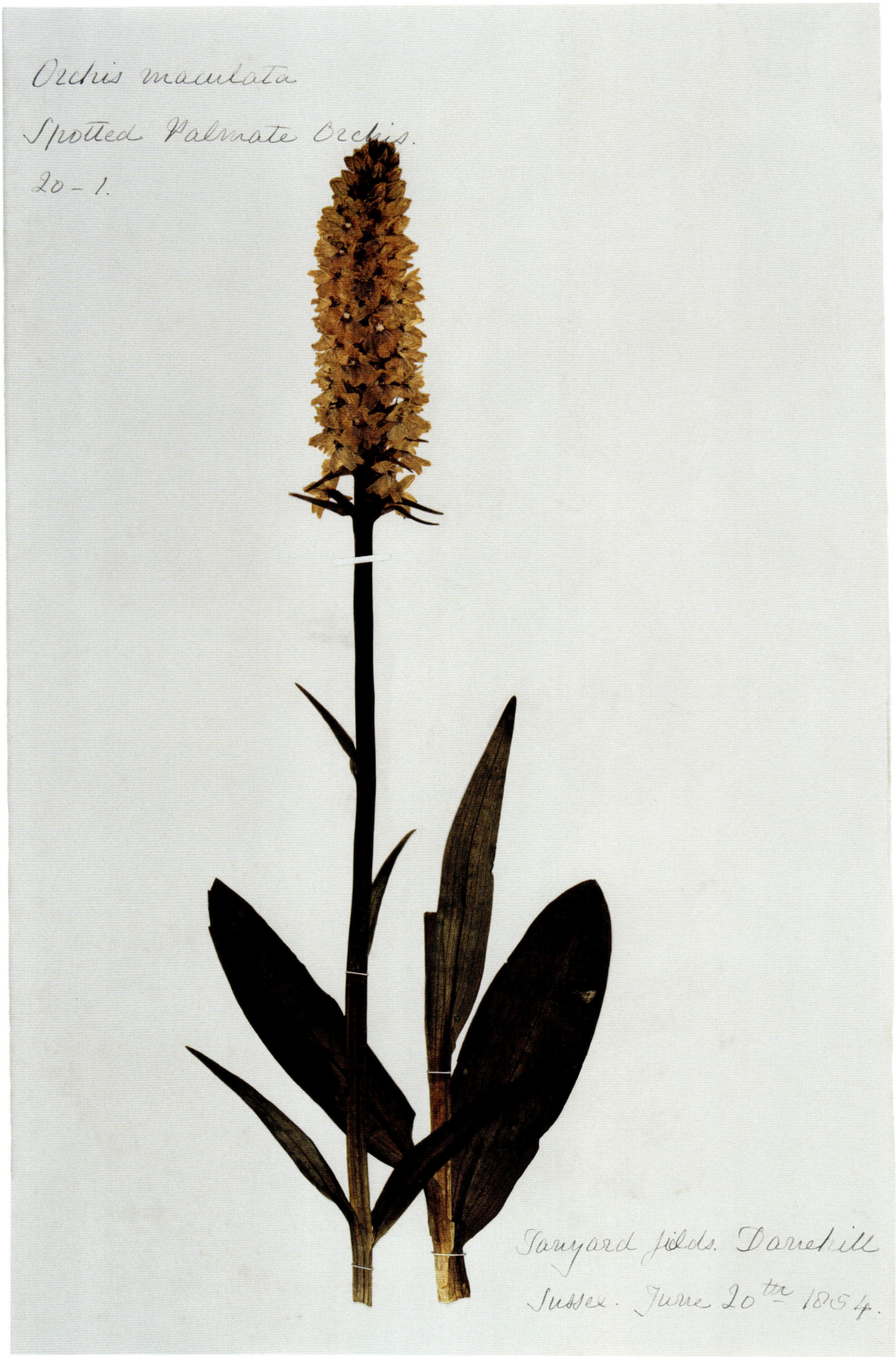

Common Spotted Orchid
Collected 1854

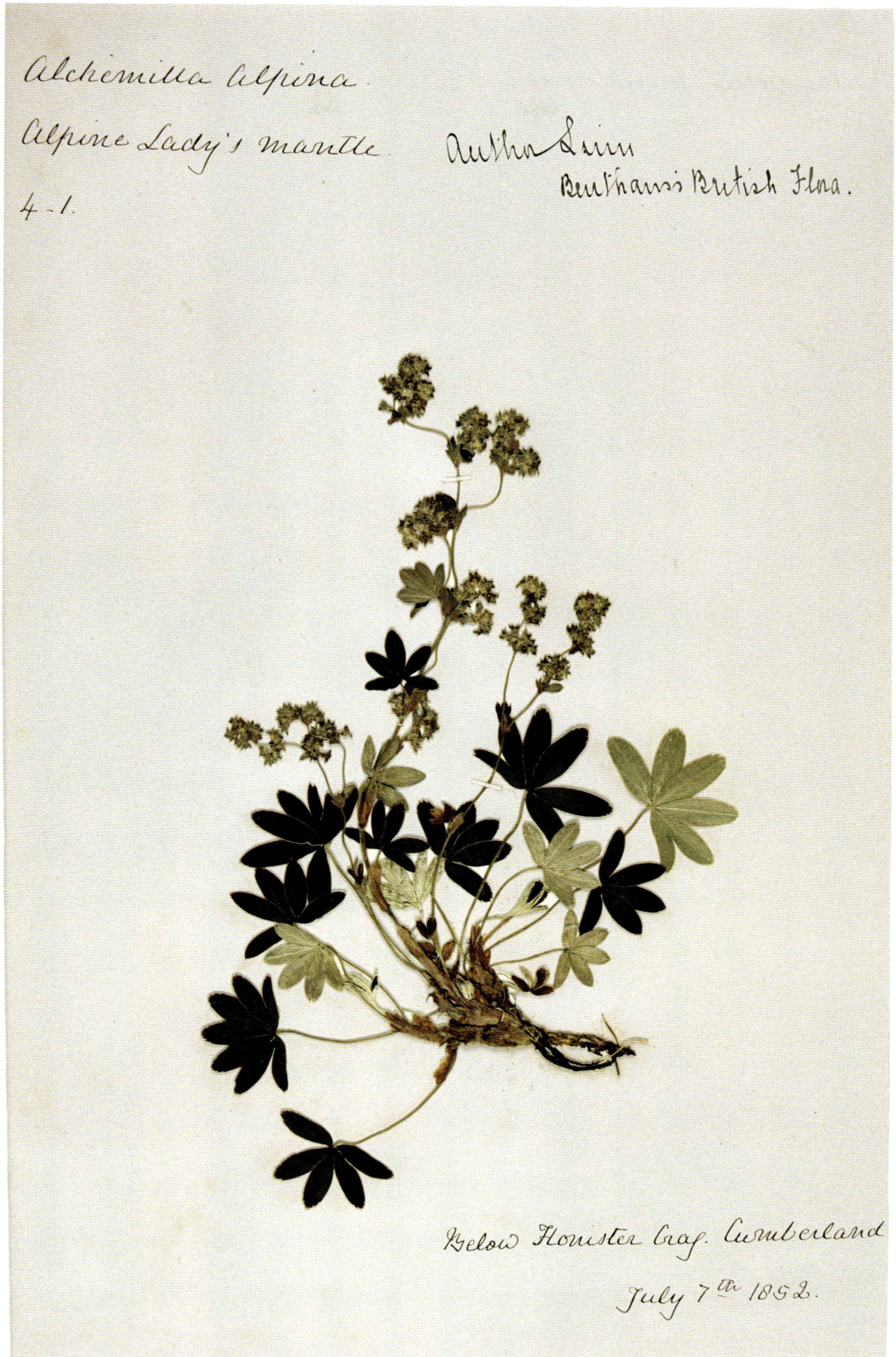

Alpine Lady's Mantle
Collected 1852

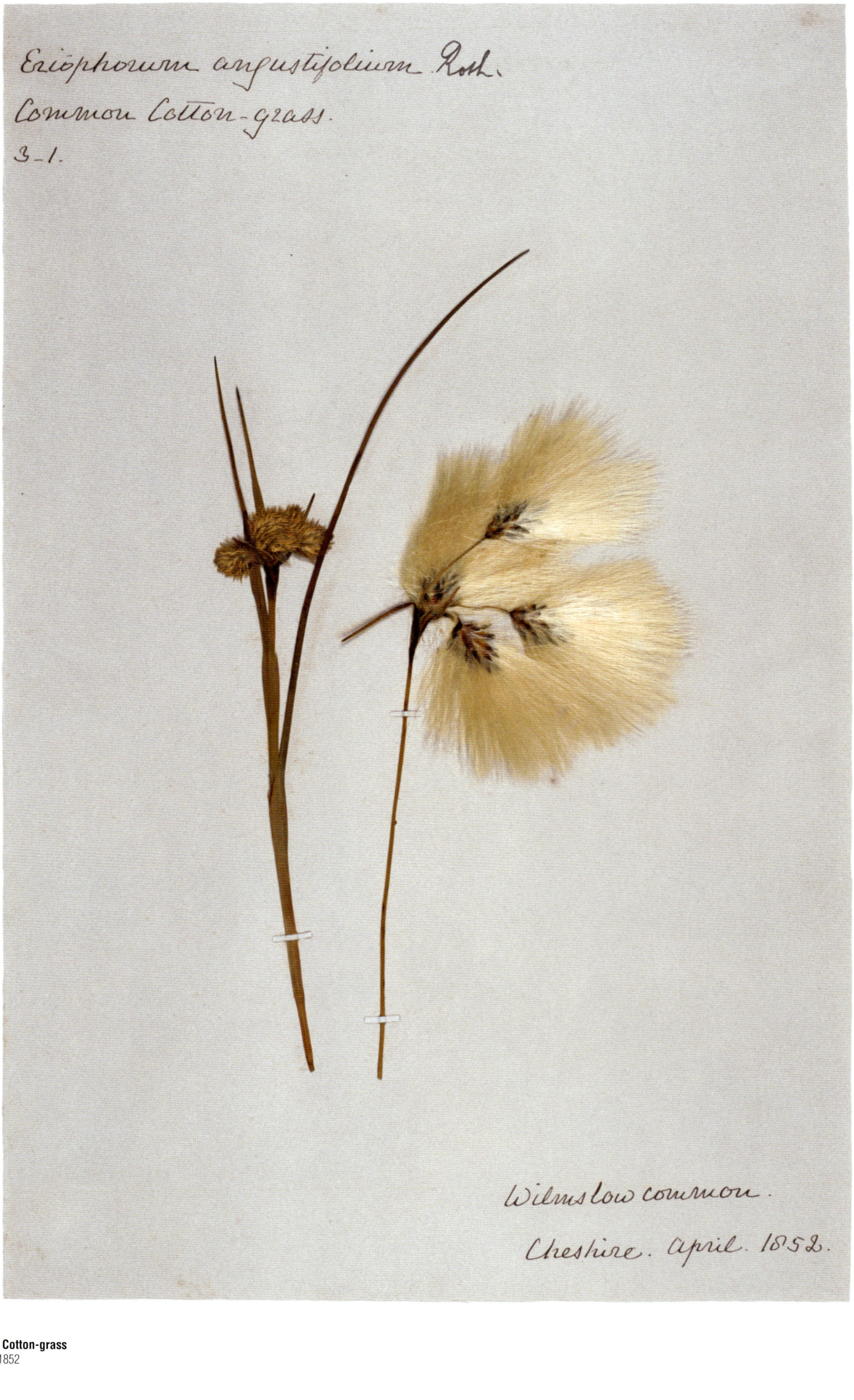

Common Cotton-grass
Collected 1852

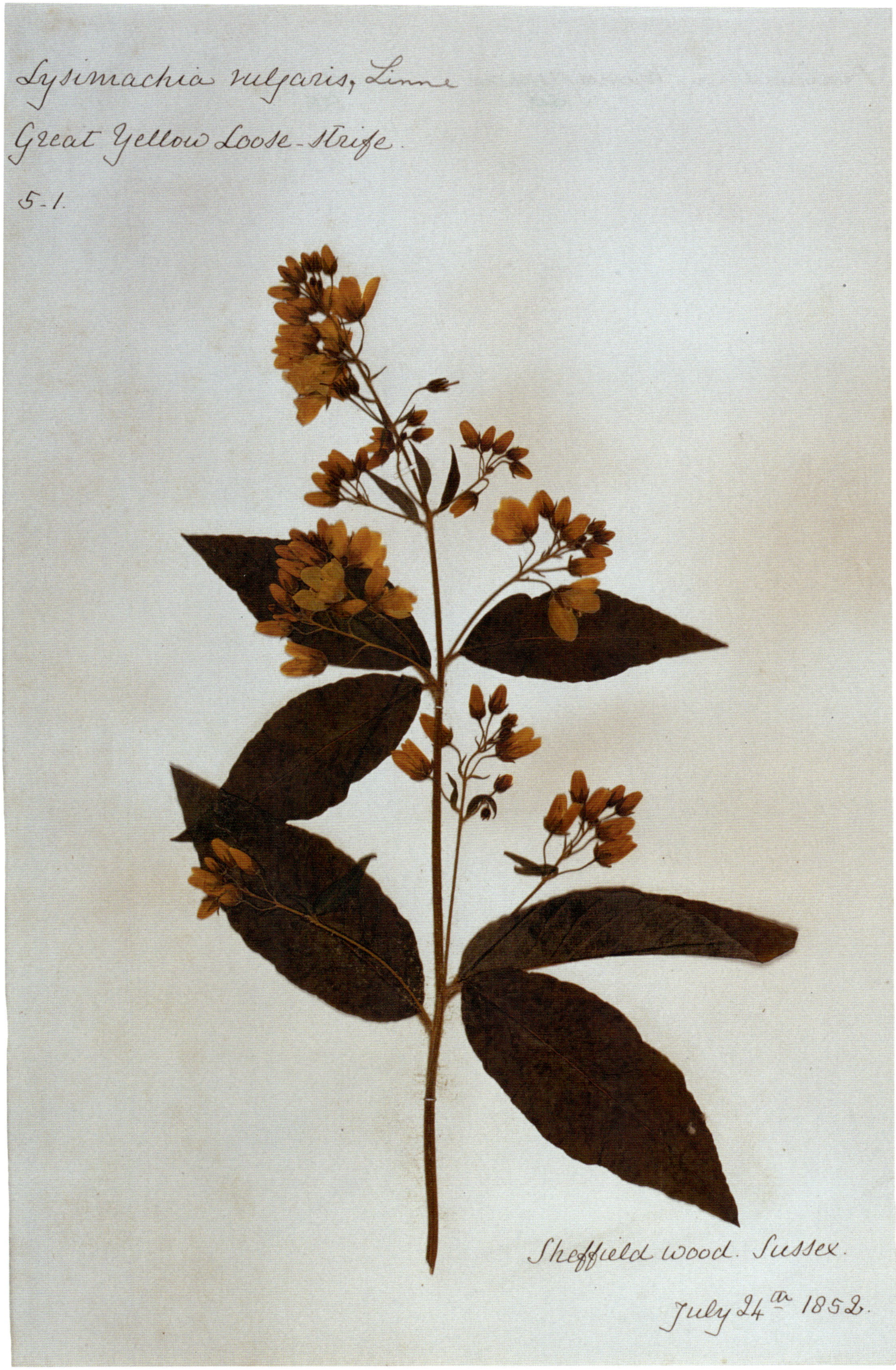

Yellow Loosestrife
Collected 1852

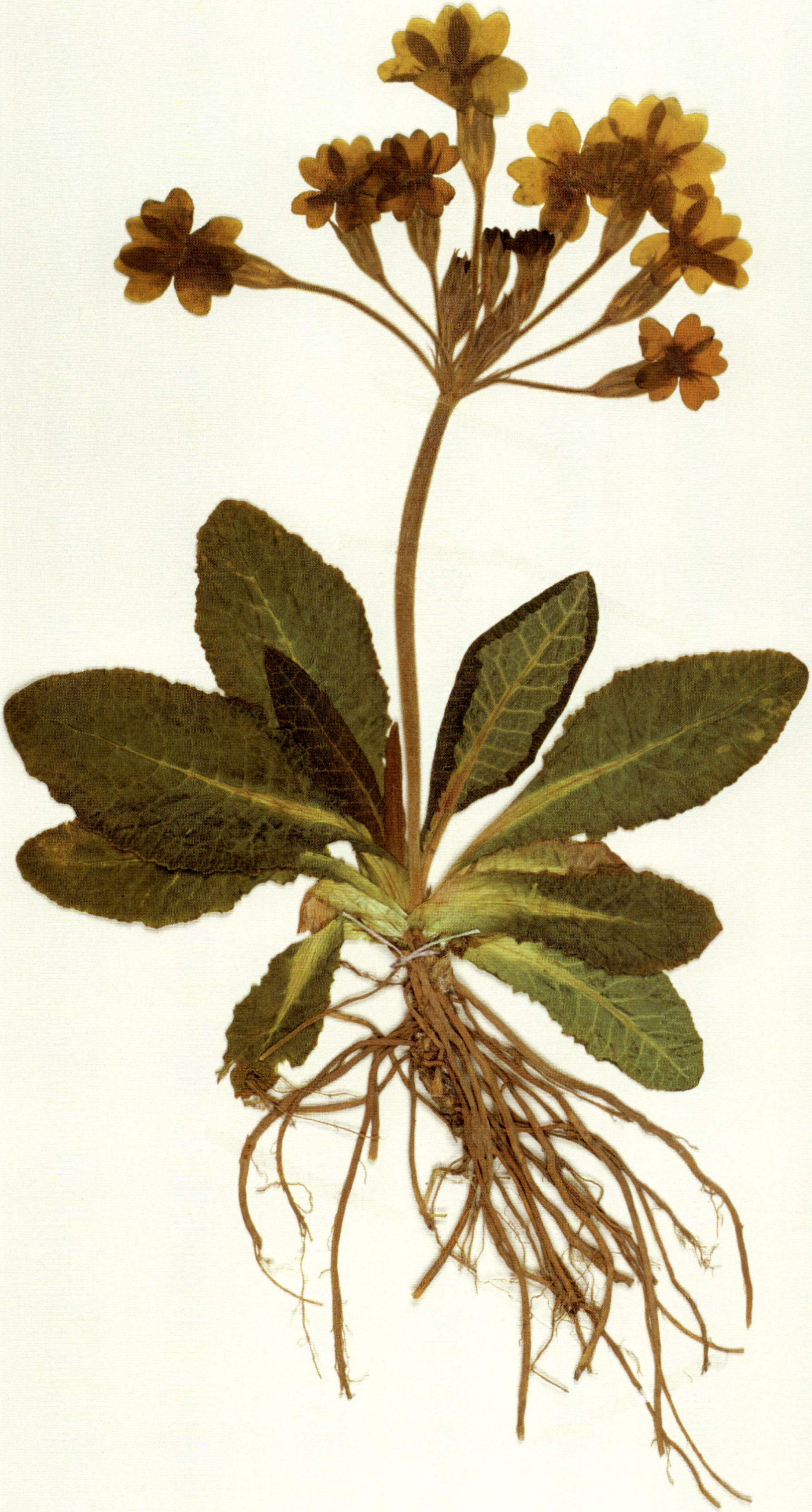

Primrose
Collected 1864

Botanical Notes

THE NEW WORLD

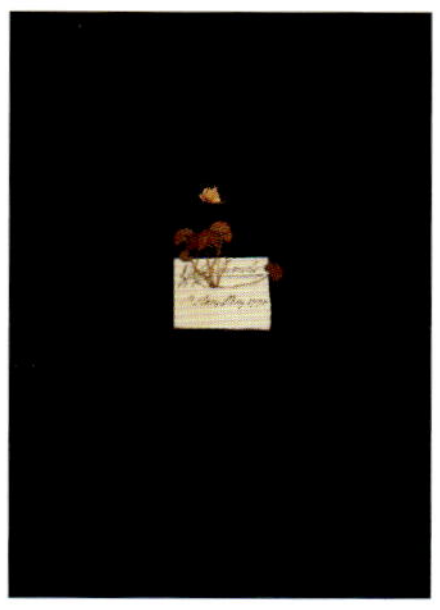

Viola banksii K. R. Thiele & Prober
A Native Violet*
family VIOLACEAE

Collection details: Collected by Joseph Banks and Daniel Solander at Botany Bay, NSW, in 1770 (NSW 171055).

This specimen has remained remarkably intact: surviving the rigours of the voyage back to England, storage there for more than a century, and then being packed up and returned to Sydney and stored there for nearly a century. It would have been common at Botany Bay in 1770 and can still be seen there near Cooks Rivulet at Kurnell. This is a recently described native species, named after Joseph Banks. It was called Viola hederacea, but recent research showed that there were about eight different species hiding under that name, including this. – LJM & KLW

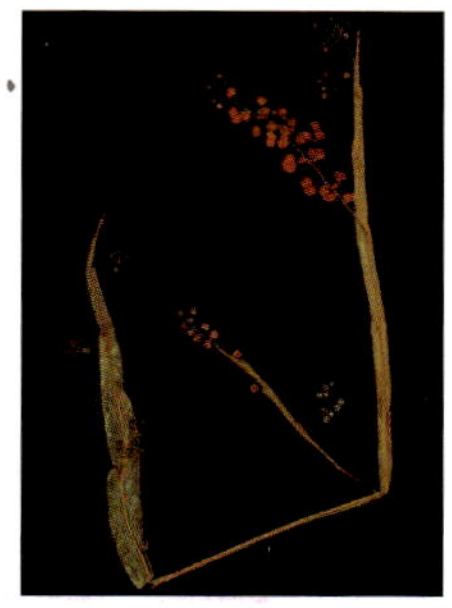

Acacia willdenowiana H.L. Wendl.
Grass Wattle, Two-winged Acacia
family FABACEAE subfamily MIMOSOIDEAE

Collection details: Collected from Wooroloo, Western Australia, by Max Koch in September 1906 (NSW 475690).

Acacia willdenowiana is endemic to parts of Western Australia, growing often in winter-wet depressions in woodland. It is a rush-like shrub with scrambling or erect stems to 1 m high. The unusual grey-green phyllodes are continuous with the branches, forming 'wings' to 15 mm wide. Named after the great German botanist Carl Ludwig von Willdenow (1765–1812) who became Director of the Berlin Botanical Garden in 1801 and Professor of Botany at the University of Berlin in 1810. Max Koch (1854–1925) was born in Germany and died in Pemberton, Western Australia. He migrated to Australia in 1878, becoming a prolific plant collector. Koch discovered nine new species in South Australia and 33 in Western Australia. – PGK

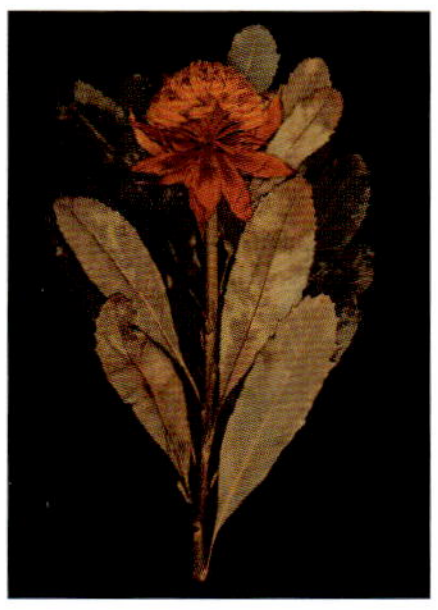

Telopea speciosissima (Sm.) R. Br.
Waratah
family PROTEACEAE

Collection details: Collected by Michael D. Crisp near Mount Victoria in the Blue Mountains, NSW in October 1984 (NSW 463647).

The waratah is the floral emblem of NSW and one of Australia's most spectacular plants. The generic name derives from the Greek tele, far, and opsis, view, i.e. 'seen from afar', while the specific name comes from the Latin speciosus, showy or handsome. T. speciosissima is native to coastal and tableland areas within about 200 km of Sydney. It is a single or few-stemmed upright shrub to 3 m high, with the red inflorescence 15 cm or more in diameter. See also p. 141. – LJM

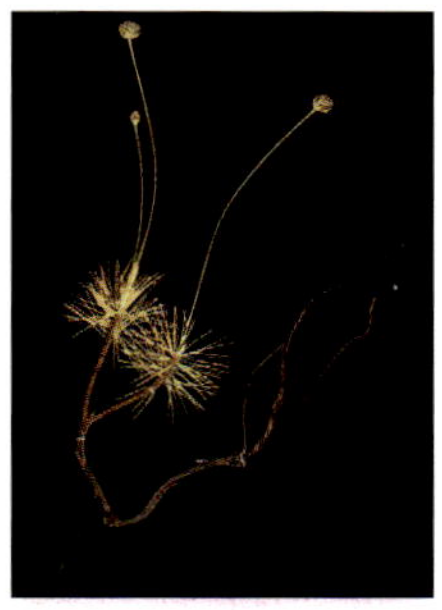

Borya septentrionalis F. Muell.
Porcupine Bush
family BORYACEAE

Collection details: Collected from Hinchinbrook Island, Queensland, by W. L. Ashburner in December 1984 (NSW 541193).

A 'resurrection' plant, whose leaves turn bright orange and appear dead in the tropical dry season but then revive and become green again after rain. This species occurs on scattered rocky mountains, usually granite, in north-eastern Australia. The plants form small clumps that look like miniature spinifex grass (Triodia spp.) hummocks, and have rounded heads of small white flowers on short stems in spring and summer. – LJM & KLW

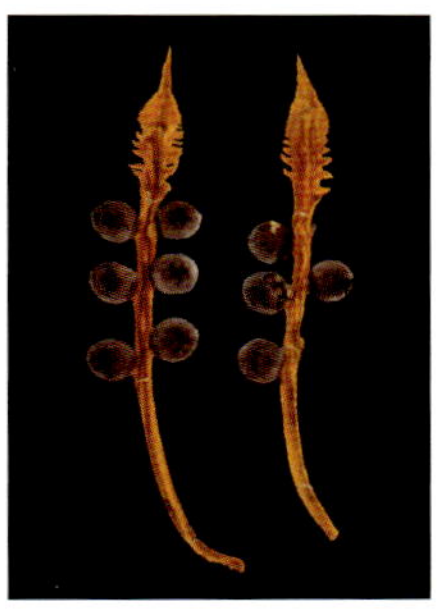

Cycas pruinosa Maconochie
A Northern Australian Cycad
division CYCADOPHYTA, family CYCADACEAE

Collection details: Collected from Spirit Hills, on 'Bullo River' station, Northern Territory, by D. T. Liddle in October 1996 (NSW 518127).

This is the first population of this species found in the Northern Territory; otherwise it is known from scattered populations in the Kimberley region of Western Australia. The generic name is derived from the Greek koikos, which was transliterated as kykas and Latinised as cycas. The specific name is from Latin pruinosus, 'covered with a waxy whitish bloom', referring to the waxy seed surface. The cycads are a small, ancient group. They superficially resemble palms or tree-ferns, particularly in their very large divided leaves, but differ greatly in almost all aspects of detailed structure and reproduction. – LS

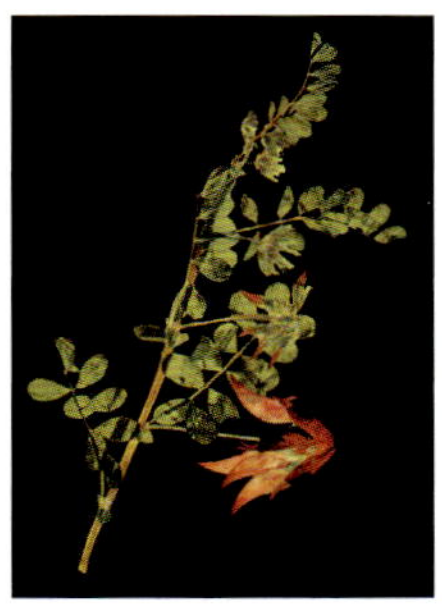

Swainsona formosa (G. Don) J. Thompson
Sturt's Desert Pea
family FABACEAE subfamily FABOIDEAE

Collection details: Collected from Granite Downs, South Australia, in September 1969 by Ian Southwell (NSW 449615).

One of the most spectacular plants found in the arid inland areas of mainland Australia, this is the floral emblem for South Australia, called Sturt's Desert Pea after the explorer Charles Sturt. The genus is named after Isaac Swainson (1746–1812), an English physician with a private botanical garden at Twickenham. The species name comes from the Latin *formosus*, beautiful. *Swainsona formosa* has long stems trailing over the ground with hairy, grey-green leaves. The standard, keel and wing petals of the flower are deep scarlet or occasionally white, with a black or dark red boss at the flower's centre. – LJM

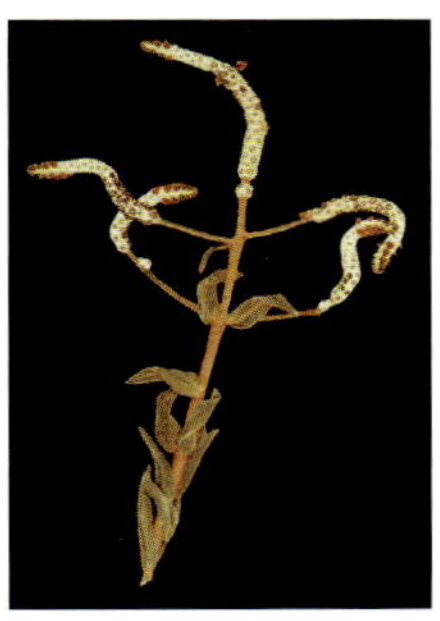

Newcastelia hexarrhena F. Muell.
A Western Australian Newcastelia
family LAMIACEAE tribe CHLOANTHEAE

Collection details: Collected by C. Hassell near Carnegie Station in September 1997, during a Gibson Desert Landscope Expedition (PERTH 5512107).

This densely hairy shrub is confined to the central, arid areas of Western Australia. The plant is covered with dense, multi-branched hairs that give the plant a woolly or felted appearance. These hairs help the plant to survive the hot and dry summers by reducing water loss. Recent, as yet unpublished, research suggests that the species of Newcastelia and those of Lachnostachys (p. 139) belong to the same genus. – BJC

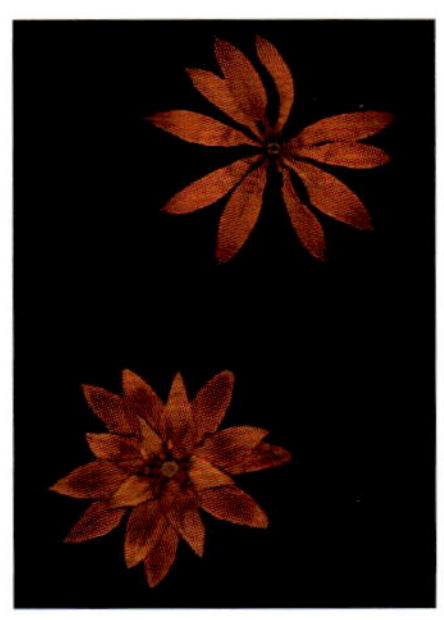

Telopea speciosissima (Sm.) R. Br.
Waratah
family PROTEACEAE

Collection details: Collected from the Blue Mountains, NSW, in October 1881 by Ernst Betche (NSW 511639).

Bulky plants like the waratah or banksias are very difficult to press, and so are often cut into two or more pieces. In this case, a flower-head has been cut across into sections to display the colourful bracts at the base of the flower-head. Ernst Betche arrived from Germany in 1881 and was first appointed as Collector for the Sydney Botanic Gardens, becoming Botanical Assistant to the Director, Joseph Henry Maiden, in 1897. See also p. 141 and above. – LJM

*Common names including the indefinite article are generic, used for a number of different plants. Common names without an article are used for a particular species only.

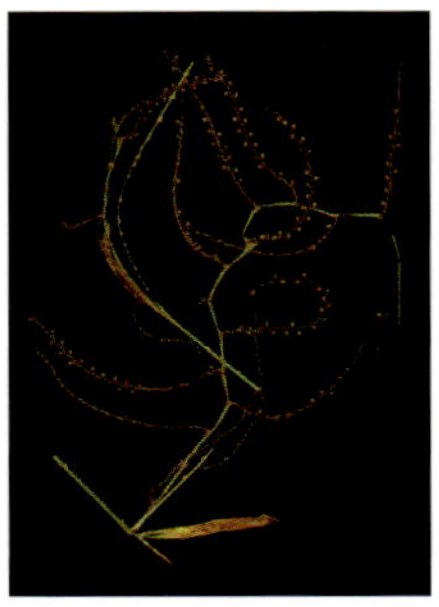

Calamus australis C. Mart.
Lawyer Vine, Wait-a-While
family ARECACEAE (also known as Palmae)

Collection details: Collected at Kuranda, Queensland, by B. Wannon and R. Jago, August 1997 (NSW 426699).

This species is a distinctive, tall, spiny-stemmed palm found in the rainforests of north-eastern Queensland. This specimen shows the young rounded fruits and the short, stout spines at the tips of the stems. These spines are the source of the common name, Wait-a-While, as you can become firmly attached to them if you are not paying attention. The generic name comes from the Greek kalamos (Latin calamus), the name for a reed, referring to the long, climbing stems of some species used for so-called cane furniture. – LJM & KLW

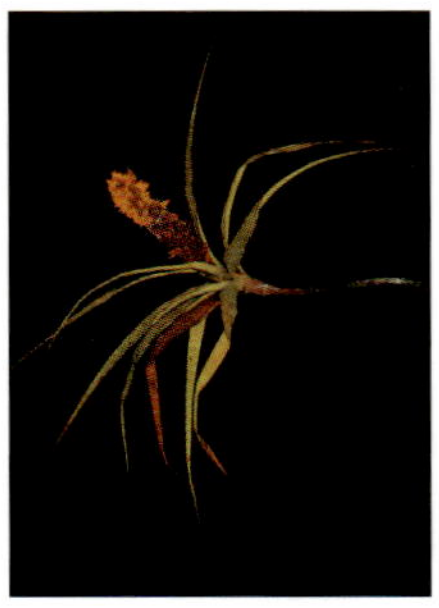

Richea dracophylla R. Br.
A Tasmanian Richea
family ERICACEAE subfamily STYPHELIOIDEAE

Collection details: Collected from South Bruny Island,Tasmania by S. Pedersen and colleagues, November 1999 (NSW 501155).

There are ten species of Richea, nine endemic to Tasmania. Their long, narrow leaves look rather like those of monocotyledon shrubs such as Dracaena, Cordyline or Pandanus. The biggest species is R. pandanifolia (called Giant Grass Tree or Pandani), which grows to 12 m. The celebrated British botanist Robert Brown named the genus after Claude Riche, a naturalist on the 1791–1794 French expedition in search of La Pérouse. The expedition was led by Joseph-Antoine Raymond Bruny d'Entrecasteaux, after whom the Bruny Islands and D'Entrecasteaux Channel were named. – LJM & KLW

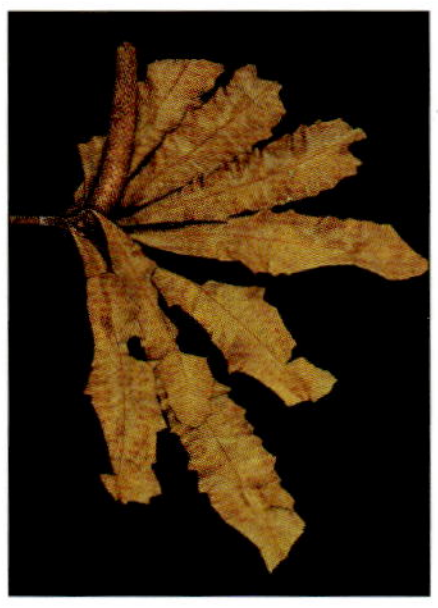

Banksia dentata L.f.
Tropical Banksia
family PROTEACEAE

Collection details: Collected by Joseph Banks and Daniel Solander in 1770 near the Endeavour River, Queensland (NSW 19609).

Banks and Solander made extensive collections along this river, named after Captain Cook's ship, the Endeavour. The ship had to be beached near the mouth of the river after it was holed on a Great Barrier Reef coral outcrop. The species grows on Cape York Peninsula, in the Northern Territory and in Western Australia's Kimberley. Linnaeus' son named the genus in 1782 in honour of Joseph Banks; the specific name derives from the Latin dentatus meaning toothed. This specimen is a duplicate (isotype) of the main type specimen (holotype) on which this name is based. – LJM

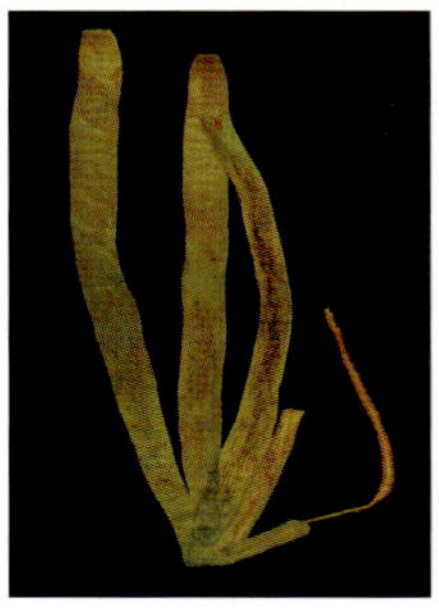

Belvisia mucronata (Fée) Copel.
A Belvisia Fern
division PTERIDOPHYTA, family POLYPODIACEAE

Collection details: Collected in what is now Papua New Guinea by the Reverend Copland King in the period 1891–1918 (NSW 618760).

This distinctive fern is an epiphyte (it grows on other plants). The long, slender tail-like section on the right is the spore-bearing part of the fertile frond. This is artistically arranged here next to what appear to be three separate ribbon-like fronds. However, the specimen is actually one sterile frond folded to fit on the sheet. The Reverend Copland King collected many plant specimens that he donated to the National Herbarium shortly before he died on 5 October 1918. See also p. 141. – LJM & KLW

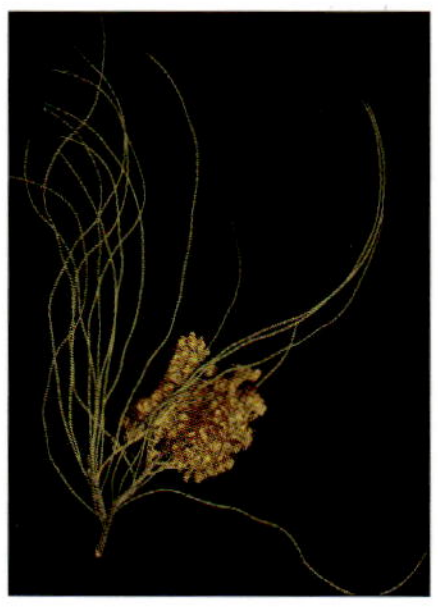

Hakea lorea (R. Br.) R. Br. subsp. *lorea*
Bootlace Tree, Cork Tree
family PROTEACEAE

Collection details: Collected by Peter Olde near Mackenzie Gorge, south of Warburton, Western Australia, in July 2000 (NSW 535643).

A widespread native tree or tall shrub of northern and central Australia, growing to 10 m with cream to greenish-white flowers. The species name derives from the Latin lorea, 'thin strips of leather', referring to the very narrow leathery leaves. The genus Hakea is very closely related to Grevillea, resembling it particularly in the flowers. In his original description of this species in 1810, even the experienced botanist Robert Brown mistakenly assigned it to Grevillea, but by 1830 he had received better specimens and correctly re-assigned the species – hence his double citation as species author. – ROM

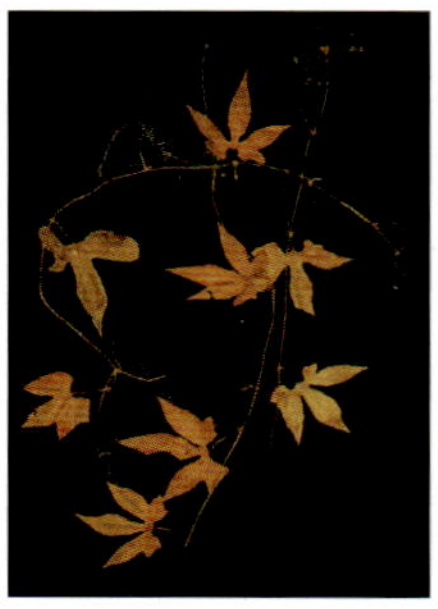

Diplocyclos palmatus (L.) C. Jeffrey
A Palmate-leaved Diplocyclos
family CUCURBITACEAE

Collection details: Collected by Joseph Banks and Daniel Solander in 1770, locality given only as 'New Holland' (NSW 133545).

Because there were so few place names in use, it can be difficult to determine exactly where the earliest explorers collected a specimen. One has to play detective and work out the likely localities by looking at the explorers' maps and journals and comparing these with the geographic distribution of the species. Banks and Solander would have collected this plant at one of their landing places in Queensland. This species is a very slender, soft-wooded climber with grasping tendrils and stems to 6 m long, forming a tangled mass over and through its supporting vegetation. – LJM

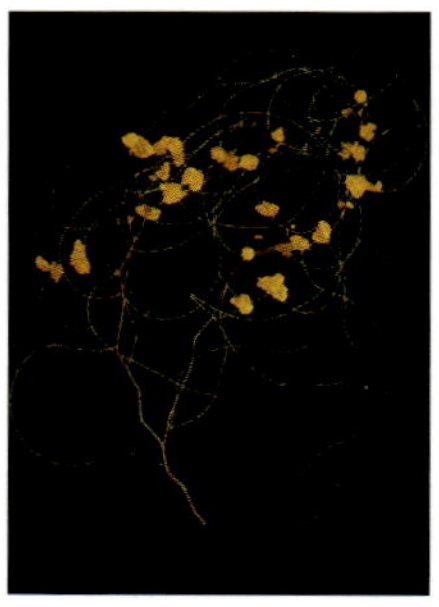

Acacia merinthophora E. Pritz.
A Western Australian Wattle
family FABACEAE subfamily MIMOSOIDEAE

Collection details: Collected from the C. A. Gardner Reserve, 14 km south of Tammin, Western Australia, in July 1980 by Michael D. Crisp (NSW 145681).

Acacia merinthophora is a shrub or small tree to 4 m high with pendulous, zig-zagged branchlet (an inverted image is shown), and golden-yellow, obloid to short-cylindrical flower heads. Its name refers to the slender, strongly curved phyllodes to 25 cm long (from the Greek merinthos, a cord or string, and phoros, carrying or bearing). It grows mostly in sandy soil in tall shrubland in southern Western Australia. – PGK

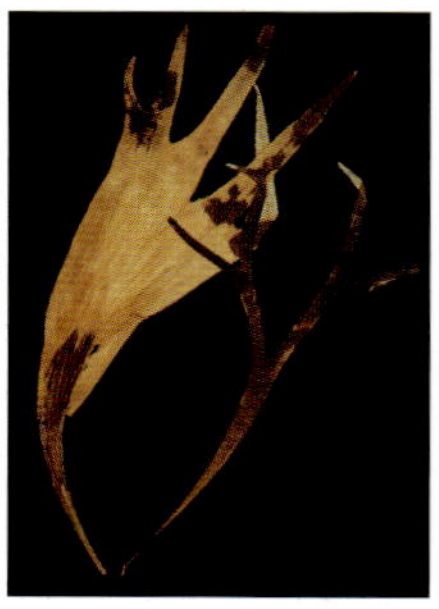

Platycerium bifurcatum (Cav.) C. Chr.
Elkhorn Fern
division PTERIDOPHYTA, family POLYPODIACEAE

Collection details: Collected by Joseph Banks and Daniel Solander in 1770 at one of their Queensland landfalls, locality given only as 'New Holland' (NSW 261602).

This epiphytic fern forms large clumps on trees and rocks. It is native to tropical and sub-tropical Australia, New Caledonia, New Guinea and Indonesia, but is widely cultivated around the world. The name is derived from Greek platys, broad, and keras, horn, referring to the antler-like fertile fronds. As with all the collections of the early explorers, this specimen went back to Europe. It was returned to Sydney in 1905 after J. H. Maiden asked for Banks and Solander duplicate specimens to be sent from the British Museum (Natural History) in London. – LJM

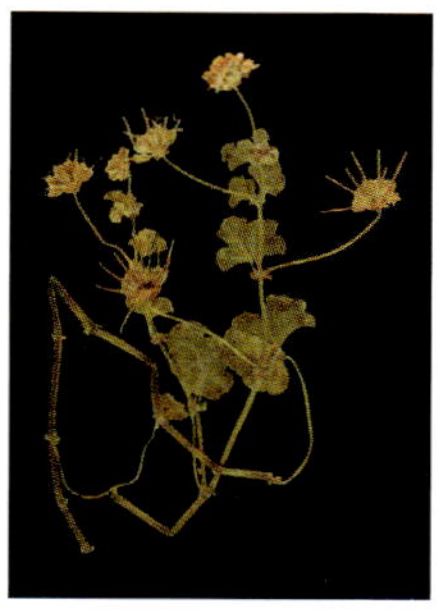

Pelargonium capitatum (L.) Aiton
Rose-scented Pelargonium
family GERANIACEAE

Collection details: Collected at Quinns Rocks, near Yanchep, Western Australia, by Barbara G. Briggs in October 1960 (NSW 52438).

This pink-flowered perennial sub-shrub is native to South Africa but is now naturalised in parts of NSW, Victoria, Tasmania, and in the south-west of Western Australia where it is a serious weed of some native coastal heaths. The generic name Pelargonium derives from the Greek pelargos, stork, from a fancied resemblance of the long fruit awn to a stork's beak. The genus includes 250 species and many artificial hybrids and varieties, selected for leaf markings, flower size and colour, and leaf aroma. – ROM & KLW

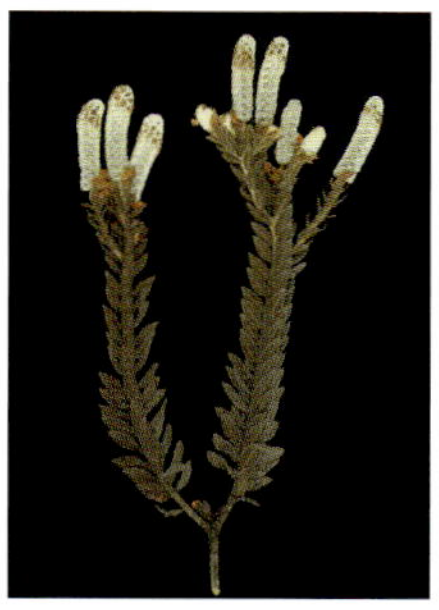

Lachnostachys albicans W.J. Hooker
A Western Australian Lachnostachys
family LAMIACEAE tribe CHLOANTHEAE

Collection details: Collected near the Brand Highway south of Moore River National Park, south-western Western Australia, by J. W. Horn in September 1999 (PERTH 5640962).

This spreading shrub grows to 1 m high. Its stems, leaves and spike-like flower-heads are densely clothed in shortly branched hairs, hence its scientific name, from the Greek lachnos, soft woolly hair, and stachys, a spike; and the Latin albicans, becoming white. Only the male stamens and female stigmas of the flowers extend beyond the very dense hairs, but the spreading purple to mauve petals of the flower are strikingly visible. – BJC & KLW

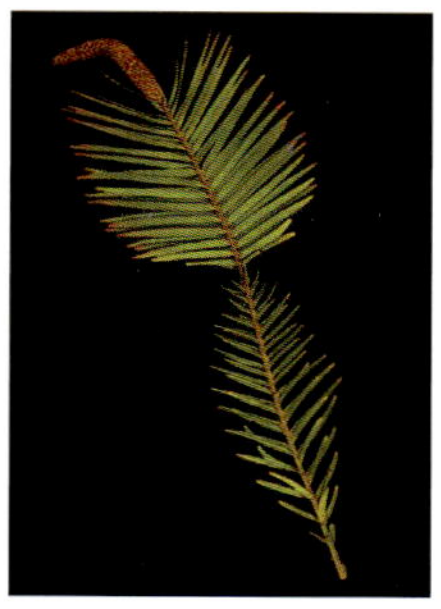

Wollemia nobilis W.G. Jones, K.D. Hill & J.M. Allen
Wollemi Pine
division CONIFEROPHYTA, family ARAUCARIACEAE

Collection details: Collected from a plant cultivated at Mount Annan Botanic Garden, near Campbelltown, NSW, in September 2002 (NSW 499861).

This cultivated plant grew from a wild seedling transplanted to Mount Annan. The slightly bent male cone was one of the first cones produced in cultivation. The wide range in leaf size results from varying seasonal growth rates. The name Wollemia recalls the Wollemi National Park where the rare species was found in late 1994, with the name originally from the Aboriginal word wollumii, 'look around you' or 'watch your step'. The specific name honours David Noble, the person who discovered it while bushwalking. The dramatic discovery of this 'living fossil' made international headlines, and was all the more remarkable because the tall, striking trees are growing within 150 km of Sydney, Australia's largest city. It is a member of the conifer family Araucariaceae, previously thought to include only two living genera, Araucaria and Agathis. – LS

SCIENTIFIC FASCINATION

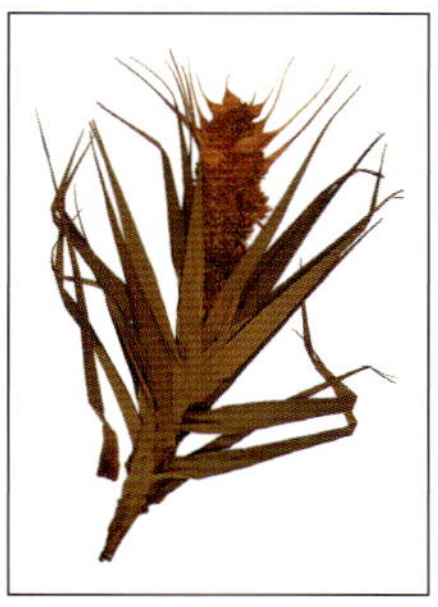

Richea dracophylla R. Br.
A Tasmanian Richea
family ERICACEAE subfamily STYPHELIOIDEAE

Collection details: Collected by W. H. Archer in Tasmania in the mid-nineteenth century (NSW 618730).

These strange-looking tall shrubs are a feature of the montane rainforests of south-west Tasmania, creating a spectacular image in early summer with their narrow, pointed leaves and spikes of white flowers. William H. Archer (1820–1874) was born in Tasmania and became an architect, naturalist and landowner there. – LJM & KLW

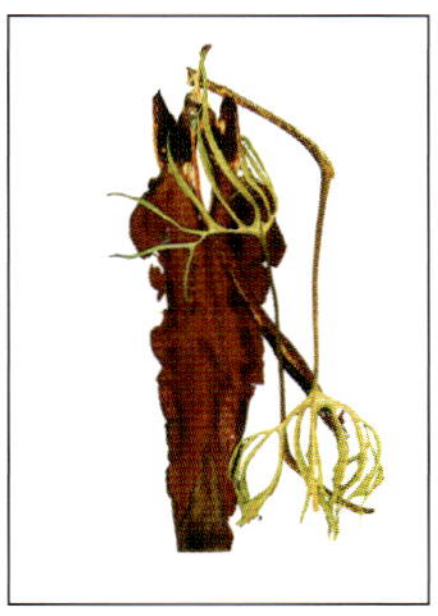

Dracunculus vulgaris Schott
Dragon Arum
family ARACEAE

Collection details: A cultivated specimen from Bowral in the NSW Southern Highlands area, collected by E. G. Knapman in November 1946 (NSW 2270).

This tuberous species is native from Corsica to Turkey, but is widely cultivated as an ornamental. The leaf stalks have dark purple spots or marbling, and the spathe (the big sheath enclosing the small male and female flowers) is dark reddish-purple inside. The live flowers have a foul odour, as do many other members of the aroid family, relying on the smell to attract carrion flies and beetles as pollinators. – PGK & KLW

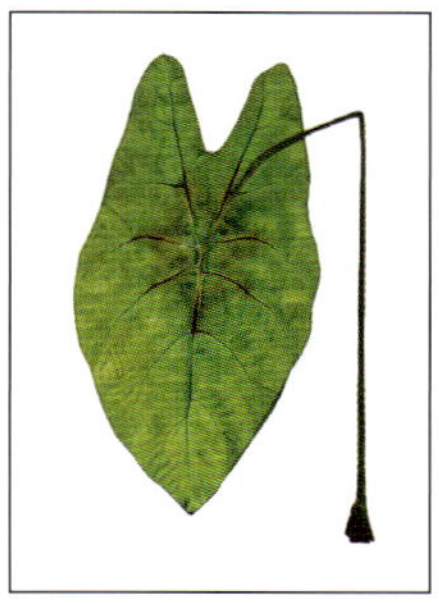

Alocasia longiloba Miq. complex, 'korthalsii' form
A Sabah Alocasia
family ARACEAE

Collection details: This specimen came from a glasshouse-cultivated plant at the Royal Botanic Gardens, Sydney (NSW 426351). It was grown from material collected in the wild, in the Sepilok Forest Reserve, Sabah, by Alistair Hay in April 1996.

Alocasia longiloba Miq. is a tall herb up to 1.5 m high with handsome peltate leaves. It varies throughout its range, but in some areas there are clearly distinguishable forms that have been given informal names. The 'korthalsii' form grows in rainforests on the island of Borneo. The Herbarium also holds a specimen of the inflorescence of this plant. It is preserved in alcohol, which provides better retention of the plant's shape and size, and allows the minute flowers to be more easily examined. – CH

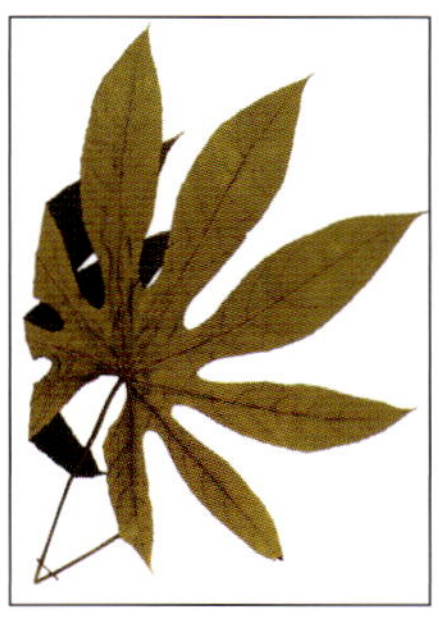

Trevesia palmata (Roxb.) Vis.
Snowflake Tree
family ARALIACEAE

Collection details: Collected in 1965 by the Gardens' Horticultural Botanist, Tony Rodd, from a young tree cultivated in the Royal Botanic Gardens, Sydney (NSW 541196).

The species is indigenous to northern India, southern China and Vietnam. It is sold as an indoor pot-plant in Europe. It grows to 8 m in the wild, with spiny stems and leaves to 60 cm wide that are deeply lobed and look like giant snowflakes. The genus is named after Enrichetta Treves de Bonfigli, an early patron of botany in Padua, Italy. – LJM

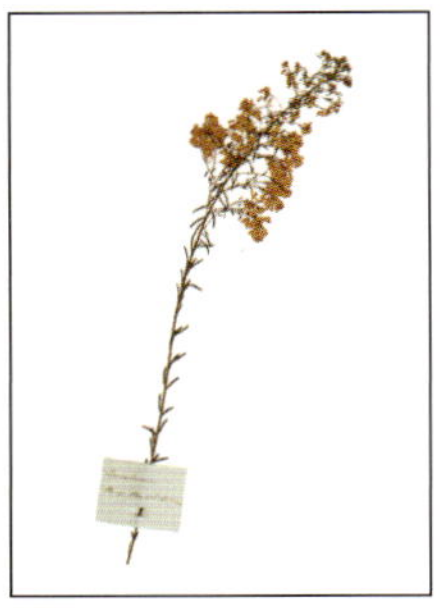

Comesperma ericinum DC.
Pink Matchheads, Heath Milkwort
family POLYGALACEAE

Collection details: Collected by William Macarthur in November 1848 at Bargo Brush, Central Coast, NSW (NSW 421500).

This small, erect shrub is found on sandstone in heath and dry eucalypt forest and becomes noticeable before its flowers open as they look like bright pink matchheads. Sir William Macarthur (1800–1882) was a horticulturist, vigneron, and agriculturalist with an interest in native plants. He was the youngest son of John and Elizabeth Macarthur, famous for introducing merino sheep to Australia. Sir William's collections are now housed at the National Herbarium. The small paper label slipped over the stem is in his hand. Such labels were used for many decades to tag specimens; today, small jeweller's tags on strings are commonly used. – LJM & KLW

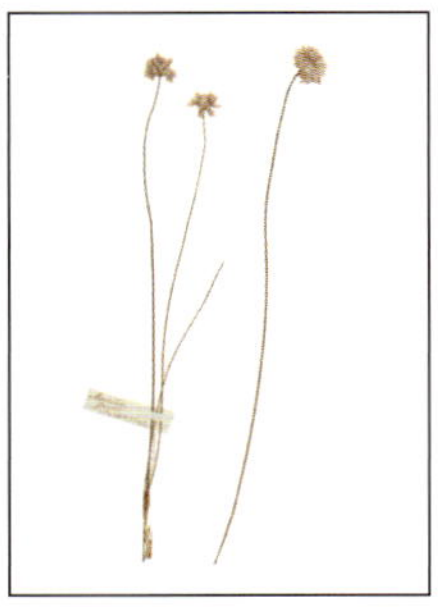

Sowerbaea juncea Sm.
Vanilla Lily
family ANTHERICACEAE

Collection details: collected in September 1802 in NSW by George Caley (NSW 618638).

The Vanilla Lily has slender leaves and an erect stem that, in spring, bears a dense cluster of lilac-pink flowers smelling of vanilla. It grows in wet soils, along the coast and mountains from southern Queensland to Tasmania. George Caley arrived in Sydney in 1800 as a botanical collector for Sir Joseph Banks. All his specimens went back to Banks in London, and are now in the Natural History Museum in South Kensington. – LJM

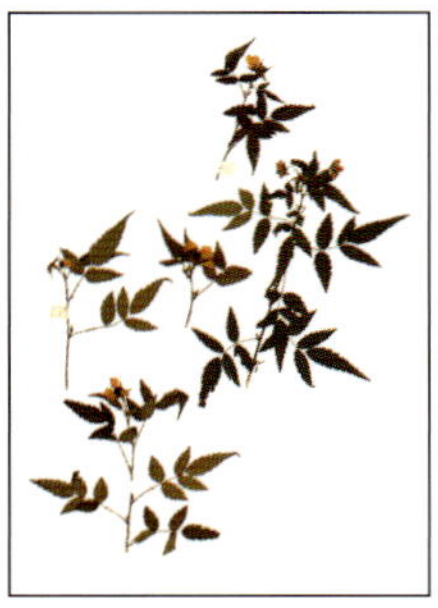

Rubus rosifolius Sm. var. *rosifolius*
Rose-leaf Bramble, Native Raspberry
family ROSACEAE

Collection details: Collected by J. H. Maiden along his route from 'George's River Swamp at Oatley's Grant, then along the ridge for ¾ mile in the direction of Penshurst' on 25 August 1888 (NSW 507538).

This specimen was originally in the Technological Museum but, with the rest of the Museum's herbarium, was donated to the National Herbarium in 1979. There are four native species of Rubus in the Sydney region besides the widely naturalised, weedy Blackberry. Joseph Henry Maiden (1859–1925) was curator of the Technological Museum in Sydney at the time of this excursion to Oatley. – LJM

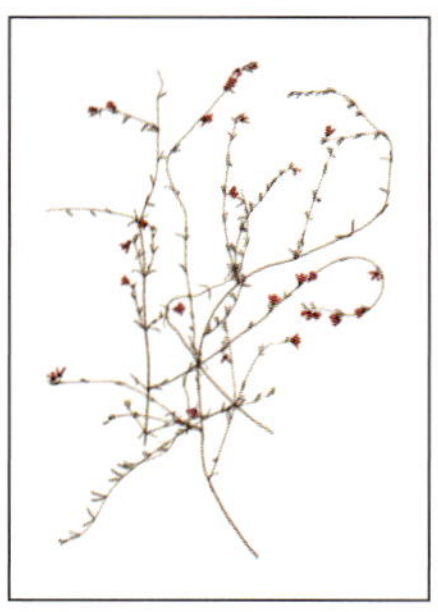

Tetratheca glandulosa Sm.
Black-eyed Susan
family TREMANDRACEAE

Collection details: Collected by D. Wotherspoon near Sackville, NSW, in October 1997 (NSW 415129).

A small, rare shrub found only from Kulnura south to Sydney Harbour and west to Sackville. This species is protected by New South Wales' Threatened Species Conservation Act and is listed as vulnerable under Federal legislation. The specific name comes from the stiff gland-tipped hairs that occur on the leaves and peduncles supporting the flowers. From late winter to early summer, the stunning, deep lilac-pink flowers can be easily spotted in the bush, hanging like little bells. – LJM

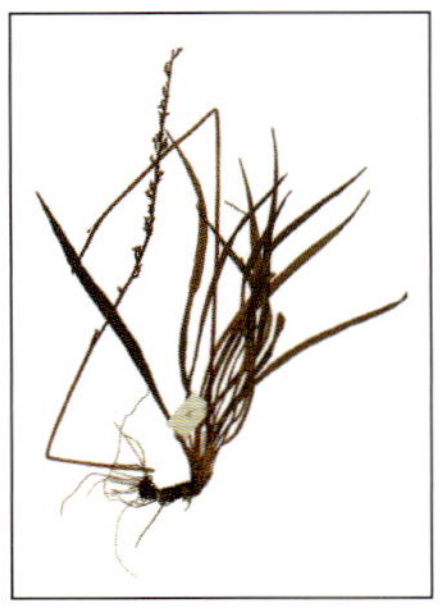

Stylidium graminifolium Sw. ex Willd.
Triggerplant
family STYLIDIACEAE

Collection details: Collected by Allan Cunningham on Boxing Day 1818 at Twofold Bay, South Coast, NSW, when he was on the first surveying voyage of the *Mermaid* under Phillip Parker King (NSW 392717).

A tufted herb with grass-like leaves and numerous small, pink flowers on long stems. The flowers have a cocked column triggered by insect visitors: when an insect probes the flower, the column springs upwards and deposits pollen on the back of the insect, which then takes the pollen to another flower.

Allan Cunningham (1791–1839) was selected by Sir Joseph Banks from among Kew Gardens staff to be an overseas collector. He arrived in Australia in December 1816 and collected widely in Australia and New Zealand. – LJM

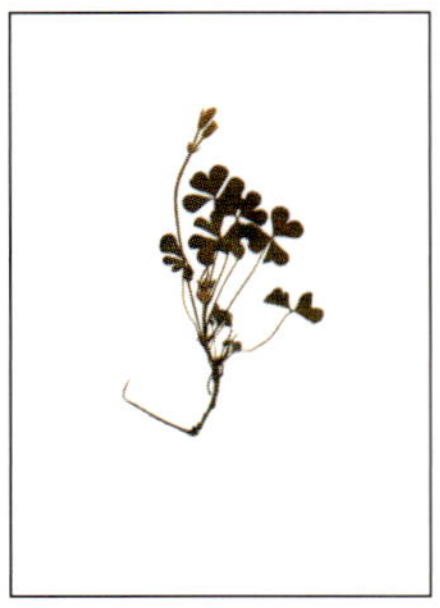

Oxalis perennans Haw.
A Native Wood Sorrel
family OXALIDACEAE

Collection details: Collected by Ludwig Leichhardt in August 1843 at Archers Station, Durundur (east of Kilcoy, Queensland), on his first northern expedition (NSW 49195).

A common character used to identify Oxalis species is the shape of the leaflets. This is a widespread, yellow-flowered native species, most common inland on heavy-textured soils. It is frequently confused with the introduced weed O. corniculata. F. W. Ludwig Leichhardt (1813–1848) was a well-educated German naturalist and explorer. He arrived in Australia in 1842 and led expeditions over large areas of the country. In 1848 he and his party disappeared during an attempt to travel from the Darling Downs in Queensland to Western Australia. – LJM

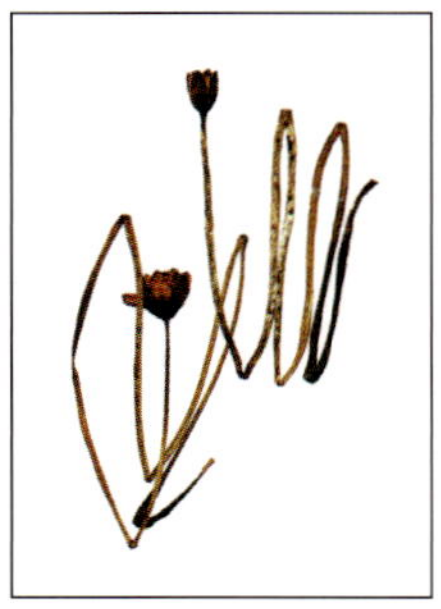

Nymphaea macrosperma Merr. & L.M. Perry
A Native Waterlily
family NYMPHAEACEAE

Collection details: Collected from Pepilum Creek, east of Aurukun, Queensland, in June 1982 by John Clarkson of the Queensland Herbarium (NSW 449953).

This waterlily grows across northern Australia and in Papua New Guinea. The specific name refers to the large seeds of this species and its allies. Its mauve, deep blue, white or pink flowers emerge above large floating leaves. It is a challenge to fit the long stems on a herbarium sheet. – PGK & KLW

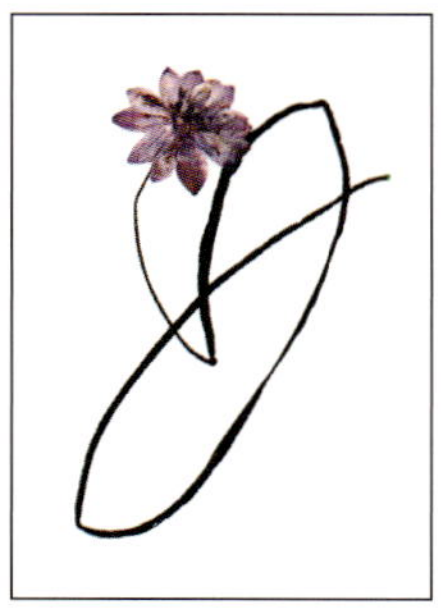

Nymphea immutabilis S.W.L. Jacobs subsp. *immutabilis*
A Native Waterlily
family NYMPHAEACEAE

Collection details: Collected south of Mutchilba, Cook District, Queensland, by Barry Conn and John DeCampo in May 1983 (NSW 339225).

This species is widespread in ephemeral or permanent waterholes in the monsoonal parts of the Australian tropics. The specific name refers to the petal colour, which does not change with age, unlike the petals of some other waterlilies that fade or darken as they age. – PGK & KLW

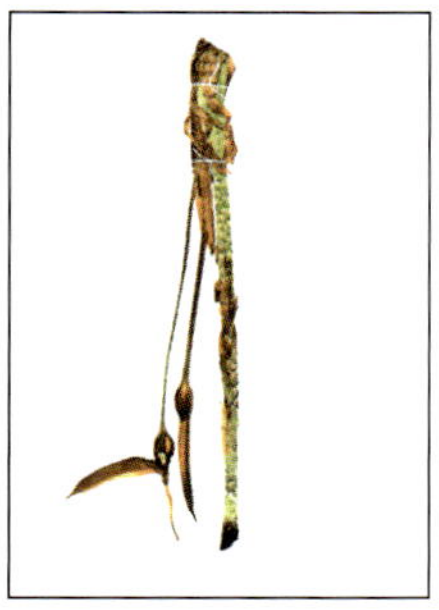

Alocasia longiloba Miq. complex, 'denudata' form
An Asian Alocasia
family ARACEAE

Collection details: A specimen collected by Clare Herscovitch in February 1996 from a glasshouse-cultivated plant at the Royal Botanic Gardens, Sydney (NSW 426450).

The plant was grown from material collected by botanist Alistair Hay in 1990 in the Singapore Botanic Gardens rainforest reserve. This is one of the variants of the more widespread species; this form is native to the Malay Peninsula and Sumatra.

For his taxonomic research on the aroid family, Dr Hay has made extensive use of plants cultivated from wild material because they provide more information about appearance and growth than pressed specimens. – PGK

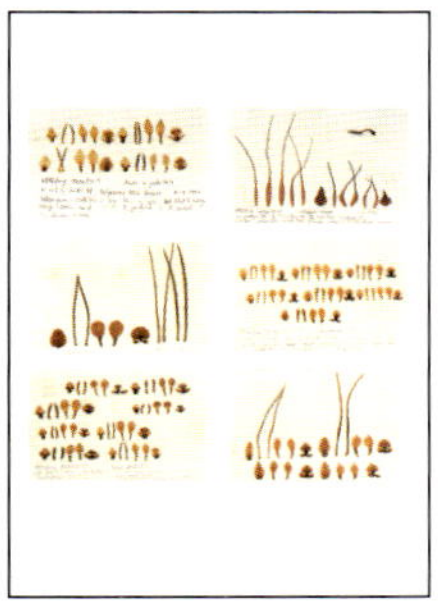

Tony Bishop's Orchid Cards

These are some of the reference cards made by Tony Bishop to study the structures in orchid flowers. These particular cards bear flower parts of Caladenia and Diuris species, each card representing samples from one population. He dissected fresh flowers and taped the parts to cards, making it easy to describe the shape of floral parts and to compare them between individuals and populations. After many years of collecting and studying orchids he wrote the Field Guide to the Orchids of New South Wales and Victoria (1996), a major contribution to the literature on Australia's native orchids. Unfortunately, he died before it was published. – LJM

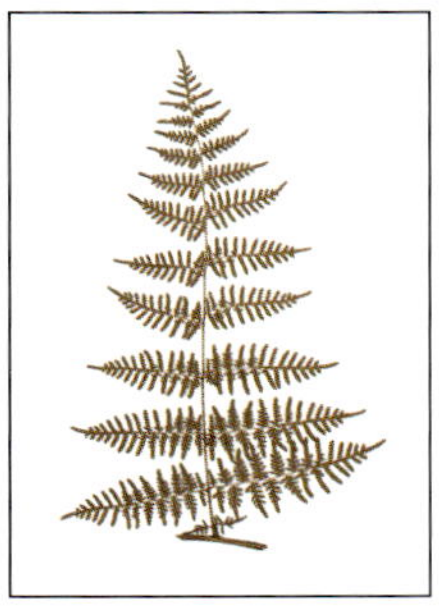

Pteridium sp.
Bracken Fern
division PTERIDOPHYTA, family DENNSTAEDTIACEAE

Collection details: Collected by the Reverend Copland King in what is now Papua New Guinea.

This specimen is possibly P. semihastatum (Wallich ex J.G. Agardh) S.B. Andrews. The genus occurs worldwide, with four or five closely related species and 14 subspecies. P. semihastatum is a tropical species occurring from northern Australia, Indonesia and Singapore to Malaysia, and arose as a natural hybrid between two Pteridium species. See also p. 138. – KLW & JT

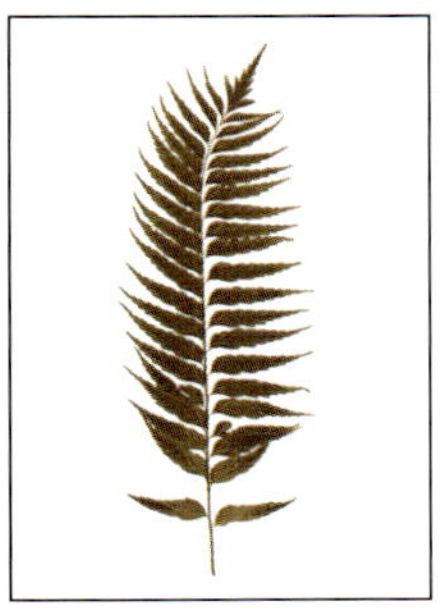

Asplenium sp.
A New Guinea Fern
division PTERIDOPHYTA, family ASPLENIACEAE

Collection details: Collected by the Reverend Copland King in what is now Papua New Guinea.

This genus includes about 700 species throughout the world; many are cultivated. The Reverend Copland King (1863–1918) was an Anglican missionary at Dogura during the period 1891–1918. He returned ill to Sydney in 1918 and died shortly afterwards. He was a great-grandson of Governor Philip Gidley King, son of the Reverend Robert Lethbridge King (a keen entomologist), and is buried in the Camden cemetery. – KLW

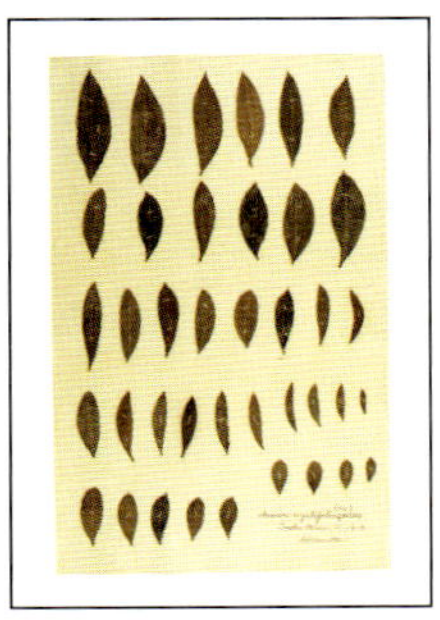

Acacia myrtifolia (Sm.) Willd.
Red-stemmed Wattle, Myrtle Wattle
family FABACEAE, subfamily MIMOSOIDEAE

Collection details: Phyllodes collected from Cooks River, Sydney, by A. A. Hamilton in November 1913 (NSW 108231).

Arthur Andrew Hamilton (1855–1929) worked at the Sydney Botanic Gardens from 1882 to 1920, holding the position of Botanical Assistant in the Herbarium from 1911. He had a keen interest in the variation between individual plants within a species. The specimens on this sheet show the variation in shape and size of A. myrtifolia phyllodes. This is a bushy shrub widespread and common in temperate southern Australia. – PGK

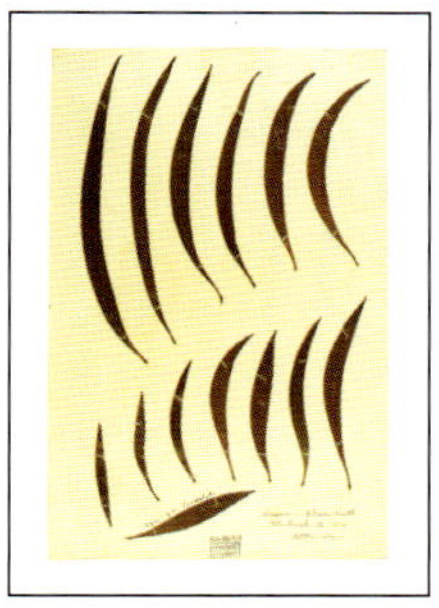

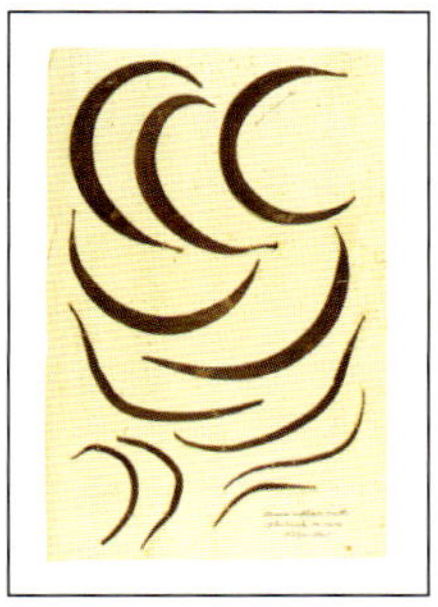

Acacia implexa Benth.
Hickory Wattle, Lightwood, Screw-pod Wattle, Bastard Myall
family FABACEAE, subfamily MIMOSOIDEAE

Collection details: Phyllodes collected from 'a small colony of plants' at Glenbrook, Blue Mountains, NSW, by Arthur Andrew Hamilton in December 1914 (NSW 486424, 613902).

A phyllode is a flattened leaf-stalk (petiole) that is leaf-like in appearance and function. The phyllodes have been mounted on four sheets (two shown here) to display the wide variation in shape and size in A. implexa, ranging from almost straight to markedly curved. A tree to 15 m high, Hickory Wattle typically has narrow-elliptic phyllodes to 20 cm long. It is widespread, occurring from south-eastern Queensland through eastern NSW to Victoria. Common names for the same species are often different in different areas, as seen with this species. These and other sheets show the use of linen thread to stitch specimens carefully to the sheet. – PGK

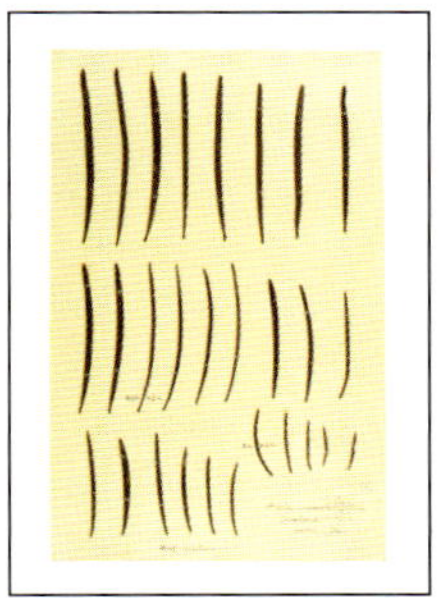

Acacia suaveolens (Sm.) Willd.
Sweet Wattle, Sweet-scented Wattle
family FABACEAE, subfamily MIMOSOIDEAE

Collection details: Phyllodes collected from plants in a population at Blaxland, Blue Mountains, NSW, by A. A. Hamilton in June 1913 (NSW 613900).

The phyllodes have been mounted on three specimen sheets (one shown here) to display their wide variation in shape and size. Sweet Wattle is a shrub to 3 m high with narrow-linear to narrow-elliptic phyllodes to 16 cm long. The species is named for the fragrant, sweet-scented flowers (Latin suaveolens, smelling fragrant) that are produced in cream to pale yellow, ball-shaped heads. It grows in coastal and near-coastal areas in Queensland, NSW, Victoria, Tasmania and South Australia. – PGK

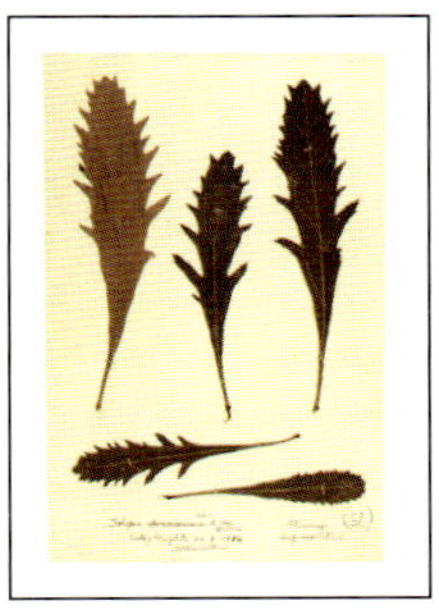

Telopea speciossissima (Sm.) R.Br.
Waratah
family PROTEACEAE

Collection details: Collected at Valley Heights, Blue Mountains, NSW, by A. A. Hamilton on Australia Day, 26 January 1914 (NSW 511606).

These sheets of waratah leaves show the variation, within a population, in leaf size, shape and type of margin, i.e. entire leaf margins to deeply serrate. The Herbarium holds many specimens collected by Hamilton. See also p. 137. – LJM

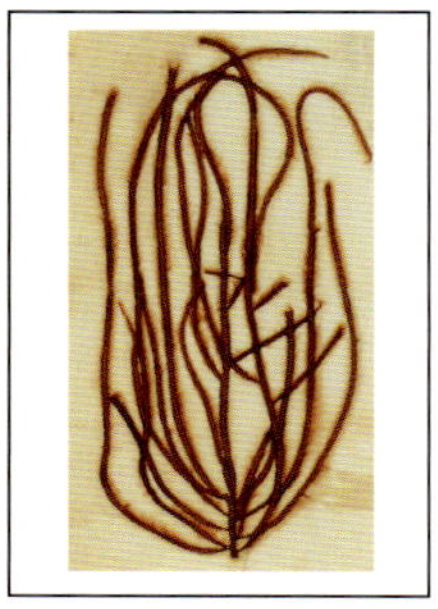

Dasya extensa Sonder ex Kuetzing
A Red Seaweed
division RHODOPHYTA, family DASYACEAE

Collection details: This specimen belongs to W. H. Harvey's exsiccatae (sets of dried specimen sheets) set 49 as species 218F – the 'F' signifies that Harvey collected it at Brighton Beach, Port Phillip Bay, Victoria, in August 1854.

This soft, furry red alga can grow to lengths of 1 m in shallow, sheltered environments such as estuaries and bays. It is found from Dongara, Western Australia, to Victoria and the north coast of Tasmania. Harvey misidentified this specimen as Dasya villosa Harvey, but its real identity is D. extensa Sonder ex Kuetzing. There are about 100 species in Dasya, found in all three major oceans. The generic name comes from the Greek dasys, hairy, shaggy. (Other Harvey specimens are on this page.) – AM

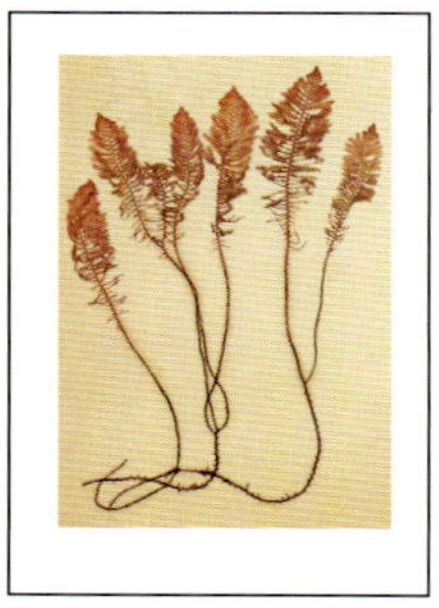

Heterosiphonia muelleri (Sonder) De Toni
A Red Seaweed
division RHODOPHYTA, family DASYACEAE

Collection details: This specimen belongs in W. H. Harvey's exsiccatae set 49 as species 208B – the 'B' indicates that Harvey collected it at King George Sound, Western Australia, early in 1854.

This marine red alga is found from Western Australia to Victoria and northern Tasmania. It is one of 20,000 specimens collected by Harvey in 1854 while he was travelling along the southern coasts of Australia from Rottnest Island off Perth to Newcastle north of Sydney. Harvey pressed and dried every specimen he collected and, on his return to Dublin, offered for sale exsiccatae containing about 250 specimens in each set at the rate of £2 5s 0d for 100 specimens – about $600 in today's values, or $1200 in today's values for each exsiccatae set. – AM

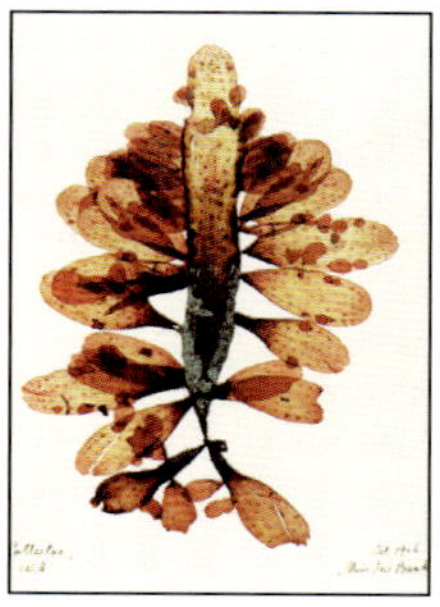

Lenormandia spectabilis Sonder
A Red Seaweed
division RHODOPHYTA, family RHODOMELACEAE

Collection details: Collected from Cottesloe, Western Australia in October 1926 by Miss Iris Banks (NSW 541192).

This specimen is a typical example of the broad-bladed red algal species that dominate many of Australia's subtidal zones. The small darker specks scattered regularly over the blades' surfaces are the reproductive bodies. Iris Banks was regarded by Arthur Lucas (see notes for image immediately to right) as one of his important correspondents and sent many specimens, such as this one, to him from Western Australia. – AM

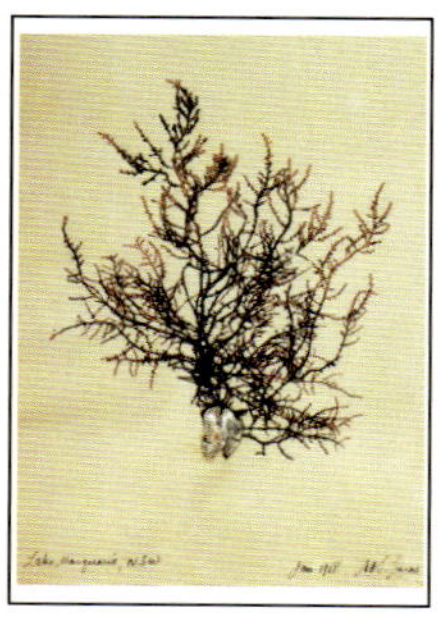

Laurencia obtusa (Hudson) Lamouroux
A Red Seaweed
division RHODOPHYTA, family RHODOMELACEAE

Collection details: Collected by Arthur Lucas from Lake Macquarie, south of Newcastle, NSW, in January 1918 (NSW 541188).

This marine red alga, attached here to a shell, is common in temperate regions of the world, found both in intertidal rock pools and in the subtidal zone. – AM

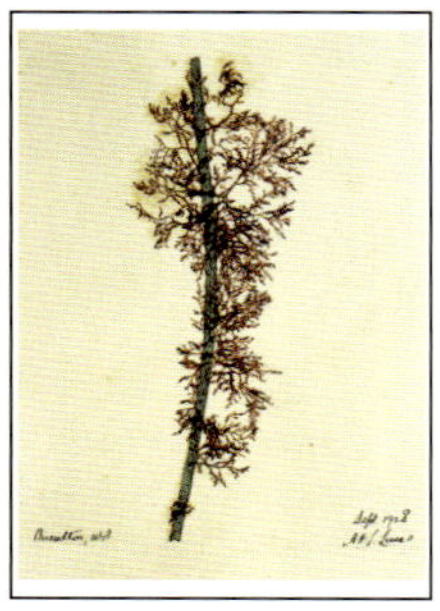

Laurencia obtusa (Hudson) Lamouroux
A Seagrass Leaf and a Red Alga
division RHODOPHYTA, family RHODOMELACEAE

Collection details: Collected from Busselton, Western Australia, by Arthur Lucas in September 1928 (NSW 541190).

This specimen includes two organisms: a ribbon-like seagrass leaf (Zostera sp.) and a small, finely divided red alga, Laurencia obtusa, attached to the leaf. Marine seaweeds are often referred to as benthic (bottom-dwelling) algae to distinguish them from pelagic algae such as phytoplankton, the single-celled algae of the oceans. Seaweeds are not pelagic (free-swimming or floating), but instead need to be attached to something. Seagrasses are marine angiosperms (that is, they have vascular tissue and flowers, albeit reduced); they are rooted in the seabed and provide a substrate for epiphytes such as this Laurencia obtusa. – AM

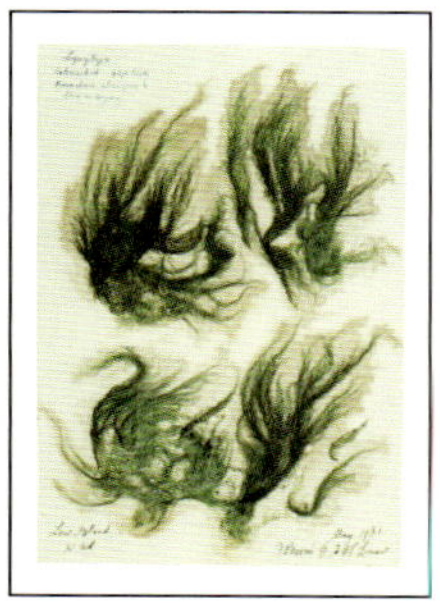

Microcoleus lyngbyaceus (Kuetz.) P. Crouan & H. Crouan
Mermaids Hair
division CYANOPHYTA, family OSCILLATORIACEAE

Collection details: Collected by F. Perrin and A. H. S. Lucas from Low Island (north-east of Port Douglas, north Queensland) in May 1931 (NSW 541197).

This filamentous blue-green alga can occur in plague proportions, causing destruction by forming dense mats covering the seabed. It is found worldwide in tropical and sub-tropical marine and estuarine environments. The Council for Scientific and Industrial Research (precursor to CSIRO) gave Lucas, when he was 78 years old, a grant to collect on Low Island. Mrs Florence Perrin, a keen phycologist, went with him as colleague and assistant collector. After his death, she saw to the 1947 publication of their joint work The Seaweeds of South Australia, Part II: The Red Seaweeds – AM, LJM & KLW

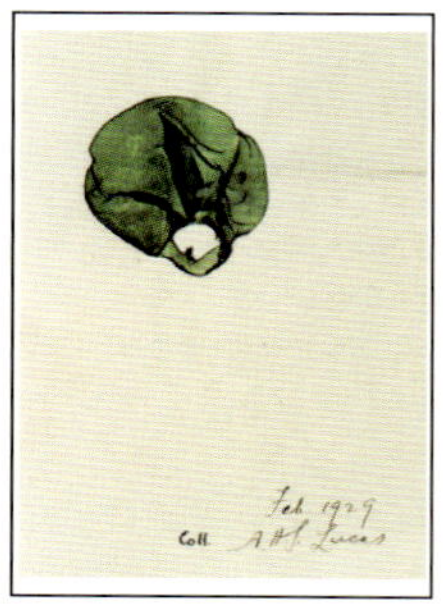

Rivularia australis (Harvey) Bornet & Flahault
A Blue-green Seaweed
division CYANOPHYTA, family RIVULARIACEAE

Collection details: Collected from San Remo in Western Port, Victoria, by Arthur Lucas in February 1929 (NSW 541191).

This blue-green alga, more correctly known as a cyanobacterium, is found on rocks in wet or damp areas. Arthur Henry Shakespeare Lucas (1853–1936) came to Australia in 1883 as a schoolteacher in Melbourne and later moved to Sydney to teach at Newington College and later Sydney Grammar School, where he was headmaster until his retirement at age 70. Besides being an excellent and much-loved teacher, he was an enthusiastic and knowledgeable naturalist, with a particular interest in algae. J. H. Maiden appointed him as honorary curator of algae in the National Herbarium in 1899. Many of his summer holidays were spent near Melbourne with his former pupil Herbert Robinson Brookes (1867–1963), the well-known businessman, pastoralist and philanthropist, and in Launceston with his phycological friends Florence and George Perrin – collecting in both areas. – AM & KLW

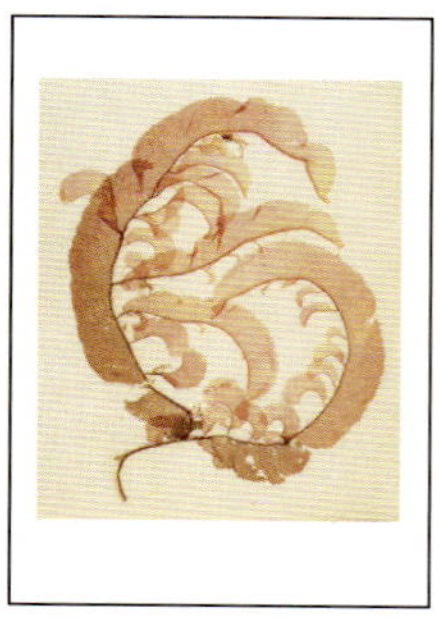

Claudea elegans Lamouroux
A Red Seaweed
division RHODOPHYTA, family DELESSERIACEAE

Collection details: This specimen is in W. H. Harvey's exsiccatae set 49 as species 109J – the 'J' indicates that Harvey collected it at Georgetown in northern Van Diemen's Land (now Tasmania) in early 1855.

This elegant red alga is found from Tasmania and Victoria to Fremantle, Western Australia. It was the first species described and illustrated in William Harvey's magnum opus Phycologia australica. This appeared as five volumes over the years 1858 to 1863, illustrating 300 marine algal species. William Henry Harvey (1811–1866) was Professor of Botany in Dublin, Ireland, with a particular interest in marine algae (seaweeds). He collected widely in Australia and the Pacific (1854–1855) and helped pay for his trip by selling sets of exsiccatae. – AM

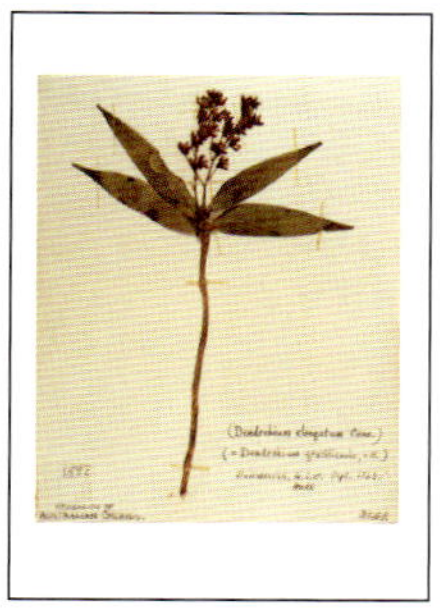

Dendrobium gracilicaule F. Muell.
A Dendrobium Orchid
family ORCHIDACEAE

Collection details: Collected in September 1923 at Bulahdelah, NSW North Coast, by the Reverend H. M. R. Rupp (NSW 618728).

Dendrobium gracilicaule grows on the edge of rainforest in NSW and Queensland. It is a spreading epiphyte (i.e. growing on other plants) or lithophyte (growing on rocks) with a few, fleshy, bell-like, yellow, perfumed flowers. Herman Montague Rucker Rupp (1872–1956) was an Anglican minister, stationed for much of his working life in country NSW. Montague Rupp, as he was known, collected widely in most plant groups, but is mainly remembered for his collections and studies of orchids. – LJM

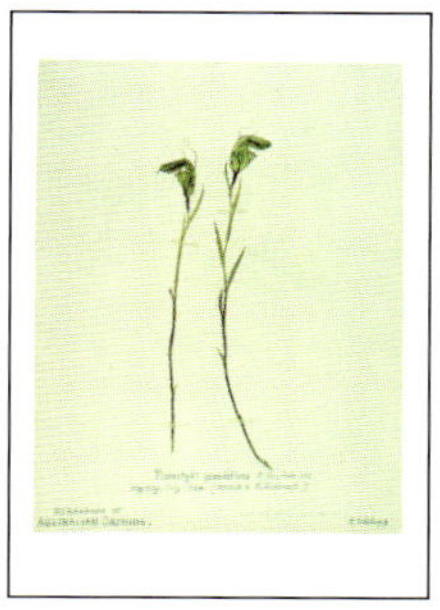

Pterostylis grandiflora R.Br.
Cobra Greenhood
family ORCHIDACEAE

Collection details: Collected by the Reverend H. M. R. Rupp at Woy Woy, NSW Central Coast, in July 1934 (NSW 582980).

This small terrestrial orchid is widespread in coastal south-eastern Australia. The species name grandiflora means large flowers; these are not actually the largest in the genus but were the largest known to Robert Brown in 1810 when he was describing this species. The generic name is derived from the Greek pteron, a wing, and stylos, a column, referring to the column with wings on either side near the flower apex. – LJM & KLW

HOBBY AND DECORATION

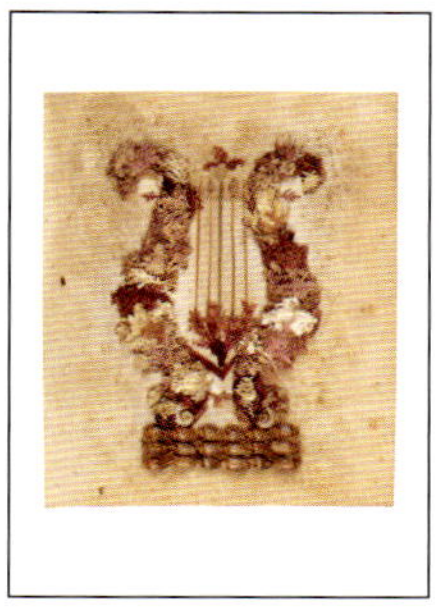

Marine Art

Collection details: A sheet in the algal collection of the National Herbarium; provenance unknown.

During the late 1800s and early 1900s, it was common for wealthy naturalists and genteel ladies to collect drift along the seashore. These seaweeds were pressed onto paper as much for their scientific curiosity as for their aesthetics, with the subsequent pressings often sent to eminent phycologists for identification. This particular collage includes animal shells and calcified coralline red algae, and is a classic example of the mixture of natural history, arts and crafts common during that period. – AM

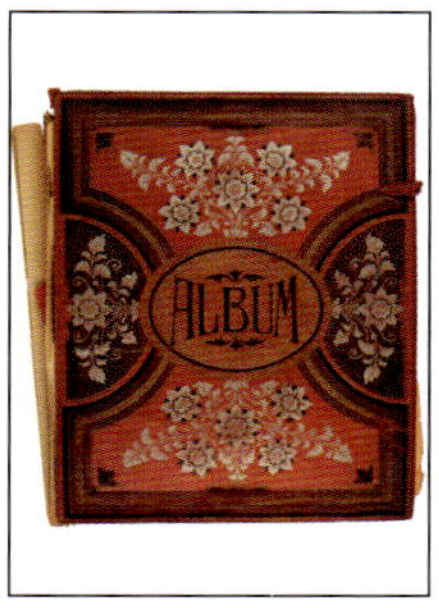

Seaweed Album

One of many albums in the phycological section of the Herbarium. These house seaweeds that have been pressed onto paper sheets, using their own natural sugars to glue them down. The seaweeds are from all over the world: the majority from Australian localities such as Sydney Harbour or Port Phillip Bay, but also from the British Isles, South Africa and the Caribbean. The frontispiece of this red-covered album bears a loving note from Charles Morrison 'To Jennie Augusta from Grandpa Dec 1901'. See also p. 144. – AM

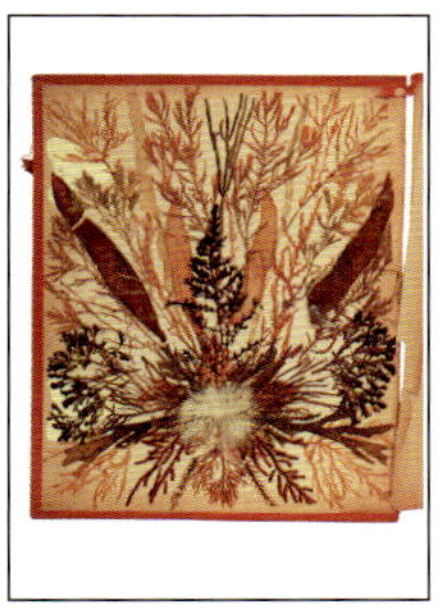

Seaweed Album Endpaper

The endpapers of seaweed albums often had an artistic arrangement of many different species of marine plants. Coincidentally, the endpapers of many old, leather-bound books were marbled (i.e. intricately patterned with many colours). This was achieved by boiling seaweeds to extract the agar, which was then poured into trays and left to set firm. Paints were then poured over the surface of that gel and mixed with various rakes to form patterns. The endpapers were then laid on the paint-strewn 'size' so that they became coloured. When dry, they were pasted onto the book covers. – AM

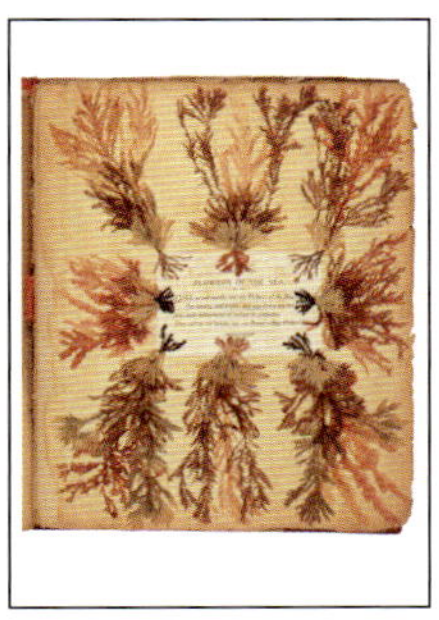

Seaweed Album Title Page

In addition to algae artistically arranged on the endpapers, the title page of these albums often had a poem printed inside a frame of radiating plant specimens. This poem is:

'Flowers of the Sea'
Call us not weeds, we are Flowers of the Sea,
For lovely, and bright, and gay-tinted are we;
And quite independent of culture or showers;
Then call us not weeds, we are Ocean's Gay Flowers.
– AM

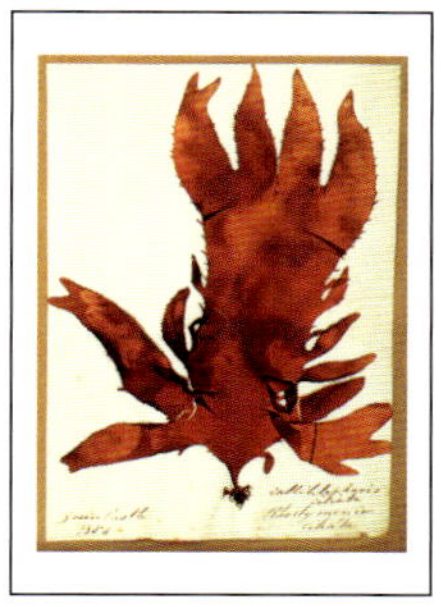

Calliblepharis ciliata (Hudson) Kuetzing
Eyelash Weed
division RHODOPHYTA, family RHODOPHYLLIDACEAE

Collection details: Collected at Green Castle, Ireland, in 1854.

This specimen is in a foolscap-size seaweed album covered in blue velvet with the title Marine Algae Rhodospermeae printed on the cover. It is in the same style as the two smaller albums covered in green velvet, one of which is annotated 'Collected by Charles Morrison Glasgow', so it is possibly another album that belonged to Charles Morrison. Most of the specimens are annotated with the name of the alga, a locality (mostly in Ireland or Scotland) and year of collection (usually in the 1850s), but without a collector indicated. See also p. 144. – AM & KLW

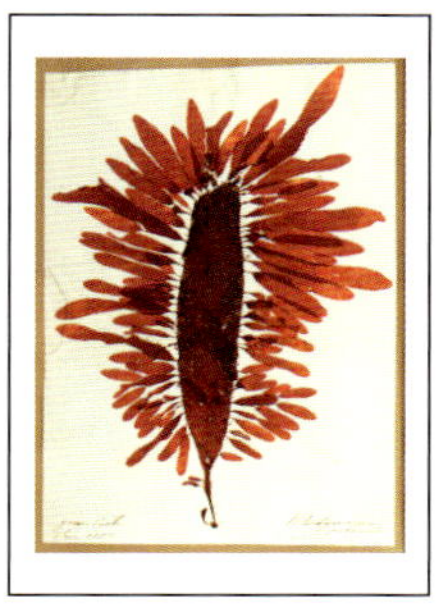

Palmaria palmata (L.) Kuntze
Dulse, Dillish, Crannach
division RHODOPHYTA, family PALMARIACEAE

Collection details: Collected from Green Castle, Ireland, in June 1854.

The species is common around the colder coastlines of the north Atlantic Ocean and north-west Pacific Ocean. It is a human and animal food in many countries. – AM & KLW

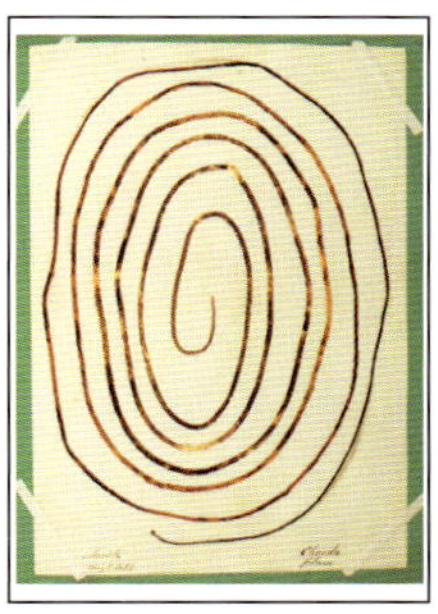

Chorda filum (L.) Stackhouse
A Brown Seaweed
division HETEROKONTOPHYTA, family CHORDACEAE

Collection details: Collected from Moville, Ireland, in August 1852; collector not named.

The form of this worm-like brown alga is reflected in the specific name, from the Latin filum, a thread. This specimen is in the Marine Algae Melanospermeae seaweed album. This and other seaweed albums in the National Herbarium are mostly from the collector Charles Morrison. One album, in particular, was donated to the Herbarium by his descendants, the Sams family. That album is inscribed 'Collected Charles Morrison Glasgow' but a few specimens in it are annotated with the name of another collector, 'Dr Dickie', who may be Dr George Dickie (1812–1882), a Scottish phycologist who was Professor of Natural History in Belfast (1849–1860). The lack of a collector's name on the other specimens may indicate that Morrison collected them himself, but this is not certain because so many specimens were swapped and sold at that time. The dates on individual specimens indicate that the collections were mostly made c. 1852–1854 in Scotland and Ireland. – AM & KLW

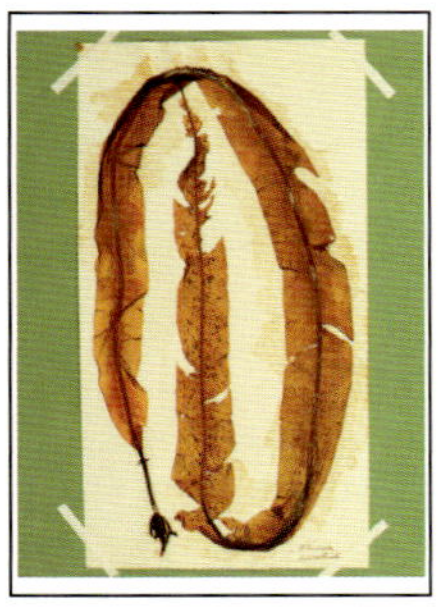

Alaria esculenta (L.) Greville
A Brown Seaweed
division HETEROKONTOPHYTA, family ALARIACEAE

Collection details: Collected c. 1852–1854, probably in the British Isles.

This brown alga is in one of Charles Morrison's seaweed albums held by the National Herbarium. The specimen is small, but this kelp species commonly has blades over 2 m long. It occurs around the coastlines of the north Atlantic and north Pacific Oceans, and is widely used for human and stock consumption, production of alginates, and as fertiliser. The species name comes from the Latin esculentus, edible. See also p. 143. – AM & KLW

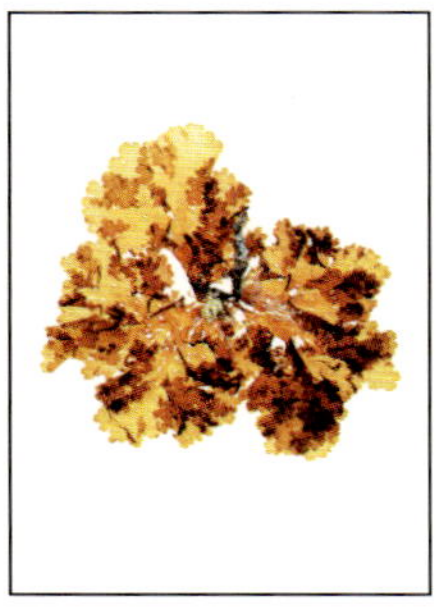

Hymenena affinis (Harvey) Kylin
A Red Seaweed
division RHODOPHYTA, family DELESSERIACEAE

Collection details: Collected by Mrs Mary Ann Barker from Cape Schanck, Mornington Peninsula, Victoria (date unknown), and given to A. H. S. Lucas for his collection (NSW 541198).

Mary Ann Elvidge (1790–1872) collected algae for William Harvey and J. Agardh. She married John Barker and they lived on the Mornington Peninsula. This red algal species was first discovered at Georgetown in northern Tasmania in 1844 by Ronald Campbell Gunn (1808–1881), a dedicated amateur botanist who sent his collections to Sir William Hooker and later his son Joseph Dalton Hooker at the Royal Botanic Gardens, Kew. Gunn's algal collections were passed to Professor William Harvey, who published the name Nitophyllum affine for this species. The Swedish phycologist Harald Kylin determined that the species belonged to a different genus, Hymenena. – AM

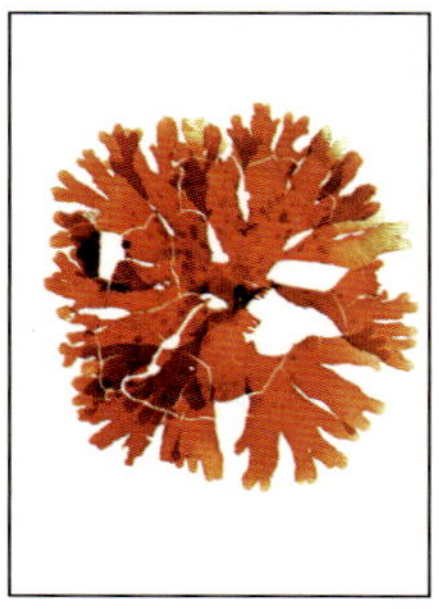

Callophyllis laciniata (Hudson) Kuetzing
A Red Seaweed
division RHODOPHYTA, family KALLYMENIACEAE

Collection details: Collected at Moville, a shipping port on the northern coast of Ireland, in June 1852.

The species grows around the coastlines of the north Atlantic Ocean and Mediterranean Sea. The generic name comes from the Greek kallos, beauty, and phyllon, a leaf, referring to the spectacular frond shapes in the species. See also p. 143. – AM & KLW

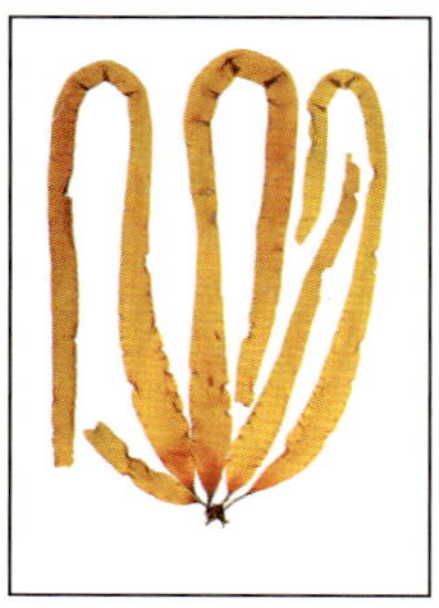

Laminaria saccharina (L.) J.V. Lamouroux
A Brown Seaweed
division HETEROKONTOPHYTA, family LAMINARIACEAE

Collection details: Collected at Dunoon, Scotland, in June 1853, with no collector indicated.

This brown alga is found around the coastlines of the north Atlantic and north Pacific Oceans. The specimen is in one of Charles Morrison's seaweed albums. See also p. 143. – AM & KLW

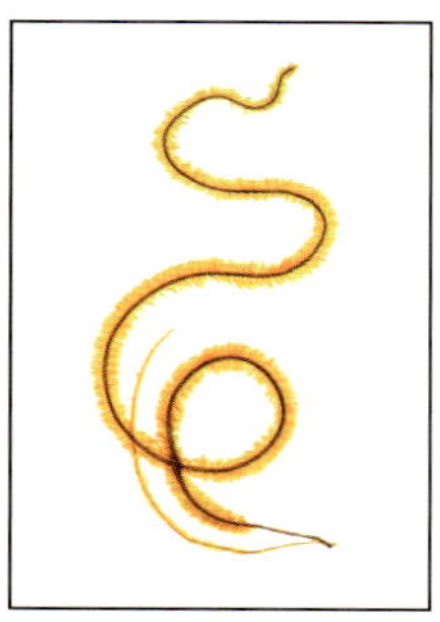

Chorda filum (L.) Stackhouse
A Brown Seaweed
division HETEROKONTOPHYTA, family CHORDACEAE

Collection details: Collected by Dr Dickie in 1852; no locality given but probably northern Ireland.

Mounted next to a specimen of Alaria esculenta in one of Charles Morrison's green velvet seaweed albums. See also p. 143. – AM & KLW

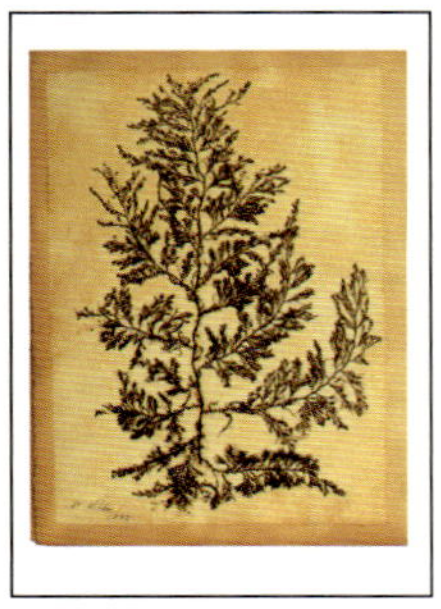

Cystoseira trinodis (Forsskål) C. Agardh
A Brown Seaweed
division HETEROKONTOPHYTA, family CYSTOSEIRACEAE

Collection details: Collected from St Kilda, Melbourne, Victoria, in April 1882, possibly by Charles Morrison (NSW).

This specimen is mounted in the red-covered seaweed album. The frontispiece bears a loving note 'To Jennie Augusta from Grandpa Dec 1901'. 'Grandpa' was Charles Morrison, who prepared similar seaweed albums for other granddaughters, and had made albums for his own interest, dating back to his time in Glasgow in the 1850s. See also p. 143. – AM

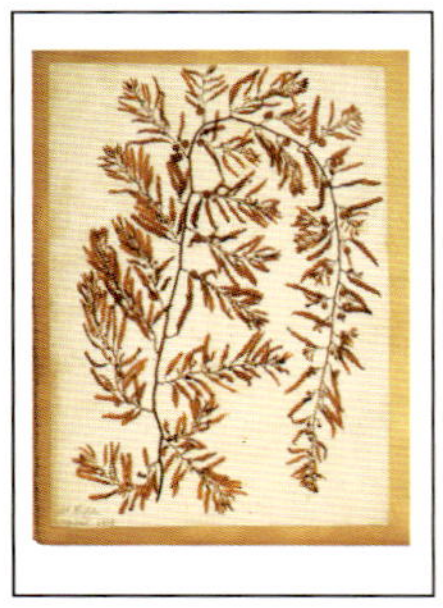

Sargassum sp.
A Brown Seaweed
division HETEROKONTOPHYTA, family SARGASSACEAE

Collection details: Collected from St Kilda, Melbourne, Victoria, in April 1882, possibly by Charles Morrison (NSW).

This is a common Australian genus, with numerous poorly known species. The genus as a whole occurs in temperate and tropical regions. It is famous for its ability to form massive floating mats that often clog the Sargasso Sea, that name actually deriving from the seaweed. – AM

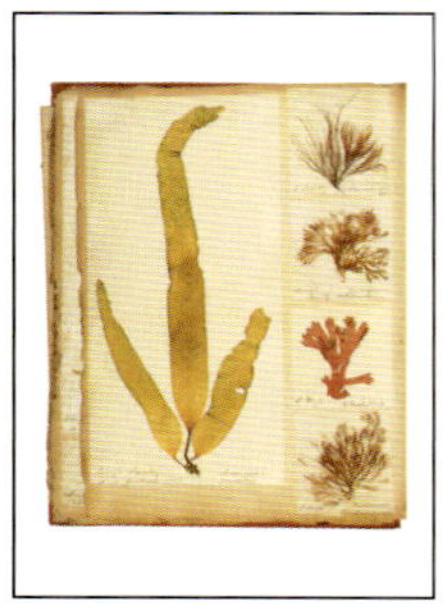

Typical Page of a Seaweed Album

One of many pages in the red-covered seaweed album (see p. 143), inscribed 'To Jennie Augusta from her Grandpa [Charles Morrison] 1901'. The large specimen on the left is a brown alga, Laminaria saccharina (L.) J.V. Lamouroux, labelled as being collected from Lough Swilly, Ireland. The four smaller Irish red algae on the right hand side of the page are (from the top): Polysiphonia atlantica Kapraun & J.N. Norris and Ceramium echionotum J. Agardh (both from Lough Swilly), Nitophyllum punctatum (Stackhouse) Greville and Ceramium deslongchampsii Chauvin ex Duby, both from Lough Foyle. – AM

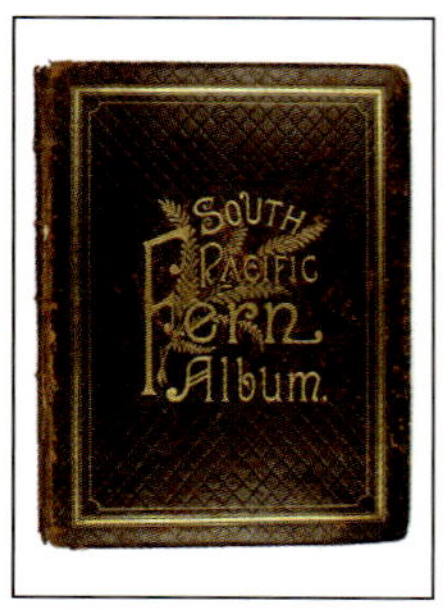

South Pacific Fern Album 'New Zealand Section'

This Victorian plant album is particularly quaint as it incorporates pressed fern specimens on each page, often with patches of moss as decoration, and two pages have hand lithographs of scenery (localities unidentified). Unlike some of the other albums held at the Herbarium, this was commercially printed and produced in Sydney: 'arranged by Mrs C. C. Armstrong and edited by J. Twomey', with specimens collected from New Zealand. The album is inscribed 'A token of esteem to E. Dixon from his sincere friend Mina Ellis 6 November 1895'. – LJM

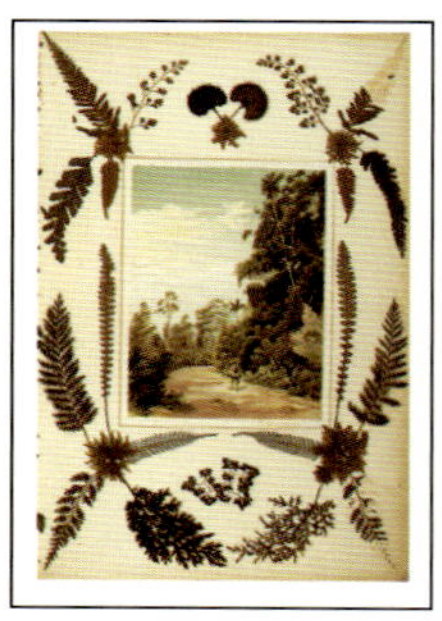

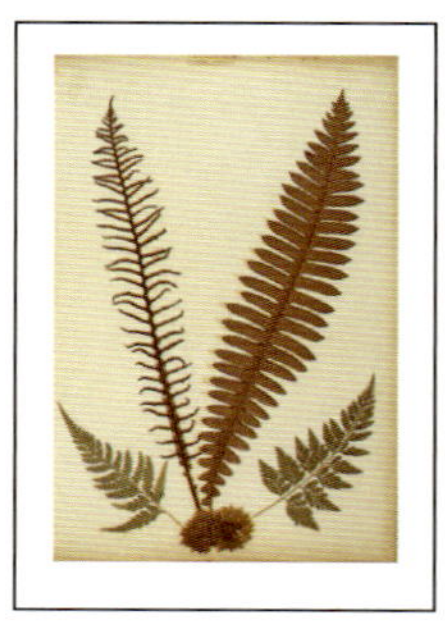

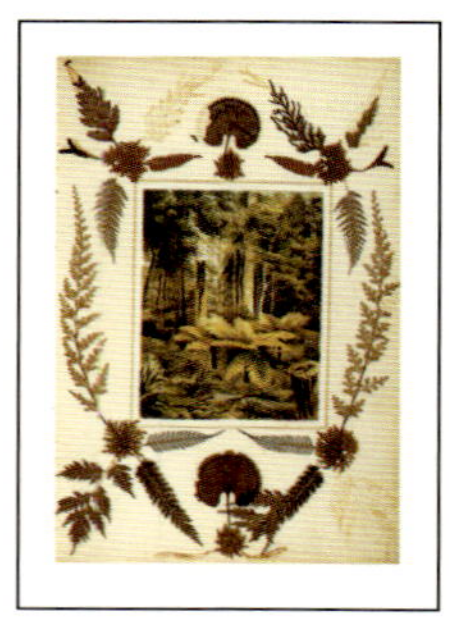

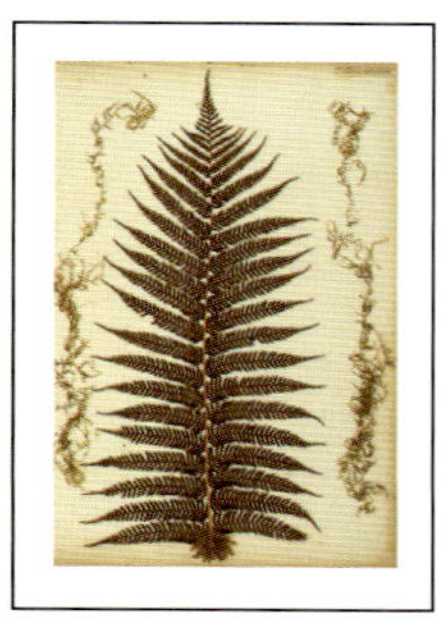

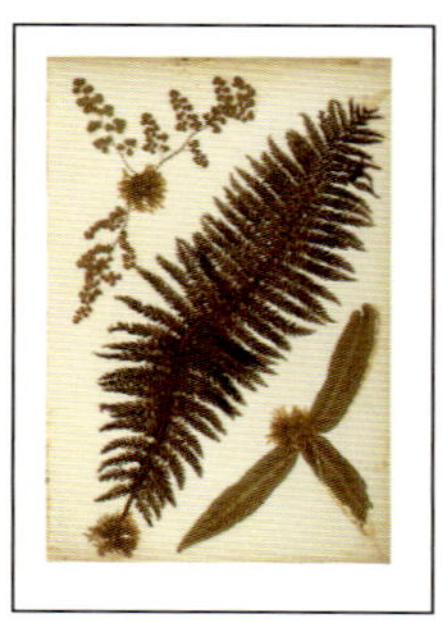

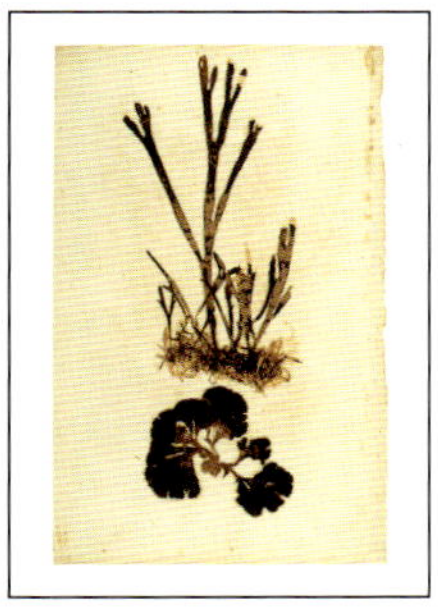

Brown Seaweeds
Division HETEROKONTOPHYTA, family DICTYOTACEAE

Collection details: Collected near Manly, NSW, in the 1860s by John Thompson.

The long, thin strap-like seaweed is a brown alga, Dilophus intermedius (Zanardini) Allender & Kraft, restricted to the waters around Australia and New Zealand. The fan-shaped plant is also a brown alga, Lobophora variegata (Lamouroux) Womersley, widespread in the coastal waters around the Atlantic, Indian and Pacific Oceans. John Thompson was born in 1828 in Scotland and migrated to Adelaide in 1842. His descendants, the Porter family, gave his album of seaweeds to the Royal Botanic Gardens, Sydney. – AM

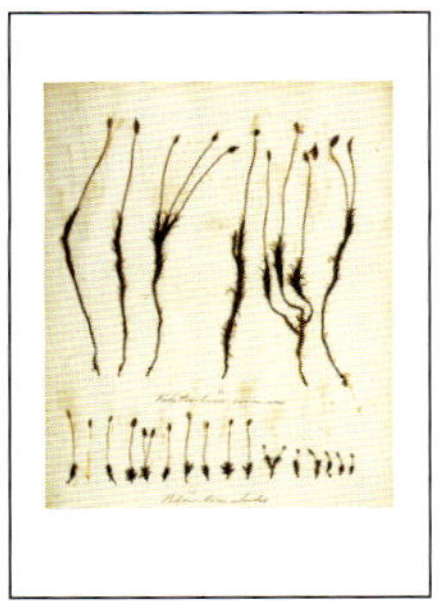

Polytrichum commune Hedwig (larger plants) and
Pogonatum aloides (Hedw.) P. Beauv. (smaller plants)
Haircap Mosses
division BRYOPHYTA, family POLYTRICHACEAE

Polytrichum commune is a cosmopolitan moss, while Pogonatum aloides is from Asia and possibly New Zealand. The generic name Polytrichum refers to the hairy calyptra (cap) that covers the capsule before it is fully mature. The stems of this moss are rigid and erect, with the long, pointed leaves arranged spirally around the stem, and at right angles to it, giving a star-like appearance when viewed from above. Groups of the plant can look like miniature conifer forests. – LJM

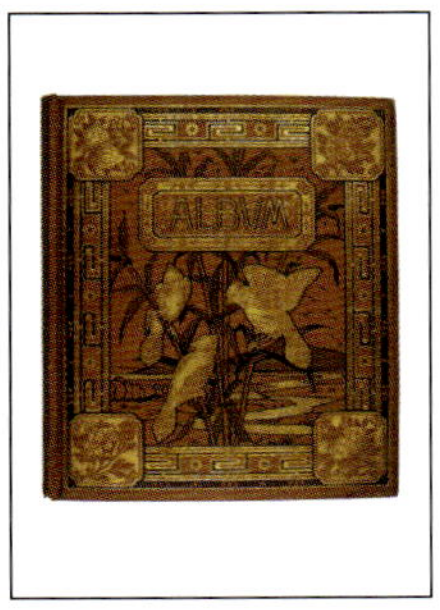

Victorian Fern Album

This typical Victorian fern album has no scientific value as there is no information about where or when the specimens were collected. However, it is interesting historically and the specimens are well-presented. – LJM & KLW

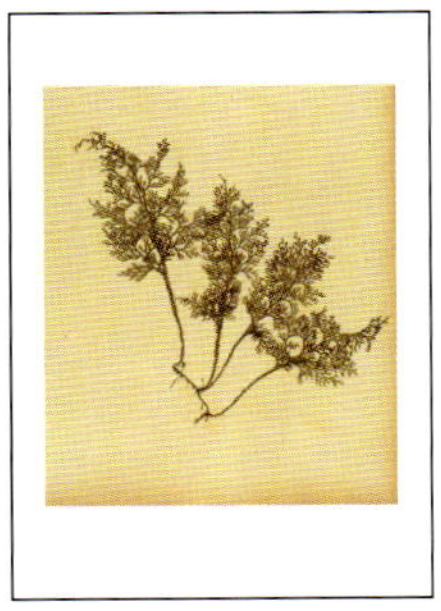

Hymenophyllum sp.
A Filmy Fern
division PTERIDOPHYTA, family HYMENOPHYLLACEAE

This genus includes about 250 species, found in most parts of the world except the Americas. – LJM & KLW

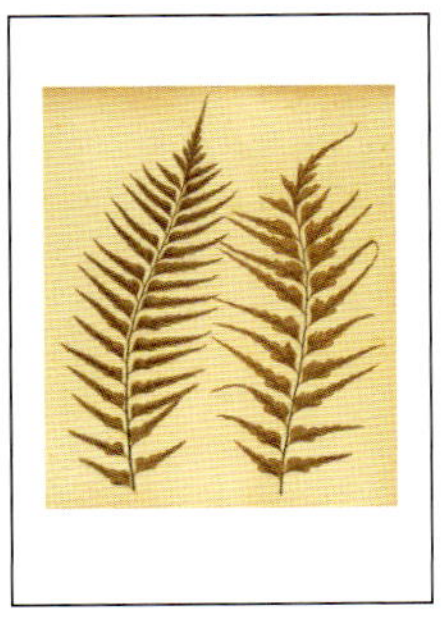

Asplenium sp.
Asplenium Ferns
division PTERIDOPHYTA, family ASPLENIACEAE

This genus includes about 720 species worldwide. – LJM & KLW

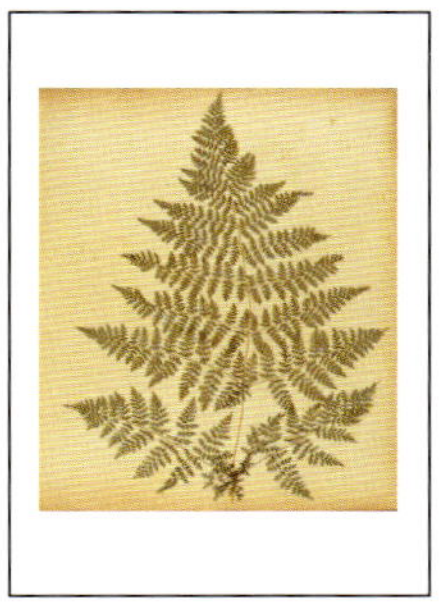

Hypolepis sp.
A Hypolepis Fern
division PTERIDOPHYTA, family DENNSTAEDTIACEAE

This genus includes about 40 species in tropical and subtropical regions of the world. – LJM & KLW

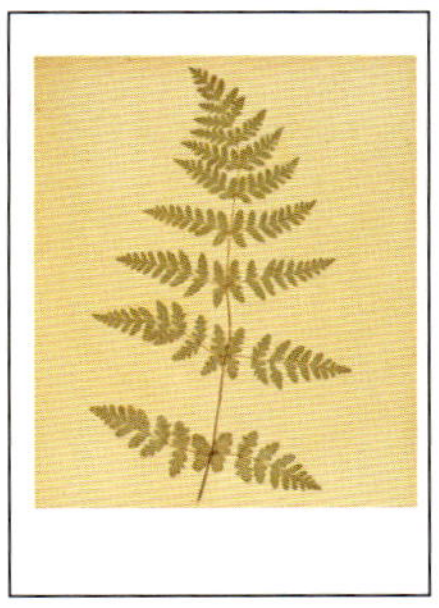

Histiopteris sp.
A Histiopteris Fern
division PTERIDOPHYTA, family DENNSTAEDTIACEAE

This genus includes eight tropical species. – LJM & KLW

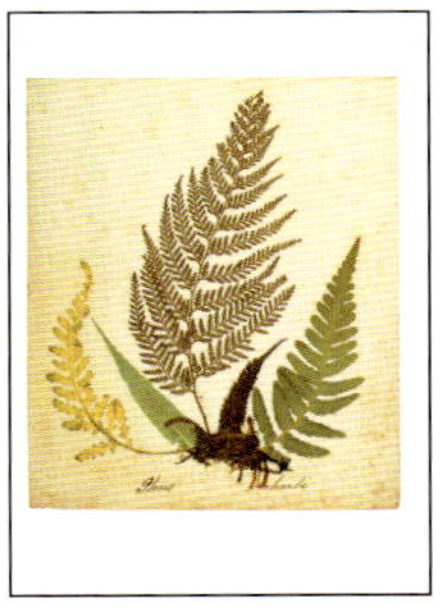

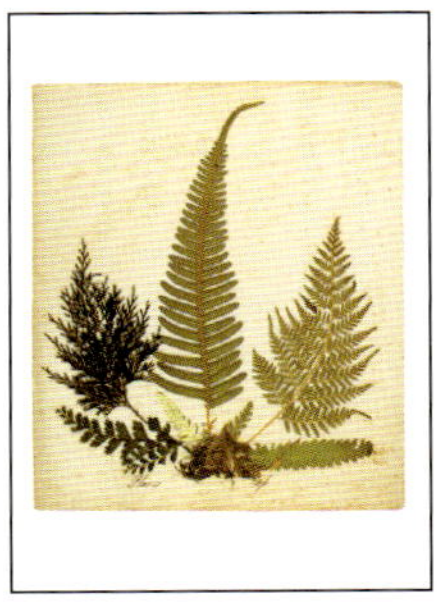

Louisa Atkinson's Fern Pressings

Collection details: Fifteen loose sheets with ferns and some mosses artistically arranged by Louisa Atkinson and presented to the National Herbarium by her granddaughter, Miss Janet Cosh of Moss Vale.

The first Australian-born woman novelist, Louisa Atkinson (1834–1872) was also an acknowledged journalist, artist, and naturalist. George Bentham, in his great Flora Australiensis, cited 116 of Louisa's specimens forwarded to him by Ferdinand von Mueller. More than 580 sheets in the National Herbarium of Victoria at the Royal Botanic Gardens, Melbourne, bear her name. Born Caroline Louisa Waring Atkinson at Sutton Forest, 120 km south of Sydney, Louisa returned to this area many times, dying nearby less than three years after marrying James Calvert. For nearly ten years Louisa and her mother lived in the Blue Mountains and from here made expeditions on horseback to Mount Tomah, Wisemans Ferry, and the Grose Valley toward Springwood. The latter were attempts to retrace the route taken by, and re-collect the plants encountered by, Allan Cunningham. – JA

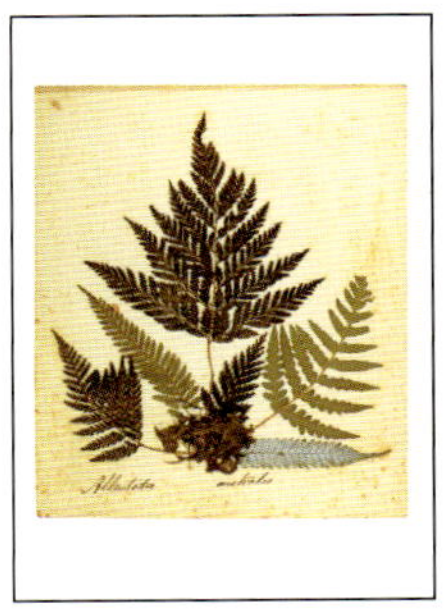

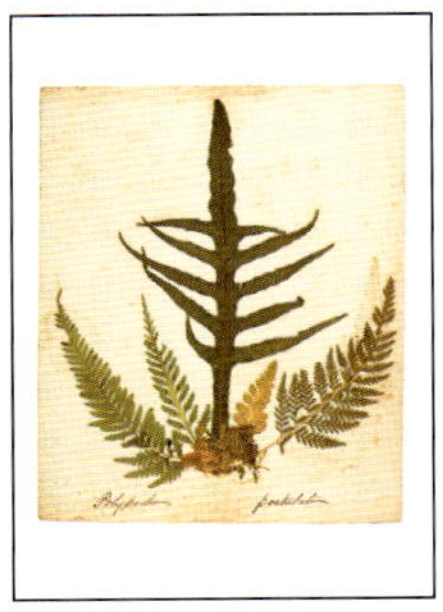

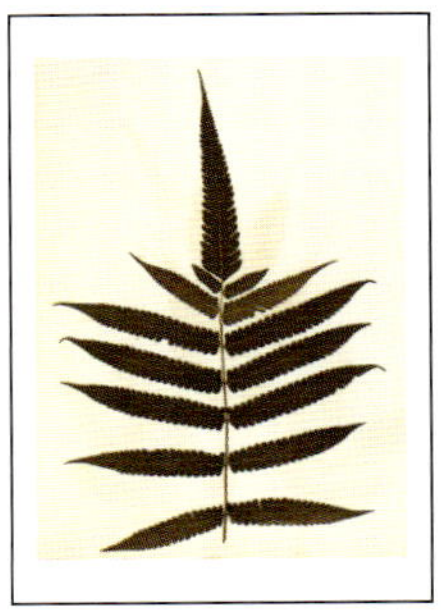

Christella subpubescens (Bl.) Holttum
A Tropical Fern
division PTERIDOPHYTA, family THELYPTERIDACEAE

This specimen is part of a donated album without annotations. The album covers are wooden, stencilled with fern frond shapes, and with the title Australian Ferns on the front. The two covers are held together by tied ribbons, enclosing homemade folders housing specimens on loose sheets of paper. This species grows in moist forests from South-East Asia to north-east Queensland, Vanuatu and Fiji. – KLW

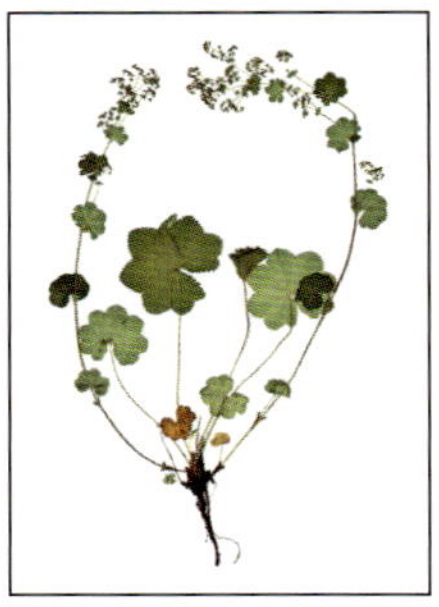

Alchemilla erectilis Juz.
A Russian Alchemilla
family ROSACEAE

Collection details: Collected from alpine meadows near Bakuaiani, Mount Tskhra-Tskharo, Georgia (former USSR) in 1953 by Sergei Vasilievich Juzepczuk (NSW 521255).

Juzepczuk subsequently officially published this species' name in 1957. (Unnamed plant species must have their new name published with a description, a type specimen and other details in accordance with the International Code of Botanical Nomenclature.) This specimen is a replicate (isotype) of the principal collection (holotype) on which the scientific name is based. The holotype is held at the V. L. Komarov Botanical Institute, St Petersburg, and another isotype is at the herbarium of the New York Botanical Garden. – BJC

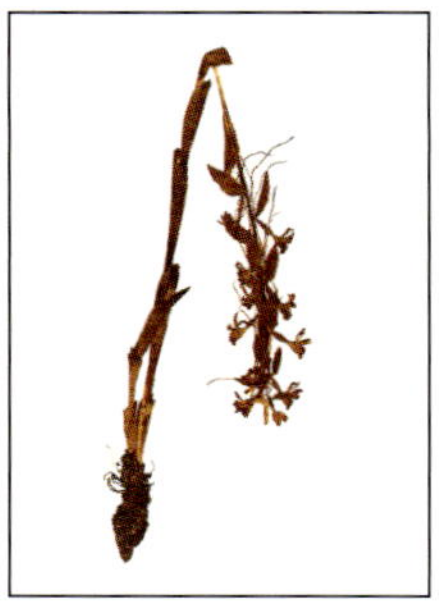

Satyrium buchananii Schlechter
An African Orchid
family ORCHIDACEAE

Collection details: Collected by Kew botanists R. K. Brummitt and R. M. Polhill in January 1975 at about 2100 m altitude near Mbeya, Iringa Region, Tanzania (NSW 541185).

This terrestrial orchid has white flowers and leaf sheaths tinged pinkish brown. The name Satyrium refers to the two-horned mythical satyr, an allusion to the two-spurred flowers. The specific name refers to the collector of the type specimen, John Buchanan (1855–1896), a European settler in what is now Malawi who grew coffee, sugar and tobacco. He also took a keen interest in the flora, sending many specimens to the Royal Botanic Gardens, Kew. – LJM & KLW

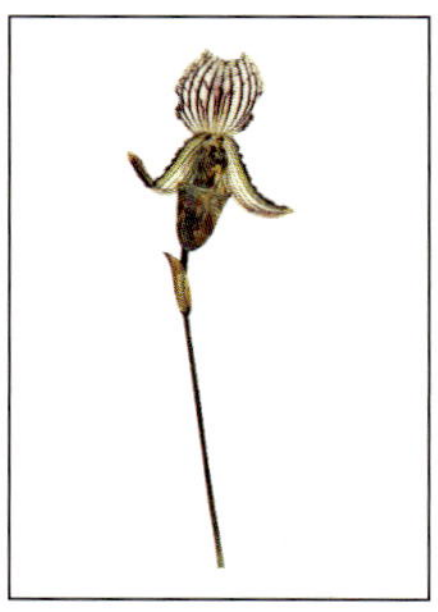

Paphiopedilum fairrieanum (Lindl.) Stein.
A Slipper Orchid
family ORCHIDACEAE

Collection details: Collected in August 1988 by technical officer Dianne C. Evans from a plant growing at the Royal Botanic Gardens, Sydney (NSW 368695).

This species is native to India and Bhutan and grows within the altitudinal range 1,200–3,000 m. It is a very distinctive and attractive species, widely cultivated since 1857. Mr Fairrie of Liverpool, for whom the species was named by English botanist John Lindley, was among the first to produce cultivated flowers of this species. – LJM

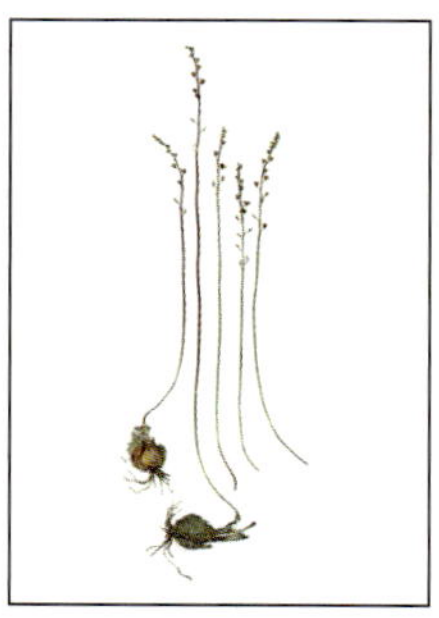

Drimia modesta (Bak.) J.P. Jessop
A South African Drimia
family HYACINTHACEAE

Collection details: Collected by W. J. Hanekom in October 1972 in Pretoria, South Africa (NSW 541186).

The specimen has insignificant yellow-green flowers and was collected from a grassland that had been burned over winter. The species was previously known as Urginea modesta Baker but the genus Urginea is now considered to belong to Drimia, a large genus of 120 species in the Mediterranean region, Africa and Asia. It includes many medicinal and poisonous species such as Sea Squill (D. maritima). – LJM & KLW

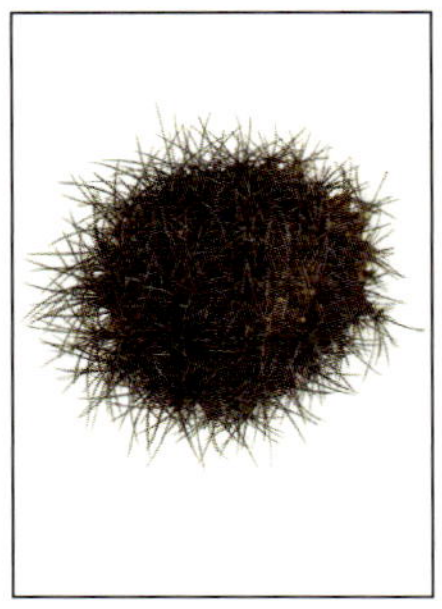

Mammillaria rhodantha subsp. *pringlei* (J.M. Coult.) D.R. Hunt
A Mexican Cactus
family CACTACEAE

Collection details: Collected by C. G. Pringle on 13 May 1901 in Mexico (where it is native) on the ledges of Tultenango Canyon at an altitude of 8,200 feet (NSW 596642).

This is a duplicate (isotype) of the main type collection (holotype) for this taxon, first published as a species, Cactus pringlei, by J. M. Coulter. Subsequent research showed that this was not very different from Mammillaria rhodantha and so was better treated as a subspecies. This is a small, barrel-shaped, green cactus up to 25 cm tall and 12 cm wide. Its bright pink flowers appear in a ring around the top. It is named after the collector: Cyrus Guernsey Pringle (1838–1911) of Vermont, USA, who collected in Mexico for the renowned American botanist Asa Gray. – LJM & KLW

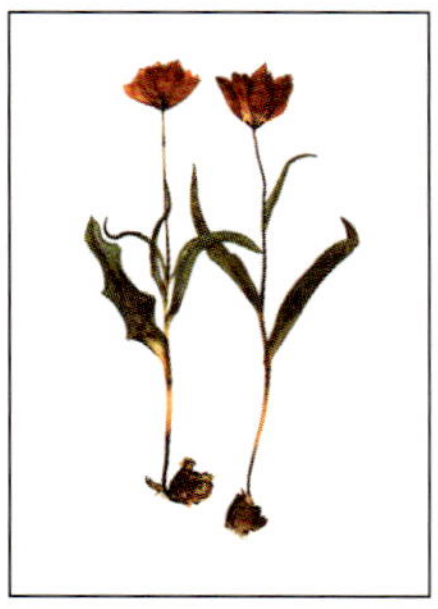

Tulipa gesneriana L.
Tulip
family LILIACEAE

Collection details: This specimen was collected in the Crimean region in 1900 by Alexis Callier (NSW 618676).

This species has vivid scarlet, orange, yellow or purplish flowers. It was originally native to Central Asia, but is now widespread in cultivation, being the source of most tulip cultivars. This species has been designated as the lectotype species for the genus Tulipa. (A lectotype is a specimen selected from original material as a representative of the taxon – chosen when the original author has not designated the type specimen.) The specific name honours the Swiss physician and botanist Conrad Gesner (1515–1565), who was the first to publish (1561) a description and drawing of a tulip flower cultivated in Europe – a plant he saw in Herwart's garden in Augsberg, Bavaria. – LJM & KLW

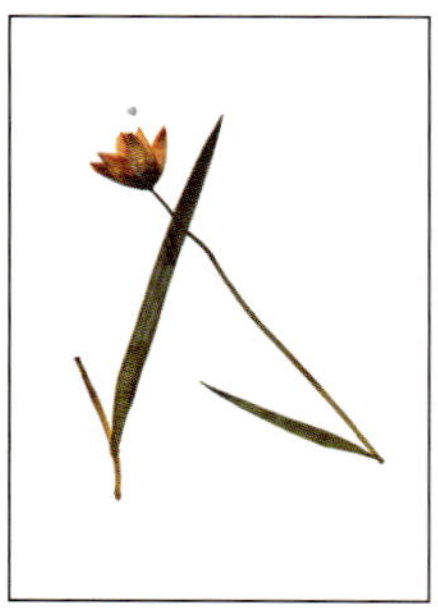

Tulipa hungarica Borb.
A Romanian Tulip
family LILIACEAE

Collection details: Collected near Orsova, Romania, by A. Richter in June 1908 (NSW 618679).

This yellow-flowered species grows on limestone rocks in the gorge of the Danube in south-west Romania and north-east Serbia. It is closely related to T. gesneriana. The specimen is stamped as received by the National Herbarium in July 1909. It may have been part of a set bought or exchanged by J. H. Maiden, who set out to ensure as wide a range of collections as possible – including at least one example of each native species – for the purposes of research and identification. – LJM & KLW

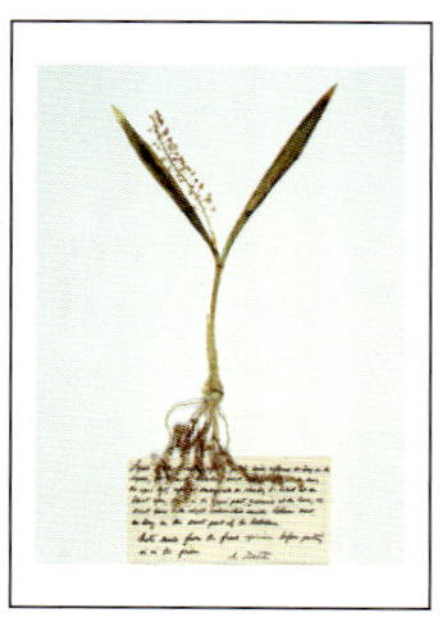

Liparis condylobulbon Rchb.f.
A Vanuatu Liparis
family ORCHIDACEAE

Collection details: Specimen taken from a living plant cultivated at the Botanic Gardens from material collected in New Hebrides (now Vanuatu) in 1897 by Dr Marriso (NSW 618644).

This widespread, epiphytic orchid has small greenish-white flowers. Ernst Betche, Botanical Assistant to J. H. Maiden, has written a note about the arrangement of the parts of the fresh specimen, e.g. 'petals linear, reflexed, as long as the sepals'. Such features are hard to study once orchids have been pressed and dried, so plants are often preserved instead in alcohol or another preservative (even whisky or rum have been used when no absolute alcohol was available). – LJM & KLW

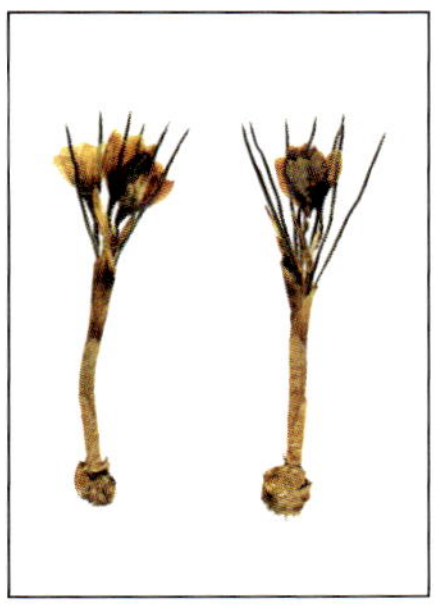

Crocus flavus Weston subsp. *flavus*
Yellow Crocus
family IRIDACEAE

Collection details: Collected in April 1844 at an indecipherable locality by an unknown collector. It is annotated as coming from Herbarium Laurer, the herbarium of the well-known collector Johann Friedrich Laurer (1798–1873) (NSW 618743).

This beautifully pressed specimen shows the structures both above- and below-ground of this commonly cultivated yellow-flowered crocus. It is not clear whether J. H. Maiden bought this from Laurer or exchanged it for some of his Australian collections.– LJM & KLW

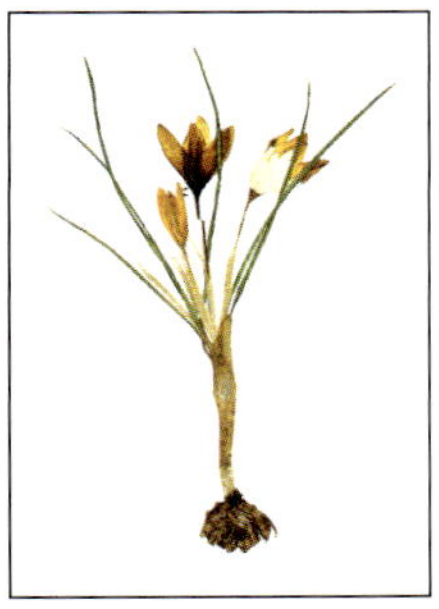

Crocus vitellinus Wahlb.
A Lebanese Crocus
family IRIDACEAE

Collection details: Collected on a plain north of Sidon, Lebanon, this specimen (NSW 618683) is part of a collection of Bible plants prepared by Sarah Stainton in 1865. The collection was donated to the National Herbarium in 1936 by Mrs Thomas, widow of Mesac Thomas (1816–1892), the first Bishop of Goulburn, NSW.

This yellow-flowered species is native to Syria, Palestine and Lebanon, but it is widely cultivated. In the nineteenth century, western European Christians took a strong interest in the plants mentioned in the Bible, and collected many of those growing in the Holy Land when touring that region. The names used by translators of the Bible are often misleading to present-day readers because the Hebrew names in the Old Testament and the Greek names in the New Testament were translated into local names recognisable by readers at that time. For example, the Hebrew word for 'crocus' was often translated as 'rose'. – LJM & KLW

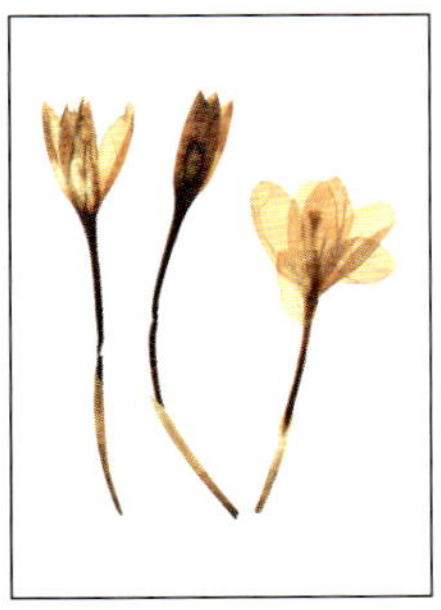

Crocus nudiflorus Sm.
Naked-flowering Crocus
family IRIDACEAE

Collection details: Collected at Adlington, Lancashire, England, in October 1858 by an unknown collector (NSW 618685).

This magnificent species produces large, vivid indigo-violet flowers in autumn. The leaves are not evident at flowering, hence both the common name and the specific epithet. It is native to the Pyrenees region of south-western Europe, but has been cultivated in England and elsewhere – possibly because it has been used as a substitute for saffron (derived from C. sativus). The central part of the C. sativus flower contains three red-orange stigmas, the female parts of the plant used to produce saffron. It takes 150,000 flowers to produce 1 kg of dried saffron. – LJM & KLW

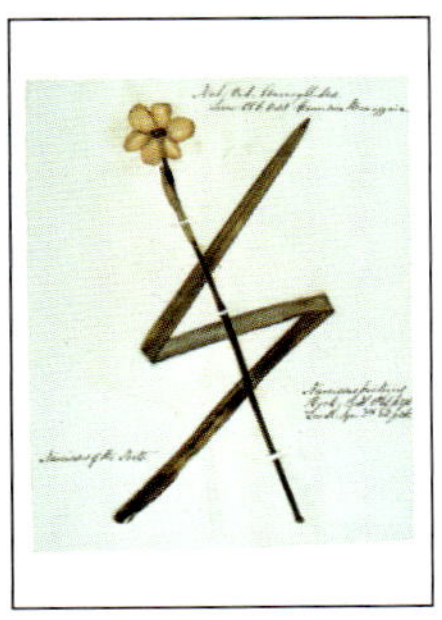

Narcissus poeticus L.
Pheasants Eye Narcissus, Narcissus of the Poets
family AMARYLLIDACEAE

Collection details: None available (NSW 618665). It was originally held in the herbarium of the Technological Museum in Sydney (now the Powerhouse Museum), which was donated to the National Herbarium in 1979.

The mounting of this specimen is interesting in that it is more like a work of art than a scientific specimen. This ornamental species is native to Europe, the Mediterranean region and western Asia but is widely cultivated. It flowers later in spring than other Narcissus species. It has an underground bulb, producing grey-green strap-like leaves and a solitary, fragrant flower 3–6 cm in diameter. The six white perianth segments of the flower surround a short central cup (the corona), which is yellow or greenish with a red or orange rim (compare the long trumpet of the common yellow daffodil). – LJM

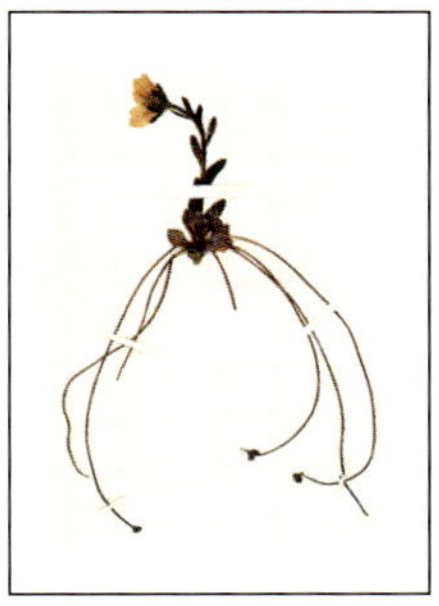

Saxifraga platysepala (Trautv.) Tolm.
Spiderplant
family SAXIFRAGACEAE

Collection details: Collected on one of the Arctic expeditions of Sir William Edward Parry (NSW 541153).

Parry (1790–1855) undertook three Arctic expeditions, unsuccessfully looking for a passage from the north-west Atlantic Ocean to the Pacific Ocean in 1819–1820, 1821–1823 and 1824–1825. In 1827 he attempted to reach the North Pole. This herbaceous species is circumpolar, growing as far as 83° north in Canada and Greenland. It is biennial or even triennial, producing bright yellow flowers. When the mother plant dies after flowering, the young rosettes at the end of the runners become detached, thus spreading the species. – LJM & KLW

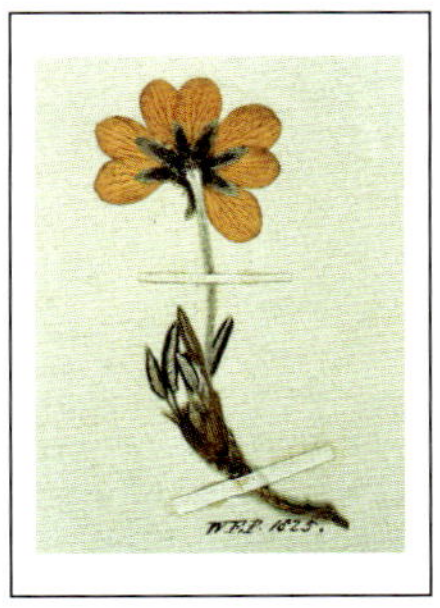

Dryas integrifolia Vahl
An Arctic Dryas
family ROSACEAE

Collection details: Collected during the third Arctic expedition of Sir William Edward Parry in 1825 (NSW 541177).

This dwarf shrub stands no more than 15 cm high, with intertwining stems forming mats to 1 m in diameter on the barren tundra of Canada, Alaska and Greenland. The flowers are white (they have turned brown in this specimen) with a yellow, central mass of 40–70 anthers. The numerous styles elongate to 2.5 cm and twist together, untwining as the fruits mature. – KLW

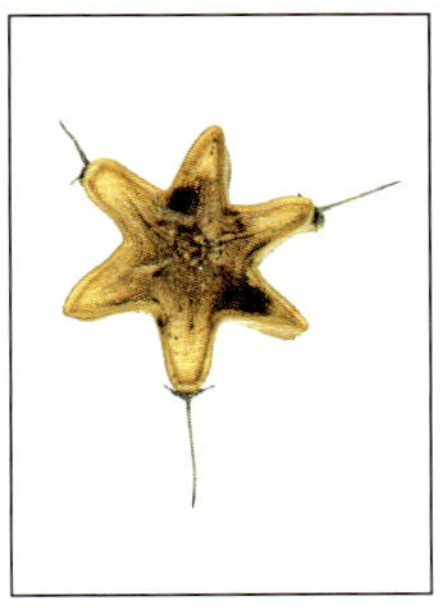

Myrtillocactus geometrizans (Mart.) Console
Blue Candle, Whortleberry Cactus
family CACTACEAE

Collection details: Collected by the American botanist C. G. Pringle in May 1891, on rocky hills near the Hacienda de Bocas, San Luis Potosi, Mexico (NSW 596641).

This succulent, candelabra-like tree, with ridged, blue-green stems 7–10 cm thick, is endemic to Mexico. This specimen includes a cross-section of a stem showing the fearsome spines on alternate ridges. The generic name apparently alludes to the similarity of the edible fruit to Vaccinium myrtillus (Bilberry); the specific name refers to the symmetry of the stems. – LJM

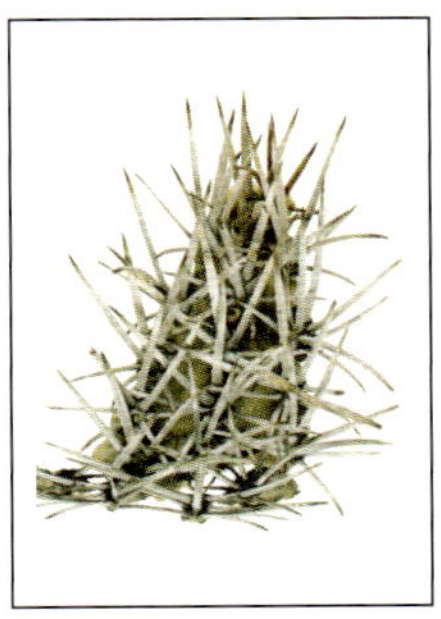

Echinocactus polycephalus Engelm. & J.M. Bigelow
A Nevada Cactus
family CACTACEAE

Collection details: Collected south of Indian Springs at 1,250 m altitude in the Charleston Mountains, Clark County, Nevada, USA, by I. W. Clokey in July 1939 (NSW 596640).

This specimen was growing in chaparral (Larrea) shrubland on low gravelly hills. Some colonies of this cactus contain over 100 spherical to short-cylindrical stems, 30–60 cm tall and 9–20 cm in diameter. Flowers are yellow and the fruits are covered with woolly hair. The spines grow to 8 cm, making the generic name an appropriate one (from Greek echinos, hedgehog or sea urchin). The specimen is stamped 'Received January 1941', indicating that exchange of herbarium specimens was still taking place despite the war in Europe. – LJM & KLW

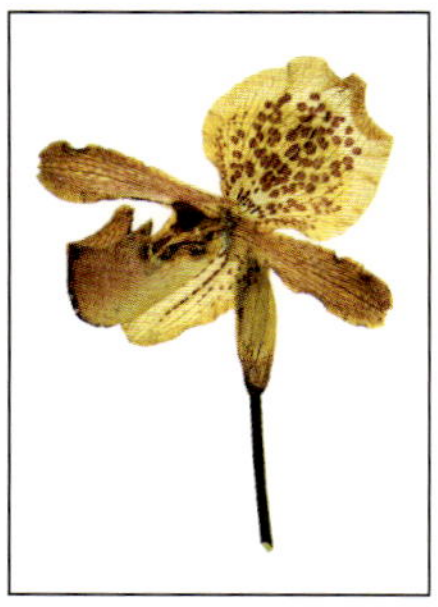

Paphiopedilum insigne Pfitz.
A Slipper Orchid
family ORCHIDACEAE

Collection details: Collected from the plant in the nursery at the Royal Botanic Gardens, Sydney, by Dianne Evans in August 1988 (NSW 368693).

This is the best-known and most widely cultivated species of Paphiopedilum. It is native to Nepal and north-eastern India, growing within the altitudinal range 1,200–2,000 metres. This flower has faded following pressing and drying, but it hints at the splendour of the live flowers: mostly yellow-green, with maroon dots and veins. These are 'trap flowers' that lure insects into the slipper (the modified third petal) to effect pollination as the insect struggles to escape. – LJM & KLW

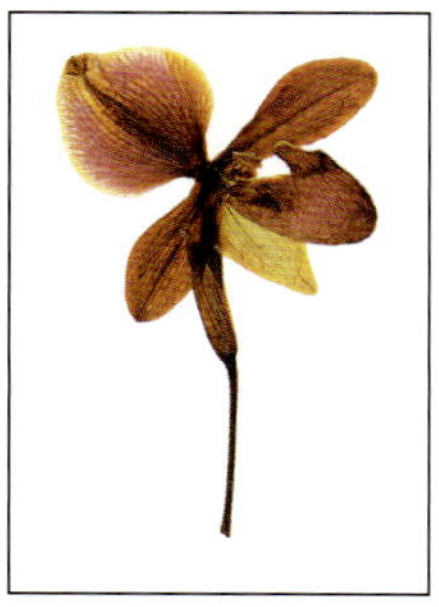

Paphiopedilum x *bingleyense*
A Slipper Orchid Hybrid
family ORCHIDACEAE

Collection details: This specimen was collected by Dianne Evans from a nursery-grown plant in the Royal Botanic Gardens, Sydney, in August 1988 (NSW 368692).

Many orchids hybridise, with the species of this genus doing so especially freely. They are native from Sikkim to New Guinea and the Solomon Islands, particularly diverse in Borneo. Plants are highly prized by collectors and many artificial hybrids have been produced. This hybrid is a cross between P. charlesworthii (female parent) and P. harrisianum (male). – LJM & KLW

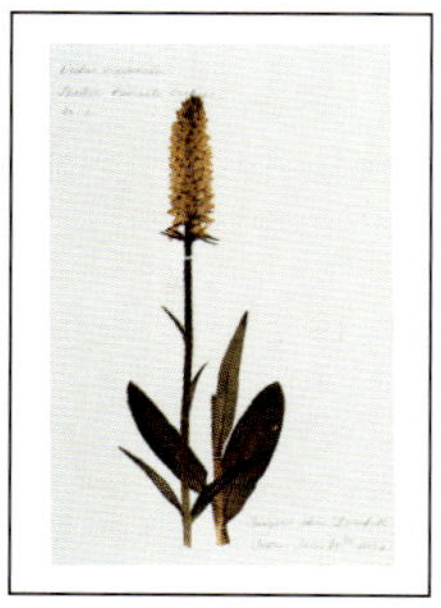

Dactylorhiza fuchsii (Druce) Soó
Common Spotted Orchid
family ORCHIDACEAE

Collection details: Collected from Tanyard Fields, Danehill, Sussex, England, in June 1854 by an unknown collector (NSW 621771).

A very attractive terrestrial orchid with purple dots on the leaves, in summer producing pale pink, mauve or white flowers with deep red or purplish dots. The species occurs in northern Europe and western Asia, mainly on calcareous soils in grassland, open scrub and woodland margins. The generic name is derived from the Greek daktylos, a finger, and rhiza, a root, referring to the forked, tuberous roots of most of the species. – LJM

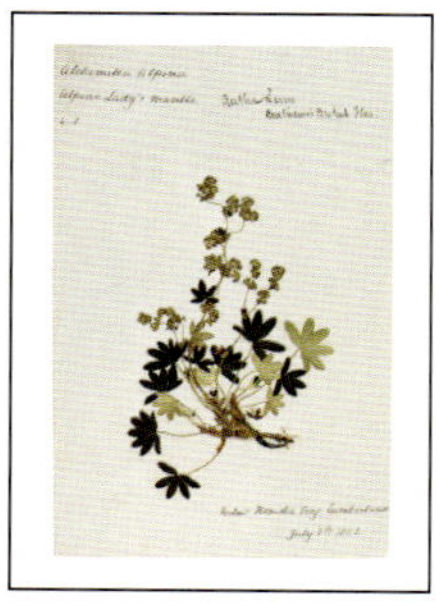

Alchemilla alpina L.
Alpine Lady's Mantle
family ROSACEAE

Collection details: Collected from Cumberland, northern England, by an unknown collector in July 1852 (NSW 541194).

This small, elegant summer-flowering herb has shining, silvery hairs on the stems and lower surfaces of the leaves, which are deeply divided like a hand (palmate). This species occurs throughout the Arctic region, from Europe and Asia to North America. In Europe, it is also found in the major mountain ranges south of the Arctic Circle. The generic name comes from the Arabic name for these species: alkemelych. – BJC & KLW

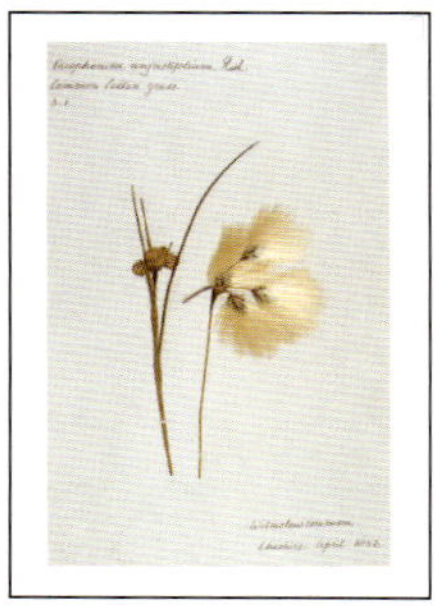

Eriophorum angustifolium Honckeny
Common Cotton-grass
family CYPERACEAE

Collection details: Collected on Wilmslow Common, Cheshire, England, in April 1852 by an unknown collector (NSW 618743).

This plant's generic name comes from the Greek erion, wool or cotton, and phoras, bearing, referring to the white cotton- or silk-like appearance of the spikelets when the perianth bristles elongate (up to 3 cm) and become conspicuous. This small, perennial herbaceous sedge is named for its narrow grass-like leaves (Latin angustus, narrow; folium, leaf). It is found in Northern Hemisphere cool-temperate regions, in and around bogs, marshes, and shallow water. – LJM & KLW

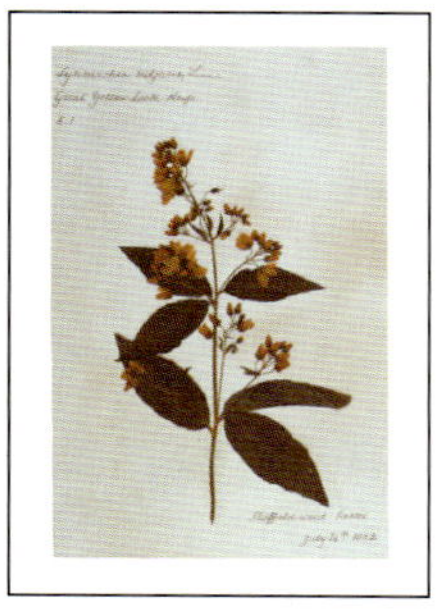

Lysimachia vulgaris L.
Yellow Loosestrife, Garden Loosestrife
family PRIMULACEAE

Collection details: Collected from Sheffield Wood in Sussex, England, in July 1852; collector unknown (NSW 541184).

Yellow Loosestrife is a perennial rhizomatous herb with erect stems to 1 m or more high and yellow, primrose-like flowers. It is a native of Eurasia, where it occurs in wetlands and other moist habitats such as fens, lake shores and river banks. Having been introduced as an ornamental, Yellow Loosestrife has become naturalised in North America where it has the potential to become an invasive weed. It has been used medicinally and as a dye in some regions. – PGK

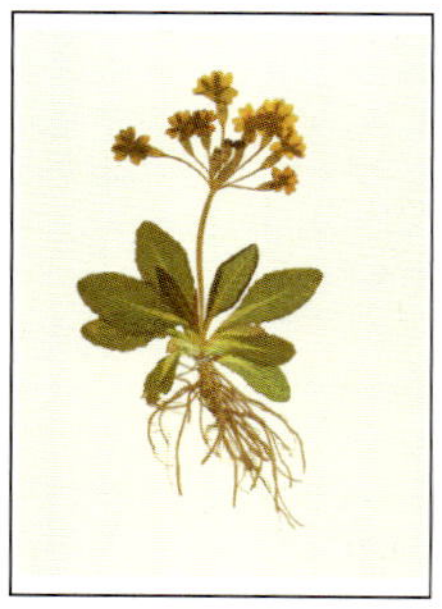

Primula vulgaris Hudson
Primrose
family PRIMULACEAE

Collection details: This specimen was collected near Geneva, Switzerland, in April 1864 by an unknown collector (NSW 598258).

This herb is native to Europe, northern Africa and western Asia but is also commonly cultivated because of the many vibrant colour-forms available. Primroses generally grow best in partial shade in rich, moist organic and well-drained soils. The petals vary from white to yellow (as in this wild specimen), orange, through pink and red to various shades of blue. In recent years more pastel colour-forms have been bred, the latest being a lime-coloured cultivar. This species has been recorded as having many herbal medicinal properties. This is a duplicate of a specimen in the Herbarium at the Royal Botanic Gardens, Kew. – BJC & KLW

Notes

Unpublished material by and relating to J. H. Maiden and the Rev. H. M. R. Rupp was consulted in the Quarto Vertical File archives of the Royal Botanic Gardens, Sydney – cited in the notes as RBGS QVF. Personnel interviewed at the Royal Botanic Gardens, Sydney, are indicated by their name and RBGS. Correspondence between, and relating to, J. H. Maiden and Wilhelm Baeuerlen was consulted in the Powerhouse Museum Archives, Sydney (formerly the Technological Museum) – cited as PHMA. Correspondence of Ferdinand Mueller was previously consulted at the Mueller Correspondence Project, Royal Botanic Gardens, Melbourne, cited as RBGM. Specimens consulted at the herbarium of the Natural History Museum, London, are indicated by the species' name and NHM, London.

p. 1 Royal Botanic Gardens, Sydney… Part of the Botanic Gardens Trust.

p. 2 'slow and roundabout' walk… Née, 1793, in Rolls, 2002, p. 31.
first land plants… Mayr, 2001, pp. 58, 63.
Twenty-five million years… Lines, 1994, pp. 6, 7.
Australia's flora… George, in Bureau of Flora and Fauna, 1981, p. 11.
Some, conservative in… Barlow, ibid, p. 52.
'consider the distinctive… Theophrastus, in Saunders, 1995, p. 17.

p. 3 the best teacher… Pliny, ibid.
'Not a hundredth… Brasavola, 1536, in Morton, 1981, p. 118.
the first herbarium specimens… Isely, 1994, pp. 20–21.
'whatsoever singularity… Bacon, 1594, in Pearce and Arnold (eds), 2000, vol. ii, p. 19.
'rare, exceptional, … Pomian, 1990, in Shelton, in Elsner and Cardinal (eds), 1994, p. 180.
he bought specimens… Stephenson, 1992, p. 7.
'Every blockhead… Banks, in Brunton (ed.), 1998, p. 4.
'No people, … Ellis, in O'Brian, 1987, p. 65.

p. 4 'Now do I wish… Banks, in Brunton (ed.), 1998, p. 6.
'well contented… Banks, 2 May 1770, ibid, p. 28.
'Our collection of Plants… Banks, 3 May 1770, ibid.
'made a voyage… *Evening Post*, 15 July 1771, in Adams, 1986, p. 125.
'which I am told… Lady Mary Coke, ibid, p. 127.
'Here is … Sheffield, in O'Brian, 1987, pp. 169–170.
'almost entirely' deprived… Linnaeus, in Gilbert, 1986, p. 6.

p. 5 'Consider my friend, … Linnaeus, in Adams, 1986, p. 140.
Why not make Kew… Ibid, pp. 136–137; Musgrave and Musgrave, 2000, p. 149.
'no doubt possess… Banks, in Hewson, 1999, p. 33.
'rare and beautiful plants… Tench, in Gilbert, in Carr and Carr (eds), 1981, p. 224; Ross, in Hughes, 1987, p. 95.
Saturday was collecting day… Gilbert, in Carr and Carr (eds), 1981, p. 226; Hughes, 1987, p. 97.

p. 6 'it is not therefore… Phillip to Banks, May 1789, in Gilbert, in Carr and Carr (eds), 1981, p. 224.
Fifteen shillings a week… Banks to Caley, 16 November 1798, in Dawson (ed.), 1958, p. 193.
'in the course of one year… Caley to Banks, 12 October 1800, in Carr and Carr (eds), 1981, p. 231.
plants he'd seen before leaving… Webb, 1995, p. 48.
'the plants grew… Brown, 9 December 1801, in Vallance *et al.* (eds), 2001, p. 92.
'remain[e]d on board… Brown, 10 December 1801, ibid, p. 93.
'more successful… Brown, 11 December 1801, ibid, p. 93.

p. 7 'Mr Brown has informed… Webb, 1995, p. 48.
'not yet begun to examine… Mabberley, 1985, p. 95.
'The kind of paper… Brown, 30 May 1802, in Vallance *et al.* (eds), 2001, p. 206.
'a single plant… Brown, 9 September 1802, ibid, p. 265.
'upon the whole disappointed… Brown to Banks, 6 August 1803, ibid, pp. 418–419.
'the loss of the garden… Brown to Banks, 16 September 1803, ibid, p. 438.
a live wombat… Mabberley, 1985, p. 129.
'provided for, … Brown, 13 October 1805, ibid, p. 127.
'by far the most excellent… Ibid, p. 130.
'employed to gather… Banks, 19 October 1805, in Vallance *et al.* (eds), 2001, p. 601.

p. 8 'in a new world… Smith, in Saunders, 1995, pp. 65–68.
'amount to about 1,600… Brown to Banks, 6 August 1803, in Vallance *et al.* (eds), 2001, p. 419.
one vicar claimed… Schiebinger, 1994, pp. 36, 37.
the *Encyclopaedia Britannica*… Fara, 2003, p. 43.
First, the ferns… Mabberley, 1985, pp. 162–164.
'the generic and specific… Ibid, p. 159.
hailed as 'most… Ibid, p. 172.
'We are indebted… J. Hooker, 1890, ibid, p. 191.
'quite as fashionable… Banks, in McMinn, 1970, p. 5.
Australian species that Brown knew… George, in Wheeler and Price (eds), 1981, p. 55.
Even his approach… Mabberley, 1985, p. 400.
'dried plants… Ibid, pp. 359, 364.
'huge collections… J. Hooker, ibid, p. 392.

p. 9 an eye to anything … Earl Bathurst, in Gilbert, 1986, p. 22.
'the highest ambition… Cunningham, ibid, p. 29.
strip off his clothes… Finney, 1984, p. 189.

p. 10 'an excellent Botanical Press… Cunningham, in Gilbert, 1986, p. 53.
'about the danger… Mitchell, ibid, p. 56.
'a kitchen garden, … *The Sydney Herald*, in McMinn, 1970, p. 110.
'a poor, decrepit… Cunningham, 16 May 1839, ibid, p. 112.
'sprang ashore… Leichhardt, in Roderick, 1988, p. 158.
'You know… Leichhardt, ibid, p. 84.
'My mind… Leichhardt, 17 May 1842, in Aurousseau (ed.), 1968, p. 470.
Leichhardt may have advocated… Finney, 1993, p. 2.
'one of those areas… Leichhardt, 6 January 1844, in Aurousseau (ed.), 1968, p. 700.

p. 11 could one tell different species… Leichhardt, 19 February 1843, ibid, p. 631.
'lift their majestic heads… Leichhardt, 9 January 1844, ibid, p. 707.
'You can perhaps form… Leichhardt, 6 February 1844, ibid, p. 729.
Unassisted, he had to… Leichhardt, ibid, p. 730.
'cook, groom, washerwoman… Leichhardt, ibid, p. 706.
'something far superior… Gilbert, 1986, p. 69.
'certainly not to be attributed… J. Macarthur, ibid, p. 69.
'nothing but the blue skies… Leichhardt, 24 January 1846, in Aurousseau (ed.), 1968, p. 839.
Our pack bullocks… Leichhardt, 7 October 1845, in Roderick, 1988, pp. 356–357.
This disastrous event… Leichhardt, 21 October 1845, ibid, p. 358.
'to render the broth… Short (ed.), 2003, p. 159.
'almost crying with vexation… Ibid.
'a saltwater and Mangrove… Leichhardt, 24 October 1845, in Aurousseau (ed.), 1968, p. 969.
'many new trees… Leichhardt, in Roderick, 1988, p. 363.

p. 12 'I hope my next expedition… Leichhardt, 20 May 1846, in Aurousseau (ed.), 1968, p. 870.
'send back a whole set… Leichhardt, 27 September 1846, ibid, pp. 906–907.
'travel down… Leichhardt, 20 May 1846, ibid, pp. 870–871.
'scarcely … possible… Leichhardt, 17 August 1846, ibid, p. 892.
'those who are not versed… Leichhardt, 24 October 1847, ibid, p. 969.
Leichhardt 'kindly showed me… Gordon, 2 April 1874, in Meacham, 2003.
'the blacks had mustered… ibid.

p. 13 'many ladies living… Mueller, *West Australian*, July 1883, RBGM.
'particularly fitted… Woolls, in Maroske, in Kelly (ed.), 1993, p. 24.
'anxious to learn… E. Macarthur, 7 March 1791, in Gilbert, 1962, p. 74.
'remarkable feature… Molloy, in Moyal, in Carr and Carr (eds), 1981, p. 335.
'who does not enter… Ibid.
'for being the cause… Ibid.
'I have almost been panting… Ibid, p. 337.

p. 14 'one of the most delightful… Molloy, in Maroske, in Kelly (ed.), 1993, p. 25.

'all my former pursuits... Molloy, ibid, p. 27.
'a bouquet... Molloy, in Hasluck, 1955, p. 199.
A Sketch of... Molloy, ibid, p. 202.
the beautiful kangaroo paw... Stephenson, 1992, p. 60.
'sent [him] everything... Molloy, in Moyal, in Carr and Carr (eds), 1981, p. 337.
'I hope this acknowledgement... Mueller, in McKay, 1997, pp. 1–2.
'very few Ladies... Mueller, in Maroske, in Kelly (ed.), 1993, p. 20.
three other taxa... Clarke, 1990, p. 136: *Doodia atkinsoniana*, *Erechittes atkinsoniae*, and *Epacris calvertiana*.
The trees, chiefly eucalypti... Atkinson, ibid, p. 125.
'nature ... as an... Atkinson, ibid, p. 127.
'Scrambling over rocks... Atkinson, ibid, p. 131.

p. 15 'My love for ... Ellis Rowan, in Moyal, in Carr and Carr (eds), 1981, p. 350.
'collections presented... Bentham, 1967 [1867], p. 14.
recognition given... Spender, 1988, p. 71.
a large portion... Martin, 1996, p. 186.
shareholders were requested... Preiss, in Ducker, in Carr and Carr (eds), 1981, pp. 132–133.

p. 16 The British Government declined... Hewson, 1999, p. 120.
'you Father... Preiss, in Mabberley, 1985, p. 362.
'I have heard you say... Harvey, 9 April 1856, in Ducker (ed.), 1988, p. 275.
'a diligent botanist... Sumner, 1993, p. 70.
If, for the majority... Schiebinger, 1994, p. 36.
Dietrich's luggage... Moyal, in Carr and Carr (eds), 1981, p. 340; McKay, 1997, p. 2.
'amongst which I disappear... Dietrich, in Moyal, in Carr and Carr (eds), 1981, p. 341.
'the loveliest plants... Dietrich, in Sumner, 1993, p. 135.
'Frau Dietrich continues... Sumner, in Voigt (ed.), 1983, p. 263.
Acacia dietrichiana... Moyal, in Carr and Carr (eds), 1981, p. 342: also *Bonamia dietrichiana*, *Carex dietrichiae*, *Cyperus dietrichiae*, *Heleocharis* (now *Eleocharis*) *dietrichiana*, and the Australian Skipper Butterfly *Cephrenes amalia*.
'badly dressed woman... Sokolowsky, in Sumner, 1993, p. 62.
Correa baeuerlenii... Adams, 1988, pp. 8–9.

p. 17 'new, rare and beautiful... Baeuerlen, 1886, p. 38.
'painstaking botanical collector'... Mueller and Maiden, in Wilson, in Short (ed.), 1990, p. 102.
'exceedingly anxious... Maiden, 15 April 1890, in Gilbert, 2001, p. 123.
'Mr Baeuerlen... Maiden, Report of the Minister for Public Instruction, 1892, MRS 117, PHMA.
'work under very... Baeuerlen to Maiden, 7 February 1891, in MRS 203, PHMA.
'require more attention... Ibid, and 14 February 1891, in MRS 203. PHMA.
'would beg and steal... Baeuerlen to Maiden, 8 April 1891, in MRS 203, PHMA.
'rather thin... Baeuerlen to Maiden, 28 February 1891, 21 March 1891, and 7 March 1891, in MRS 203, PHMA.
'They all seem... Baeuerlen to Maiden. 8 April 1891, and 15 April 1891, in MRS 203, PHMA.
Other localities... Baeuerlen to Maiden, 9 April 1891, in MRS 203, PHMA.
'*very good* value... Maiden, 15 April 1890, in MRS 14, PHMA.

p. 18 'I cannot understand... Smith to Baeuerlen, 27 April 1899, in PHM MRS 240, PHMA.
'I am the Botanical ... Baeuerlen to Baker, 16 March 1904, in Gilbert, 2001, p. 216.
the *Evening News... Evening News*, 12 September 1908, in MRS 14, PHMA.
'My service... Baeuerlen, 1912.
'Even in this... Tomlins to Maiden, 1916, and reply, in Gilbert, 2001, pp. 277–278.

p. 19 'thousand or two... Maiden, 2 August 1898, ibid., p. 173.
'proceeds of some... Mitchell, in Gilbert, 1986, p. 88.
'I do not know... Maiden, 1898, in Gilbert, 2001, p. 173.
'the care of the Herbarium... Maiden, 1897, in Gilbert, 1986, p. 118.

p. 20 'properly perform... Gilbert, 2001, p. 180.
'the mecca... Maiden, 1898, ibid, pp. 172–173.
'Over 20 years... Maiden, 1907, ibid, p. 195.
'get on the tracks... Maiden, 1907, ibid, p. 195.
'two days' blizzard... Lucas, 1898, ibid, p. 176.
'Collecting,' Maiden said... Maiden, 1898, ibid, p. 175.
'not less than 15,000... Maiden, 1899, 1901, ibid, p. 179; Gilbert, 1986, p. 128.
97 per cent... Betche, 1903, in Gilbert, 1986, p. 129.
'all the Australian... Maiden, 1903, ibid, p. 129.

p. 21 this banksia... *Banksia dentata*, NHM, London.
that eucalypt... *Eucalyptus alba*, NHM, London.
'the return half... Britten, 1900, in Gilbert, 2001, p. 183.
'a set of 586... Maiden, in *Sydney Morning Herald*, 31 May 1905, ibid, p. 199.
'absolutely the oldest... Ibid.
'Dried specimens... Maiden, 1907, in Gilbert, 2001, p. 229.
'one of the best herbaria... Maiden, 1905, ibid, p. 213.
'wild, untamed things... Maiden, 1919, in Gilbert, 2001, p. 275.
'I pride myself... Maiden, 1925, ibid, p. 302.
18-and-a-half inches... Maiden, 1921, ibid, p. 331.

p. 22 a combination of economic depression... Briggs, 2001, p. 224.
'The taste for collecting... Rheims, in Baudrillard, in Elsner and Cardinal, (eds), 1994, p. 9.
'full of genial... Rupp, in Gilbert, 2001, p. 307.
'Whatever little contributions... Rupp, in Gilbert, 1992, p. 110.

p. 23 a purple orchid whose 'tuber... www.beautifulorchids.com
'a thrilling serial... Rupp, in Gilbert, 1992, p. 29.
'a bit dotty... Rupp, ibid, p. 106.
'small pay'... Mueller, ibid, p. 45.
'I am merely... Rupp, ibid, p. 179.
'[It's] a good... Rupp, 27 April 1933, ibid, p. 47.
'Some folks object... Rupp, ibid, p. 82.
'the non-orchidaceous ... Rupp, ibid, p. 55.

p. 24 'Are [New South Wales' orchids ... Rupp, 1925.
'discovered and named... Rupp, in Gilbert, 1992, p. 124.
'clearing away... Rupp to Rodway, 9 December 1931, RBGS QVF.
'an orchid which... Gilbert, 1992, p. 71.

p. 25 'the National Herbarium... Rupp to Nichols, 30 January 1946, RBGS QVF.
'I am no pioneer... Rupp, in Gilbert, 1992, p. 95.
'continue my botanical researches... Rupp to Nichols, 9 July 1956, RBGS QVF.

p. 26 On a bush block... P. Weston, RBGS, 2003, pers. comm.
Like one red alga ... *Psilosiphon*, T. Entwisle, RBGS, 2004, pers. comm.
that small packet of violets ... Thiele and Prober, 2003, p. 17; L. Murray, RBGS, 2004, pers. comm.
The proudly painstaking portraits... See Saunders, 1995.
Some of Borneo's Dayak... Jordan, 1993, pp. 32, 53; Fontana, 1993

Bibliography

Adams, Brian (1986), *The Flowering of the Pacific: Being an Account of Joseph Banks' Travels in the South Seas and the Story of his Florilegium*. Sydney: Collins, and the British Museum (Natural History).

Adams, L. (1988), The orthography of Mueller's epithets commemorating Wilhelm Baeuerlen. *Australian Systematic Botany Society Newsletter* 24: 8–9.

Aurousseau, M. (ed.) (1968), *The Letters of F. W. Ludwig Leichhardt, vols i–iii*. Cambridge: Cambridge University Press.

Baeuerlen, William (1886), *The Voyage of the Bonito: An Account of the Fly River Expedition to New Guinea*. Sydney: Gibbs, Shallard and Co.

Baeuerlen, William (1912), Letter, *Sydney Morning Herald*, 22 April 1912.

Benedict, Barbara M. (2001), *Curiosity: A Cultural History of Early Modern Inquiry*. Chicago: University of Chicago Press.

Bentham, George (1967) [1867], Flora Australiensis: *A Description of the Plants of the Australian Territory*. Ashford: A. Asher and L. Reeve & Cos.

Bernhardt, Peter (1990), *Wily Violets and Underground Orchids: Revelations of a Botanist*. Sydney: Allen & Unwin.

Blunt, Wilfred, and William T. Stearn (1994), *The Art of Botanical Illustration*. London: The Antique Collectors' Club in association with the Royal Botanic Gardens, Kew.

Briggs, Barbara (2001), The National Herbarium of New South Wales – 100 years. *Telopea* 9(2): 223–226.

Brunton, Paul (ed.) (1998), *The Endeavour Journal of Joseph Banks: The Australian Journal*. Sydney: Angus & Robertson in association with the State Library of New South Wales.

Bureau of Flora and Fauna, Canberra (1981), *Flora of Australia: vol. i*. Canberra: AGPS.

Campbell-Culver, Maggie (2001), *The Origin of Plants: The People and Plants that have Shaped Britain's Garden History Since the Year 1000*. London: Headline.

Carr, D. J., and S. G. M. Carr (eds) (1981), *People and Plants in Australia*. Sydney: Academic Press.

Chippendale, G. M. (1988), Eucalyptus, Angophora *(Myrtaceae). Flora of Australia, vol. xix*. Canberra: AGPS.

Clarke, Patricia (1990), *Pioneer Writer: The Life of Louisa Atkinson, Novelist, Journalist, Naturalist*. Sydney: Allen & Unwin.

Coats, Alice M. (1969), *The Quest for Plants: A History of Horticultural Explorers*. London: Studio Vista.

Dawson, Warren R. (ed.) (1958), *The Banks Letters: A Calendar of the Manuscript Correspondence of Sir Joseph Banks Preserved in the British Museum, the British Museum (Natural History), and Other Collections in Great Britain*. London: British Museum.

Ducker, Sophie (ed.) (1988), *The Contented Botanist: Letters of W. H. Harvey about Australia and the Pacific*. Melbourne: Miegunyah Press.

Ebes, Hank (1988), *The Florilegium of Captain Cook's First Voyage to Australia 1768–1771*. Melbourne: Ebes Douwma Antique Prints and Maps; Sydney: Sotheby's; and London: British Museum (Natural History).

Elsner, John, and Roger Cardinal (eds) (1994), *The Cultures of Collecting: From Elvis to Antiques – Why Do We Collect Things?* Melbourne: Melbourne University Press.

Fara, Patricia (2003), *Sex, Botany and Empire: The Story of Carl Linnaeus and Joseph Banks*. Sydney: Allen & Unwin.

Finney, C. M. (1984), *To Sail Beyond the Sunset: Natural History and Australia 1699–1829*. Sydney: Rigby.

Finney, Colin (1993), *Paradise Revealed: Natural History in Nineteenth-Century Australia*. Melbourne: Museum of Victoria.

Flannery, Tim (ed.) (1999), *The Birth of Sydney*. Melbourne: Text.

Fontana, David (1993), *The Secret Language of Symbols: A Visual Key to Symbols and Their Meanings*. London: Pavilion.

Gilbert, Lionel (1962), *Botanical Investigation of Eastern Seaboard Australia 1788–1810: An Examination of the Means by Which Botanical Knowledge of the Seaboard of Eastern Mainland Australia was Gained So Rapidly During the First 22 Years of Settlement*. Thesis submitted as part of the requirements for History Honours (BA), University of New England, Armidale (unpublished).

Gilbert, Lionel (1985), *William Woolls: 1814–1893 – A Most Useful Colonist*. Canberra: Mulini Press.

Gilbert, Lionel A. (1986), *The Royal Botanic Gardens, Sydney*. Melbourne: Oxford University Press.

Gilbert, Lionel (1992), *The Orchid Man: The Life, Work and Memoirs of the Rev. H. M. R. Rupp, 1872–1956*. Kenthurst, NSW: Kangaroo Press.

Gilbert, Lionel (2001), *The Little Giant: The Life and Work of Joseph Henry Maiden, 1859–1925*. Sydney: Kardoorair Press with the Royal Botanic Gardens, Sydney.

Griffiths, Tom (1996), *Hunters and Collectors: The Antiquarian Imagination in Australia*. Cambridge: Cambridge University Press.

Hall, Norman (1978), *Botanists of the Eucalypts*. Melbourne: CSIRO.

Hamilton, Jill, Duchess of, and Julia Bruce (1998), *The Flower Chain: The Early Discovery of Australian Plants*. Kenthurst, NSW: Kangaroo Press.

Hasluck, Alexandra (1955), *Portrait with Background: A Life of Georgiana Molloy*. Melbourne: Oxford University Press.

Hewson, Helen (1999), *Australia: 300 Years of Botanical Illustration*. Collingwood, Vic.: CSIRO Publishing.

Home, R., A. M. Lucas, Sara Maroske, D. M. Sinkora, and J. H. Voigt (eds) (1998, 2002), *Regardfully Yours: Selected Correspondence of Ferdinand Mueller, vols i–ii*. Bern: Peter Lang AG.

Hughes, Joy (1984), *The Journals and Letters of Elizabeth Macarthur, 1789–1798*. Sydney: Historic Houses Trust of NSW.

Hughes, Robert (1987), *The Fatal Shore: A History of the Transportation of Convicts to Australia, 1787–1868*. London: Collins Harvill.

Isely, Duane (1994), *One Hundred and One Botanists*. Ames: Iowa State University Press.

Jordan, Michael (1993), *Myths of the World: A Thematic Encyclopedia*. London: Kyle Cathie.

Kelly, Farley (ed.) (1993), *On the Edge of Discovery: Australian Women in Science*. Melbourne: Text.

Le Plastrier, Constance M. (1933), *The Story of Our Plants: First Steps in Australian Botany*. Sydney: Shakespeare Head Press.

Lines, William (1994), *An All-Consuming Passion: Origins, Modernity and the Australian Life of Georgiana Molloy*. Sydney: Allen & Unwin.

Mabberley, D. J. (1985), *Jupiter Botanicus: Robert Brown of the British Museum*. Braunschweig: Verlag von J. Cramer; and London: British Museum (Natural History).

McKay, Judith (1997), *Brilliant Careers: Women Collectors and Illustrators in Queensland*. Brisbane: Queensland Museum.

McMinn, W. G. (1970), *Allan Cunningham: Botanist and Explorer*. Melbourne: Melbourne University Press.

Maiden, J. H. (1902, 1903), *The Sydney Botanic Gardens: Biographical Notes Concerning the Officers in Charge*. Sydney: S. T. Leigh & Co..

Maiden, J. H. (1908), Records of Australian botanists – (a) general; (b) New South Wales, in *Journal of the Royal Society of New South Wales*, vol. xlii.

Maiden, J. H. (1909), *Records of West Australian Botanists and Australian Botanists who have Dealt with the Flora of Western Australia*. Perth: K. Jones.

Maiden, J. H. (1909), *Records of Queensland Botanists*. Brisbane: Government Printer.

Maiden, J. H. (1912), *Records of Australian Botanists – First Supplement*. Sydney: W. E. Smith, 1912.

Martin, Susan K. (1996), Baron Ferdinand von Mueller and his 'lady' correspondents. *The Victorian Naturalist* 113(4): 184–187.

Mayr, Ernst (2001), *What Evolution Is*. New York: Basic Books.

Meacham, Steve (2003), Looking for Lud in the dead heart of darkness. *Sydney Morning Herald*, 26 September 2003.

Morton, A. G. (1981), *History of Botanical Science: An Account of the Development of Botany from Ancient Times to the Present Day*. London: Academic Press.

Musgrave, Anthony (1936), With the orchid hunters in the Mullumbimby District. *Australian Museum Magazine*, 20 November 1936.

Musgrave, Toby, and Will Musgrave (2000), *An Empire of Plants: People and Plants that Changed the World*. New York: Cassell and Co.

O'Brian, Patrick (1987), *Joseph Banks: A Life*. London: Collins Harvill.

Olde, Peter, and Neil Marriott (1994), *The Grevillea Book, vol. ii*. Kenthurst, NSW: Kangaroo Press.

Orchard, Anthony E., and Helen S. Thompson (eds) (1999), *Flora of Australia, vol. i*, Second edition. Melbourne: Australian Biological Resources Study/CSIRO.

Pearce, Susan, and Ken Arnold (eds) (2000), *The Collector's Voice: Critical Readings in the Practice of Collecting*. Burlington, USA: Ashgate.

Reinikka, Merle A. (1995), *A History of the Orchid*. Portland: Timber Press.

Roderick, Colin (1988), *Leichhardt: The Dauntless Explorer*. Sydney: Angus and Robertson.

Rolls, Eric (2002), *Visions of Australia: Impressions of the Landscape 1642–1910*. South Melbourne: Lothian.

Rupp, H. M. R. (1925), The cult of the orchid. *Sydney Morning Herald*, 1 December 1925.
Rupp, H. M. R. (1930), *Guide to the Orchids of New South Wales*. Sydney: Angus & Robertson.
Rupp, H. M. R. (1969) [1943], *The Orchids of New South Wales*. Sydney: Government Printer of NSW.
Saunders, Gill (1995), *Picturing Plants: An Analytical History of Botanical Illustration*. Los Angeles: University of California Press in association with the Victoria and Albert Museum, London.
Schiebinger, Londa (1991), *The Mind has No Sex? Women in the Origins of Modern Science*. Cambridge, MA: Harvard University Press.
Schiebinger, Londa (1994), *Nature's Body: Sexual Politics and the Making of Modern Science*. London: Pandora.
Short, P. S. (ed.) (1990), *History of Systematic Botany in Australasia: Proceedings of a Symposium Held at the University of Melbourne, 25–27 May, 1988*. Melbourne: Australian Systematic Botany Society.
Short, Philip (ed.) (2003), *In Pursuit of Plants: Experiences of Nineteenth and Early Twentieth Century Plant Collectors*. Crawley, WA: University of Western Australia Press.
Spender, Dale (1988), *Writing a New World: Two Centuries of Australian Women Writers*. London: Pandora.
Stephenson, Len (1992), *What's in a Name? Derivation of Generic Names of Australian Native Plants*. Adelaide: Society for Growing Australian Plants.
Stuessy, Tod F., and S. H. Sohmer (eds) (1996), *Sampling the Green World: Innovative Concepts of Collection, Preservation and Storage of Plant Diversity*. New York: Columbia University Press.
Sumner, Ray (1993), *A Woman in the Wilderness: The Story of Amalie Dietrich in Australia*. Kensington, NSW: New South Wales University Press.
Taylor, J. E. (ed.) (1876), *Notes on Collecting and Preserving Natural History Objects*. London: Hardwicke & Bogue.
Thiele, K., and S. Prober, (2003), *Viola hederacea* species complex. *Muelleria* 18: 14–18.
Vallance, T. G., D. T. Moore, and E. W. Groves (eds) (2001), *Nature's Investigator: The Diary of Robert Brown in Australia, 1801–1805*. Canberra: Australian Biological Resources Study.
Voigt, Johannes M. (ed.) (1983), *New Beginnings: The Germans in New South Wales and Queensland*. Stuttgart: Institute for Foreign Cultural Relations.
Walker, David, and Jürgen Tampke (eds) (1991), *From Berlin to the Burdekin: The German Contribution to the Development of Australian Science, Exploration and the Arts*. Kensington, NSW: New South Wales University Press.
Webb, Joan (1995), *George Caley: Nineteenth Century Naturalist*. Chipping Norton, NSW: Surrey Beatty and Sons.
Webster, E. M. (1980), *Whirlwinds in the Plain: Ludwig Leichhardt – Friends, Foes, and History*. Melbourne: Melbourne University Press.
Wheeler, Alwynne, and James H. Price (eds) (1981), *History in the Service of Systematics: Papers from the Conference to Celebrate the Centenary of the British Museum (Natural History) 13–16 April, 1981*. London: The Society for the Bibliography of Natural History.
White, John (1790), *Journal of a Voyage to New South Wales*. London: [no publisher].
Willis, J. H., D. Pearson, and M. T. Davis (1984), *Australian Plants: Collectors and Delineators 1780s–1980s*. Melbourne: [no publisher].
Wilson, K. (2003), *Arrangement of Specimens in the National Herbarium of New South Wales (NSW) at the Royal Botanic Gardens Sydney*. Sydney: unpublished.
Woodford, James (2000), *The Wollemi Pine: The Incredible Discovery of a Living Fossil from the Age of the Dinosaurs*. Melbourne: Text.

Index

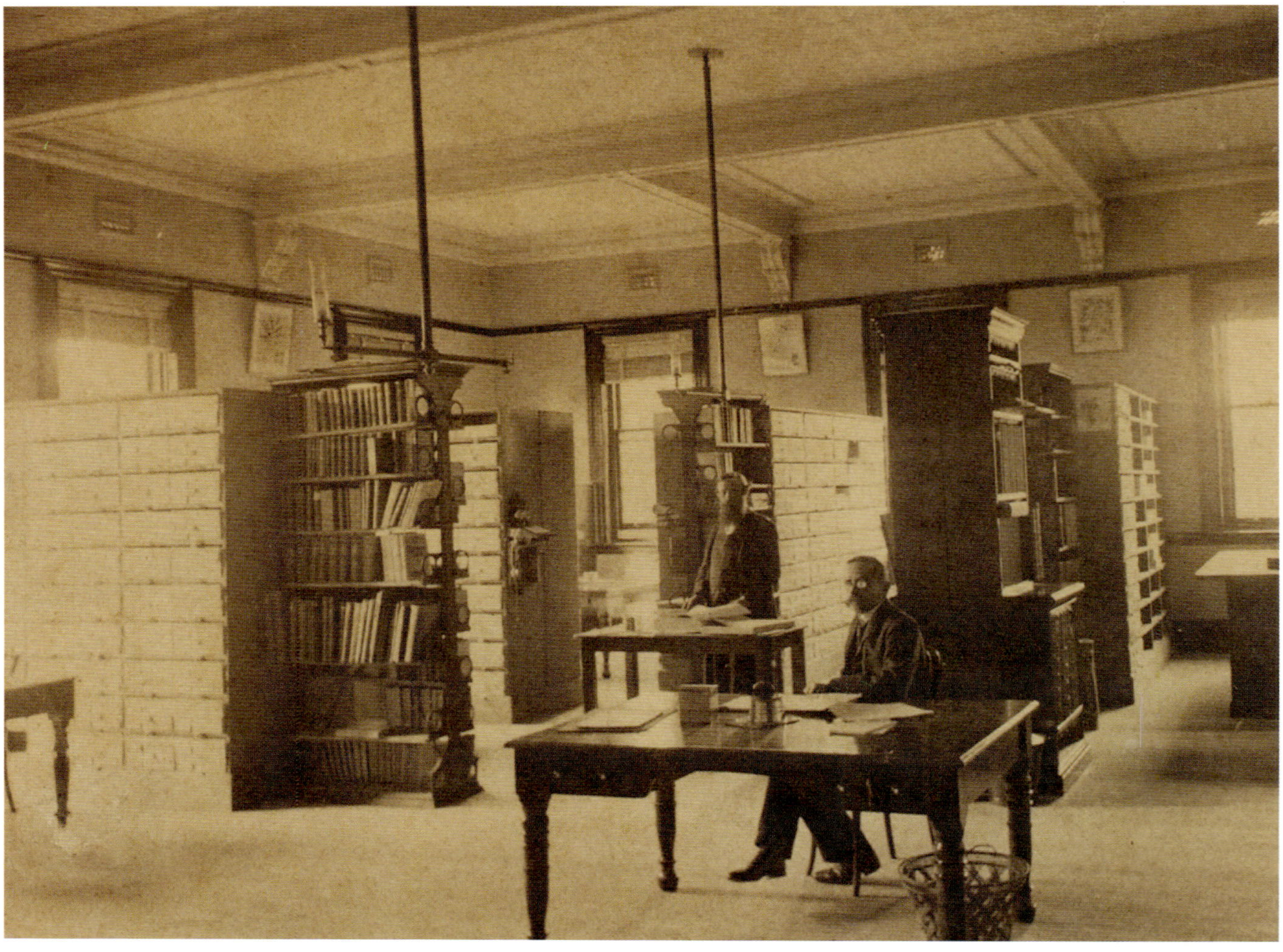

The NE corner of the old Herbarium (now known as the Anderson Building), c. 1910

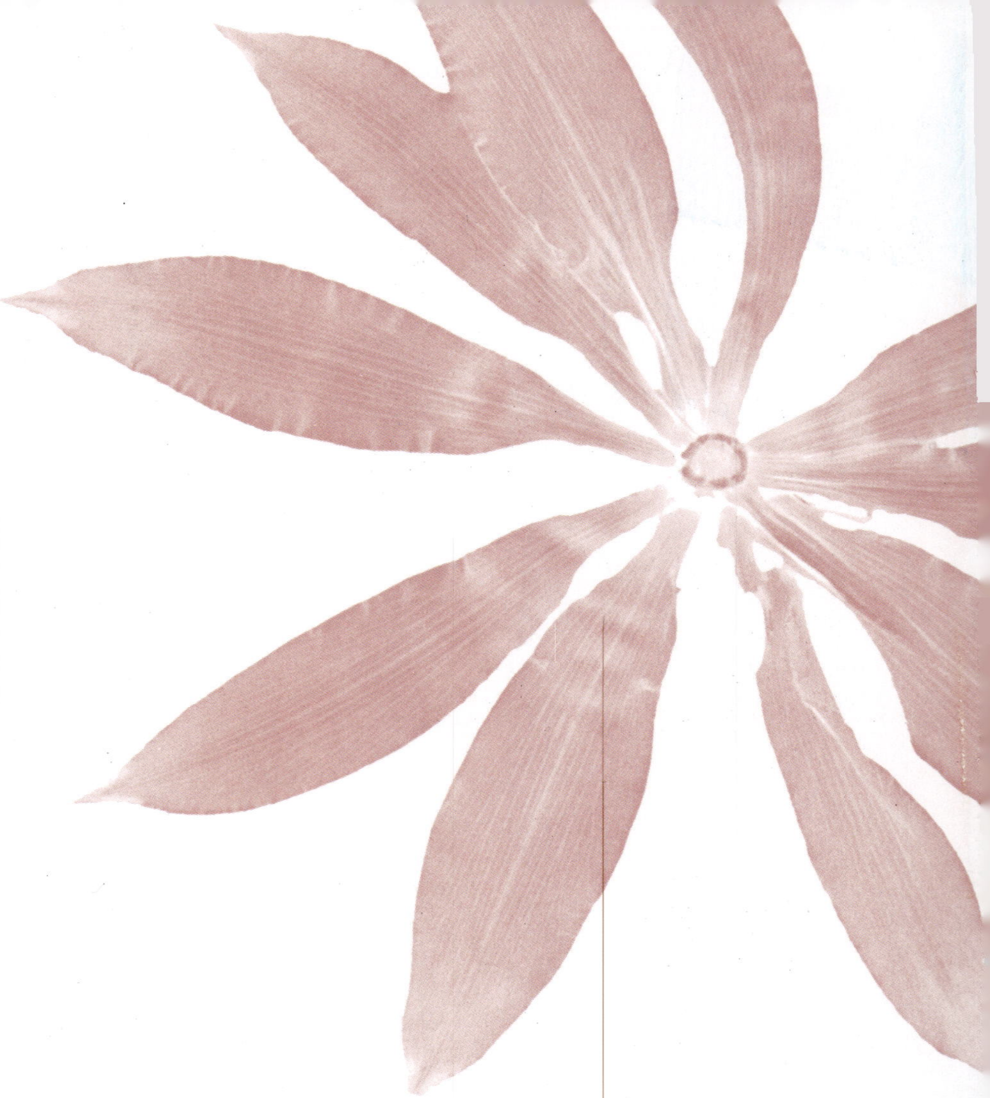

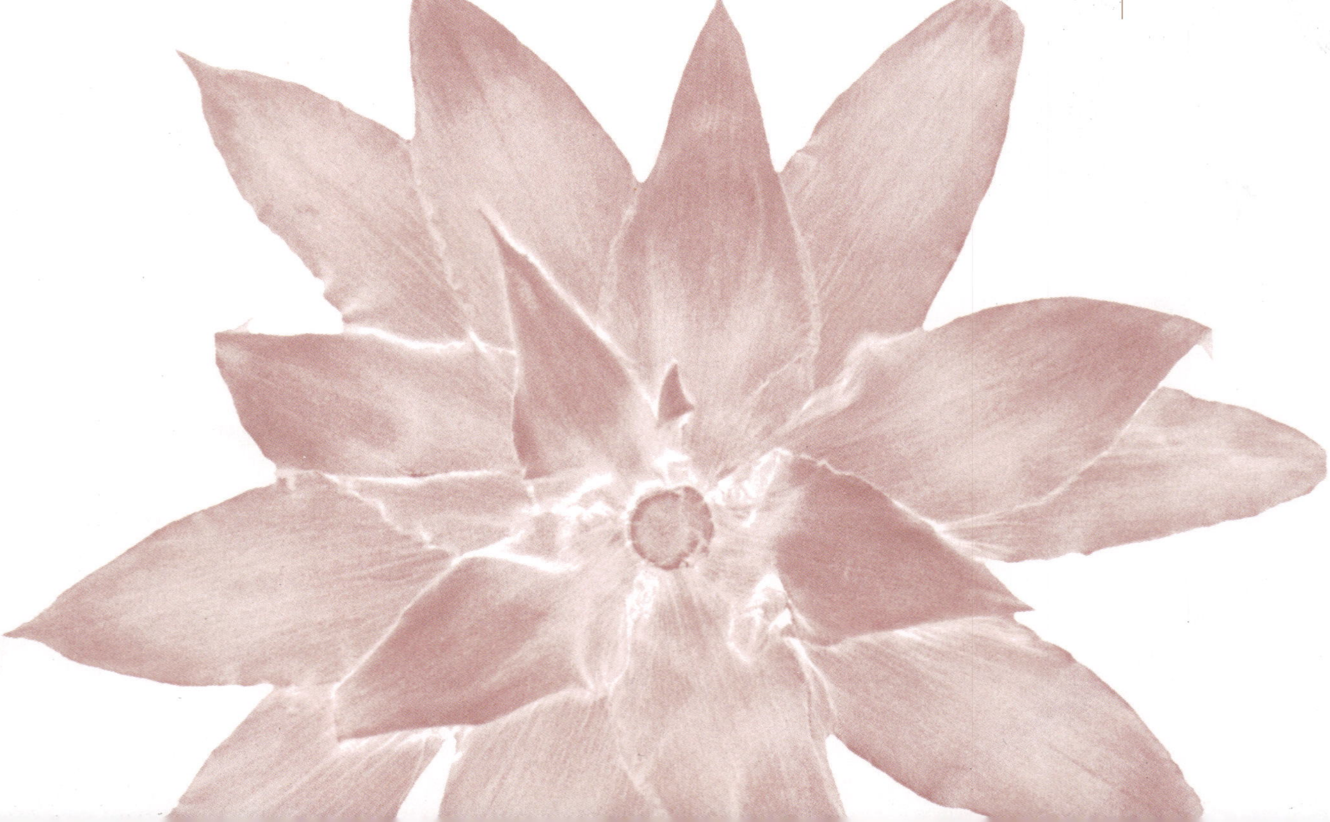